DOES SUCCESS BREED SUCCESS?

1

DOES SUCCESS BREED SUCCESS?

Effects of news and advertising on corporate reputation

MAY-MAY MEIJER

aksant
2004

ISBN 90 5260 145 3

© 2004, Aksant Academic Publishers, Amsterdam

Cover illustration: Rijkman Groenink chairman of the managing board of ABN AMRO
                      © ANP 2001/Foto: Marcel Antonisse
Cover design: Room for ID's
Lay out: BoekVorm, Amsterdam

Aksant Academic Publishers, Cruquiusweg 31, NL-1019 AT Amsterdam, www.aksant.nl

To my parents

Contents

Preface

Corporate reputation has fascinated me ever since I wrote my master's thesis, which was why I found a study on the effects of media coverage and advertising intensity on corporate reputation so appealing. There is relatively little literature on the effects of media coverage on corporate reputation, which made this study an interesting scientific journey across the different fields of political communication, corporate reputation, marketing, and advertising.

This book has three key characteristics: content analysis of the news about eight companies and two sectors, public opinion polls to assess the reputation of the companies, and data on the advertising expenditures of the focal companies and organizations. I would like to thank Patrick Blokzijl, Bram Büscher, Annemarie van Elfrinkhof, Peter Meijer, Laura Scheffer, Rens Vliegenthart, and Leonique Vis for coding the newspaper articles and the television news. I really appreciated your help. Or to put it in NET-terms:

May-May/appreciated/+1/AFF/coders. Nezet Fares, Indira Sam-Sin and Liobe Wouters, thank you for your help with gathering the newspaper articles.

The Dutch public broadcasting company NOS and the commercial broadcasting company RTL kindly let me use their electronic news archives, which saved a great deal of time. In addition, FactLANE, which is currently part of LexisNexis Benelux, let us use their news archives of daily newspapers. We really appreciated this, especially bearing in mind that university resources are always very limited. Another issue in the present study was how to weight the different newscasts of the television channels. Meinte Rebel of Intomart GFK provided me with the ratings and made it possible to construct a weighting factor based on the ratings of the different newscasts.

The second key characteristic of this book is the use of public opinion polling to measure corporate reputation. The Dutch Institute for Public Opinion and Market Research (TNS NIPO) let us use their panel of respondents for three consecutive years. The sample of respondents represents the population of Dutch households, which contributed enormously to the quality of the present study. I am particularly thankful to Henk Foekema (research director), Remco Frerichs (opinion researcher), and Patrice Weijer (programmer).

The third key characteristic of this book is the effects of paid publicity. I am very thankful to Berry Punt of BBC The Media and Advertising Bank in Amsterdam, because he provided me with data on the advertising expenditures of the ten focal organizations per medium title. This made it possible to assign the advertising expenditures to the respondents on an individual level in a real-life setting.

In addition to the people who helped me with the three key characteristics of this book, there were other people who were very important to making it a reality.

For the analysis of the pooled data and general discussions about research methods, I consulted several experts in research methods: Marcello Galluci (Department of Social Psychology at the Vrije Universiteit), Dick de Gilder (Department of Public Administration and Organization Science at the Vrije Universiteit), Jacques Hagenaars (Department of Research Methods and Statistics at Tilburg University), Adriaan Hoogendoorn (Department of Research Methods and Statistics at the Vrije Universiteit), Bas van der Klaauw (Department of Economics at the Vrije Universiteit), and Remco Peters (Department of Science at the University of Amsterdam). I thought it was very special you were willing to give your opinions about the methodological part of the present study, and some of you did not even know me. Although I did not follow all your recommendations, I really appreciated your comments.

During my time as a Ph.D. student, I joined the educational program for Ph.D. students at the Netherlands School of Communication Research. I found the courses offered very informative and I would like to thank my fellow Ph.D. students and the supervisor of our Ph.D. club, Klaus Schönbach, for their useful advice. I would also like to thank my fellow board members on the Dutch Ph.D. council, Pauline, Raoul, Astrid, Rogier, and Miranda. Being allies in promoting the interests of Dutch Ph.D. "students" and trying to make a contribution to academic culture and management of the universities meant that my research was stimulated in a much broader fashion.

I also wish to thank my esteemed colleagues of the Department of Communication Science. Cees and Dirk, I really think the final version of this book benefited greatly from your valuable comments on the draft version. Peter Kerkhof, thank you very much for your useful advice on Chapter 2 of this book.

Anita, Bob, Charles, Elly, Enny, Guda, Hans, Lonneke, Luuk, Paula, Peter 't Lam and Tinca, thank you for always being prepared to share your thoughts with me, whether about the operationalization of reputation as attitude towards the organization, or about Eagly and Chaiken's interpretation of the experiments of Cacioppo and Petty. I am also thankful to my former roommate Inge for her editorial suggestions concerning the explanation of the NET-method. Sabine, Antoinette and Margriet, thank you for all your secretarial work and for using your organizational abilities on my behalf. I want to thank the employees of our IT department for their flexibility with regard to the size of my data. Colleen Higgins, thank you very much for correcting my English.

In addition, I would like to thank Theo, Barbara, Brigitte, Corine, Elly, Latifa, and Pamala at the Department of Philanthropic Studies for being great colleagues. It's a pleasure to work with you!

The project was supervised by Jan Kleinnijenhuis, Professor of Communication Science at the Department of Communication Science at the Vrije Universiteit. In 1998, Jan wrote the book chapter "Effects of free publicity on image" that formed the start of the present study. Jan, I am especially grateful for the wealth of knowledge you were willing to share with me, and for your infectious passion for science, science, and science. You were so closely involved in the media monitor project that it was difficult to keep myself from writing phrases like "our project" and "in our opinion," instead of "my project" and "in my opinion." Nevertheless, any shortcomings in this book are, of course, entirely my own responsibility.

Of another dimension was the pressure from my friends, family and Professor Keman, who kept asking if my dissertation was finished yet. Mom and Dad and my darling sister Femke, you backed me up. Dad, thank you for sharing your insights about academia with me. Thierry, love of my life, you have supported me despite the crazy working hours, and Saturdays and Sundays at the Vrije Universiteit. Thank you for being my beloved husband and wonderful friend.

1

INTRODUCTION

> "We must distinguish carefully between the image and the messages that reach it.
> The messages consist of *information* in the sense that they are structured experiences.
> *The meaning of a message is the change which it produces in the image.*"
> Boulding (1956, p. 7)

1.1 INTRODUCTION

Corporations spend the largest part of their total communications budget on marketing communication (Van Riel, 1997). For example, Dutch supermarket chain Albert Heijn spent 438,976 euros on television advertising during a week chosen at random in January 2004 (*BBC*, 2004). In some cases, however, a single negative press report can be enough to damage a reputation and to cause share prices to plummet. The American IT company Emulex lost 2.5 billion dollars of its market value due to a fabricated report issued to the press by a former employee seeking revenge (*NRC Handelsblad*, September 1, 2000, p. 13).

In 2002 Royal Ahold, the parent company of Albert Heijn, placed thirty-fifth on the *Fortune* 500 list of the world's largest corporations. News about the accounting scandal at Ahold, which came to light in February 2003, shook the foundations of the once so authoritative supermarket concern. Ahold was compared with Enron,[1] and its bankruptcy was discussed in the Dutch national daily newspapers with such headlines as "Enron, WorldCom and now also Ahold" (*Trouw*, February 25, 2003, p. 5), and "Fifty-fifty chance of bankruptcy for Ahold" (*NRC Handelsblad*, March 6, 2003, p. 11). Although monitoring media coverage cannot prevent disasters like this, one would expect that companies would want to know more about the effects of media coverage on corporate reputation, in view of the capital that companies have tied up in this.

Although scholars have studied corporate reputation and related concepts for decades, there have been few attempts to develop and empirically test theories about *the effects of media coverage on corporate reputation*. That is why this will be the main theme of this book. In addition, the effects of paid publicity on corporate reputation will also be taken into account.

This chapter is structured as follows: In the next subsection, 1.2, possible explanations will be given for why the field of the effects of media coverage on corporate reputation is relatively new. The aim of this study and the research question will be formulated in subsection 1.3. Subsequently, in subsection 1.4, the definition of corporate reputation and its relationship to other concepts such as "image" and "identity" will be discussed. The consequences of a positive reputation on financial performance will be focused upon in subsection 1.5. Finally, the outline of the book will be presented in subsection 1.6.

1.2   POSSIBLE EXPLANATIONS FOR THE RELATIVE NEWNESS OF THIS TOPIC

In the section that follows, possible explanations will be given for why research on the effects of media coverage on reputation is relatively new. This will be done by first outlining the prevailing *effects of mass media paradigm* during the period in which publications appeared about corporate image, the precursor of corporate reputation. Second, the early publications on corporate image will be discussed in more detail.

Early publications about corporate image appeared in the second half of the 1950s (Boulding, 1956; Eells, 1959; Martineau, 1958). McQuail characterized research on media effect during this period as *the theory of powerful media put to the test*. There were doubts about the theory of the powerful media, which was the prevailing paradigm in the preceding phase. The phase of the theory of powerful media put to the test began in the early 1930s and continued until the early 1960s (McQuail, 2000).

The work of Lazarsfeld, Berelson, and Gaudet (1944) gave reason to doubt the strength of the effects of mass communication. During the American presidential election campaign in 1940, the mass media turned out to be less capable of changing people's preferences for presidential candidates than expected. Lazarsfeld et al. had the impression there seemed to be a great deal of person-to-person interaction, particularly during the critical months. They alluded to the concept of a *two-step flow of communications*, the idea being that the mass media often reaches its audiences in two stages. In the first stage, the messages in the newspapers and broadcasts are picked up by the opinion makers. In the second stage the opinion makers give their interpretation of the news, which they communicate to the less active audience. Lazarsfeld et al. were early users of a now-accepted method of social research: the panel technique. In the case of a change in vote intention, the respondents were asked the reason for their change. Some respondents attributed the reason for the change in vote intention to a certain article in the newspaper or to a radio report. Lazarsfeld et al.[2] paid scarcely any attention to the *content* of the media coverage, however, as will be examined further in the next subsection on the minimal effect theory.

In addition, in the 1960s the still-influential summary of early research by Klapper (1960) – often referred to as the "limited effects model" – gained ground. Based on the effects of mass communication studies, Klapper stated that the dominant media effect he found in his meta-analysis was a reinforcement of existing attitudes, and that conversion of attitudes rarely happened. It was not that the media had been found to be without effects, but rather to operate amid other factors, which seemed to mediate its influence in such a way as to render it an agent of information rather than change.[3] Klapper argued that mass media have a greater capability of creating opinions about new issues than in the conversion of already existing attitudes.

Lang and Lang (1981) referred to the "minimal effect" conclusion as the "no effect" myth that closed off certain avenues of research temporarily. Although Klapper (1960) did not conclude he had found no effects,[4] scholars despaired of finding effects.

It was in this scientific climate of the effects of mass media that publications appeared about corporate image, the precursor of corporate reputation.

Another reason media content did not really play a role in the early publications about image may be that these publications originated in the marketing and management field. Studies were published in journals like the *Harvard Business Review*, *California Management Review*, and the *Journal of Marketing* (Eells, 1959; Martineau, 1958; Spector, 1961). These journals did not have a tradition of publishing on media coverage.[5]

In addition, in the early writings on corporate image, most scholars were busy exploring the corporate image concept itself. Spector (1961) used factor-analysis to obtain image dimensions. Martineau (1958) argued that different groups like stockholders, employees, vendors, and buyers will see different aspects of the corporate image. In one of the earliest publications about image, however, Boulding (1956) had already described three possible effects of a message on an image. The first possibility is that the message leaves the image unaffected. According to Boulding, this is the most likely one, due to the great number of messages a person receives. The second possible effect is that a message may change the image by adding knowledge, for example, when you hear that Dutch bank ABN AMRO is offering a new way to pay off mortgage payments. This knowledge will be added to the original image of ABN AMRO without fundamentally revising it. A third type of change of image is what Boulding (1956) described as a revolutionary change such as conversion. The example is given of a man who thinks he is a pretty good guy until he meets a preacher who convinces him that his life as he is presently living it is in fact worthless and shallow. These examples show that with the *message* concept, Boulding did not necessarily mean a message from the mass media.

Nevertheless, these early thoughts about the effects of messages on image did not seem to be picked up until a decade later when, in their book about content analysis, Stone, Dunphey, and Bernstein (1966) suggested monitoring the mass media consistently together with public opinion data.

Meanwhile, various scholars challenged Klapper's (1960) minimal effect theory. Bandura (1994) stated that Klapper's (1960) view is at variance with a substantial body of evidence. De Boer and Brennecke (1999) mentioned some revealing titles that attack the paradigm of limited effects, like Noelle-Neumann's publication *Return to the Concept of Powerful Mass Media* (1973) and Zaller's publication *The Myth of Massive Media Impact Revived: New Support for a Discredited Idea* (1996).

Kleinnijenhuis (1998) remarked that Klapper's (1960) summary of previous research relied heavily on the American election research by Lazarsfeld et al. (1944) and on Berelson, Lazarsfeld, and McPhee (1954), in which the content of media coverage was not taken into account. Kleinnijenhuis (1998) stated that Klapper's (1960) reasoning is as follows: Newspaper readers do not change their opinion and radio listeners do; that means that newspapers do not have any effect and the radio does. In order to investigate the effects of media coverage on public opinion, media content should be studied as well (see also Shoemaker and Reese 1990). If newspapers report more or less the same news every day, newspaper readers who do not change their opinions are consistent with the hypothesis that newspapers do indeed have an impact. That is why two methodologies will be applied in the present study: content analysis and survey research. But this does not mean that Klapper's (1960) ideas will be completely ignored. One of the theories that will be used in this study is the agenda-setting theory. This theory was described by McCombs (1994) as a theory of limited effects. Contemporary research is trying to identify the conditions under which this agenda-setting influence of the news media does and does not occur (McCombs, 1994).

After Stone et al. (1966) suggested investigating the effects of media coverage on corporate image, it took nearly 25 years for the first empirical research (Fombrun & Shanley, 1990) to be carried out. For the last decade, investigations into the impact of media coverage on corporate reputation seem to be getting onto the corporate reputation research agenda. The following studies appeared on this topic: Carroll and McCombs (2003), Fombrun and Shanley (1990), Verčič (2000), and Wartick (1992). These studies will be discussed in Chapter 2. At the Corporate Reputation Identity & Competitiveness Congress in Paris in 2001 (Deephouse, Carroll, Meijer, Kleinnijenhuis, Verčič & McCombs, 2001), a symposium was held on the effects of media coverage on reputation.

1.3 CENTRAL RESEARCH QUESTION AND AIM OF THE STUDY

As outlined in the sections above, studies on the effects of media coverage on reputation are rare. Moreover, very few studies examine both the effects of media coverage and the effects of paid publicity on reputation. The study by Fombrun and Shanley (1990) is one of the exceptions, and leads to the following central question:

What are the effects of media coverage and paid publicity on corporate and sector reputation?

The aim of the present study is to contribute to the building of a theory about the effects of media coverage and paid publicity on corporate reputation. Since the study of the effects of media coverage on political and other attitudes has a long tradition in the field of political communication, theories from this field contribute to knowledge about the effects of media coverage on reputation. Political communication theories together with social psychological theories will be used in the present study to try to explain the mixed results of the previous empirical "media effect on reputation" studies that will be elaborated upon in Chapter 2.

In order to do so, distinctions need to be made between *different types of news*. Another aspect that may be important to explain the mixed results of the previous empirical studies is the media use of individual respondents, which will be taken into account in the present study. Moreover, in order to empirically test the effects of media coverage on public opinion, media content and public opinion will be measured at different points in time in order to trace shifts in corporate reputation.

Working out the research problem was chiefly restricted to the situation in the Netherlands. Dutch newspapers and Dutch television news were investigated, and the companies and sectors investigated were mainly Dutch. This study was originally designed as a media effects study only. Since it is likely that paid publicity influences corporate reputation as well (Brown, 1998; Fombrun & Shanley, 1990), the effects of advertising intensity were also taken into account in the present study.

Practical relevance of this study

For a communication team, systematic media analysis can be a useful tool when developing a communication strategy. Without media analysis, old newspaper cuttings will just yellow with age – there will be no consequences for policy and superficial notions will prevail, such as "the press is always negative" and "journalist X is against us" (see Kleinnijenhuis, 2001). With this study, it is our intention to contribute to the development of a media monitor that can be used as an aid by communication professionals in order to monitor media coverage systematically.

Since this is the practical aim of this study, it is necessary not only to register the frequency and the tone of news about the focal company, but also to examine the issues about which the company was in the news and to investigate which actors criticized or supported the company. As will become clear in the next chapter, the issues the company is linked to in the media may influence its reputation, and not all the negative news necessarily has a negative impact on reputation. The insights resulting from a media monitor combined with knowledge about the effects of media coverage on reputation may influence the approach chosen by the communication team.

In the present study, we talk about a media monitor as an instrument used to systematically analyze news about an individual organization, its competitors, and relevant issues over a longer period of time (McKeone, 1995; Meijer, Oegema, & Kleinnijenhuis, 2000; Kleinnijenhuis, 2001). The foundation of the media monitor is laid by the content analysis used to create a media profile.

1.4   THE DEFINITION OF CORPORATE REPUTATION AND ITS RELATION TO IMAGE AND IDENTITY

Corporate reputation and image
Although in the present study a distinction is made between the two terms, in early publications about corporate reputation the terms corporate image and corporate reputation were used alternately and as synonyms, for example by Riley and Livey, who stated the following:

> 'We might have used the term *corporate reputation* as carrying essentially the same meaning as the corporate *image*. In this sense, contrary to the popular impression of *the corporate image* as a recent Madison Avenue invention, the concept is probably as old as the corporation itself although it has not been called by this name' (Riley & Levy, 1963, pp. 176-177).

The corporate reputation concept has become popular in the last decade. In the previously mentioned publication by Fombrun and Shanley (1990), reputation appeared in the title of their article "What's in a name? Reputation-building and corporate strategy." Dowling (1994) and Fombrun (1996) mentioned the term *corporate reputation* in the main title of their books. In addition, a new journal for scholars and practitioners, *Corporate Reputation Review,* was launched in 1997. According to Van Riel (2001), the increasing influence of American scholars in the corporate reputation field stipulates the use of the concept of corporate reputation rather than the corporate image concept. It should be noted that in the marketing field the related concept of "brand image" is still frequently used (Timmerman, 2001).

Recently, some scholars in the corporate reputation field have made a distinction between the two terms (Bromley, 2000; Davies, Chun, Da Silva, & Roper, 2001; Fombrun, 1996; Maathuis, 1999). Several authors remarked that the terms image and reputation are not used consistently due to their multiple meanings (Abratt, 1989; Brown, 1998). Let's examine some of the various definitions of (corporate) *reputation*:

Box 1.1.   Definitions of (corporate) reputation.

"The overall estimation in which a company is held by its constituents" (Fombrun, 1996, p. 37).

"The evaluation (respect, esteem, estimation) in which an organisation's image is held by people" (Dowling, 1994, p. 8).

"We use the term *corporate reputation* to refer to the overall evaluation of these associations [the set of associations with a company an individual holds in mind], in terms of the relative preference position of a company within a peer group" (Maathuis, 1999, p. 31).

"A collective term referring to all stakeholders' views of corporate reputation, including identity and image" (Davies et al., 2001, p. 114).

These definitions share a common element,[6] namely the overall evaluation. There is also a difference, however, in the way the corporate reputation concept is used by Fombrun, Maathuis, Davies et al. on one hand and Dowling on the other. Fombrun (1996) argued that corporate reputation is an aggregated concept: the overall evaluation of the several corporate images that exists among the different stakeholders. To put it more specifically, Fombrun portrayed corporate reputation as the overall evaluation of the customer image, community image, investor image, and employee image. In contrast to Fombrun, Maathuis, and Davies et al., Dowling (1994) focused on the level of the individual. He used the terms corporate reputation and corporate image to describe the total beliefs and feelings a person may hold about an organization.

Dowling's view is employed in this dissertation, which means that the corporate reputation and the corporate image concept are applied on the level of the *individual*. The term *corporate image* is used "to refer to the set of associations with a company an individual holds in mind" (cf. Maathuis, 1999, p. 31). The term *corporate reputation* is used to refer to the overall evaluation (usually in terms of good or bad) of the associations with a company an individual holds in mind.

According to Bromley (1993), the adjective *corporate* can be used in at least two ways. It can be employed to refer to a particular organization (the Dutch airport Schiphol), to an organized system of firms (Albert Heijn, Super de Boer, the Dutch Railways), or to multinationals (Shell, BP, Rabobank, ABN AMRO bank). This will be done in the present study. A second way of using *corporate* is to refer to the public image of the higher-level organization – the body that attends to the business of co-coordinating the diverse interests of its various departments or subsidiaries: the holding company (Bromley, 1993).[7] The term *sector reputation* is used in this study to refer to the reputation of a sector such as that of the agricultural or police sector. For reasons of readability, however, the term sector reputation will not always be mentioned.

Delineation of corporate reputation: corporate image versus corporate identity
In this dissertation, the concept of corporate or organizational identity is characterized as the personality of the organization expressed either through its behavior, communication, or symbolism (Birkigt & Stadler, 1995), or what the organization is: its character (Balmer, 1995). This means that corporate identity is mainly characterized by what the organization is (how its employees behave and which symbols represent the organization), whereas corporate image and corporate reputation are mainly characterized by how the organization is *perceived*. Or, to put it in mass communication terms: identity is a concept that belongs to the sender side of communication, while image and reputation are concepts that belong to the receiver side. For more extensive discussions of the corporate identity concept, see Albert and Whetten (1985), Balmer (2001), Christensen (2001), Gioia, Schultz and Corley (2000), Hatch and Schultz (2000), Moingeon and Ramanantsoa (1997), Van Rekom (1998), and Van Riel and Balmer (1997).

Nevertheless, there are different and even opposing ways in which the relationship between corporate image and corporate identity is defined. Bromley (2000, p. 241), for example, defines corporate *image* as "the way an organization presents itself to its publics, especially visually." Note that this definition of Bromley is closer to the corporate *identity* concept as employed by Birkigt and Stadler (in which communication and symbolism are part of corporate identity) than to the definition of the corporate image concept of Maathuis. Because of this confusion, we found it necessary to explain our perception of the distinction between the terms, although the concept of corporate identity will not be used further in this book.

1.5 THE CONSEQUENCES OF A POSITIVE CORPORATE REPUTATION

The introduction to this chapter provided two examples of the disastrous effects media coverage can have on the reputation and market value of a corporation. In the subsection below, the effects of reputation on financial performance will be discussed, based on three recent empirical studies and a study that conducted a literature survey on this topic. The three empirical studies are based on research designs that include at least two measurement points in time, therefore allowing testing for longitudinal cause-and-effect relationships.[8]

However, the relationship between reputation and financial performance is not the only reason why the reputation field is such an interesting area of research for both academics and those working in the field. Approximately 30 years ago, Markham (1972) argued that corporations with a good reputation will sell more products, attract better applicants (see also Stigler, 1962 for more on this aspect), and attract more investors to buy shares. Moreover, a good reputation is essential for the growth and security of a corporation. These relatively early ideas

about the consequences of a positive corporate reputation are still being empirically tested in recent publications on corporate reputation. For an elaborated
overview of empirical research about the consequences of a positive reputation
on different stakeholder groups (including consumers, employees, and competitors), see Brown (1998) and Maathuis (1999).

Consequences of a positive corporate reputation on financial performance
Jones, Jones, and Little (2000) examined the question of whether corporate reputation serves as an intangible asset that can help protect the organization in times
of crises – the reservoir of goodwill presumption. Crises were operationalized by
two significant one-day stock market declines: October 19, 1987 and October 13,
1989. Corporate reputation was measured by *Fortune* magazine's annual survey
of corporate reputations.[9] They conducted a regression analysis in which the dependent variable was the post-crisis closing stock market price. The independent
variables were: prior-day stock price, corporate reputation rating, a control variable for the systematic rise or fall in stock price associated with the stocks, size,
and an interaction term of year reputation. When the market took an unexpected
downturn in 1989, the stock prices of companies with better reputations dropped
significantly less than those of companies with poor reputations.

In a similar fashion Black, Carnes, and Richardson (2000) investigated
whether market value (the expected future cash flows) is influenced by corporate
reputation score. Corporate reputation scores were obtained of all of the firms
rated by *Fortune* magazine from 1982 through 1996. In addition, financial data
were gathered. This resulted in a sample size of 2,769 firm-year observations.
Reputation score was corrected for a financial performance halo, since individual
raters appear to be heavily influenced by previous financial performance. The significant positive coefficient on the non-financial components of corporate reputation score provides evidence of influencing the market value of a firm.

Dunbar and Schwalbach (2000) explored the reputations of 63 German
firms during the period 1988 to 1998. They used data from a survey by *Manager
Magazin*, similar to *Fortune* magazine's annual survey of corporate reputations.
Five attributes were used: management quality, innovativeness, communication
ability, environmental communication, and financial and economic stability. An
overall financial performance score was calculated from the sum of an accounting score and a market performance score. A pooled regression analysis showed
that corporate reputation predicts future financial performance (Beta = .17). Furthermore, financial performance also influences future firm reputation (Beta =
.65). The authors did not specifically address the matter of feedback or bilateral
causality, which could be due to the fact that the attribute "financial and economic stability" of the corporate reputation variable closely resembles the variable "financial performance."[10]

De la Fuente Sabate and De Quevedo Puente (2003) surveyed 18 different empirical studies analyzing the relationship between corporate reputation and financial performance. Special attention was paid to the direction of causality between corporate reputation and financial performance. De la Fuente Sabate and De Quevedo Puente discussed studies that focused solely on the effects of corporate reputation on financial performance, studies that focused solely on the effects of financial performance on reputation, and studies that tried to test both directions of the relationship in a single study. Some of the studies were longitudinal, but there was a great variety in the lags selected and in the measures of financial performance and reputation (De la Fuente Sabate & De Quevedo Puente, 2003). Nevertheless, they concluded that there is empirical support for both directions of the relationship between corporate and financial performance: corporate reputation positively influences financial performance, and vice versa. They used a theoretical framework (the contractual perspective) to explain the bilateral causality.

In short, according to the contractual perspective, a firm with a good reputation is placed in a privileged market position, which enables it to capture better resources and will help to generate value. In addition, the contractual view also supports the inverse relationship. The greater the value creation of a firm, the more friendly the atmosphere among stakeholders will be, as there is a better chance of meeting the claims of every single stakeholder (De la Fuente Sabate & De Quevedo Puento, 2003).

1.6 OUTLINE OF THE BOOK

This chapter argues that research on the effects of media coverage has not yet been well developed because in the late 1950s – when the first concepts of corporate image appeared – the media were assumed to have only minimal effects. McQuail (2000) called this period "the theory of powerful media put to the test." For years, the role of the mass media in the formation of public opinion was ignored (Lang & Lang, 1981). Another reason why media content did not really play a role in the early publications about image could be that the publications originated in the marketing and management fields. At that time, these journals published very little about media content and content analysis. Moreover, the corporate reputation field is a relatively young discipline and at the moment, attention is growing for media coverage as an antecedent for corporate reputation. In this book, the effects of paid publicity on reputation will be studied along with the effects of media coverage. The importance of this research for companies and sectors is stressed, bearing in mind the capital involved to maintain a good corporate or sector reputation.

In Chapter 2, a theoretical model of the effects of media coverage on corporate reputation will be elaborated upon. Previous studies on the effects of media coverage on corporate reputation will be discussed, as well as theories coming from the field of political communication. Theoretical explanations will be given in order to try to explain the sometimes seemingly contradictory results of empirical investigations. In addition, a distinction will be made in different types of news. Each of these types of news can be assigned to several theories that explain the effects of the type of news on corporate reputation. Following the discussion of the effects of media coverage, studies on the effects of advertising intensity will be described. At the end of the chapter a conceptual model is presented that combines previously discussed studies and theories. Hypotheses are formulated to test whether the research model holds up.

The research design is focused upon in Chapter 3. As mentioned by McQuail (1994; 2000), in order to prove there is a causal connection between the media coverage and the public agenda, content analysis plus evidence of opinion change over time in a given section of the public (preferably with panel data) is needed. Moreover, an indication of relevant media use by the public concerned is needed. This is the approach taken in this study. Media coverage is analyzed over a period of three years, and public opinion is measured at three points in time. One of the public opinion variables is measured at two points in time. Also, panel data are used in which all respondents are asked about media use. The coding process and the selection of newspaper articles and television news are outlined in Chapter 3, as well as the method used to measure corporate reputation.

Chapter 4 gives a general description of the data. It describes how the companies and sectors were portrayed in the media, and will elaborate on the amount and different types of news about the companies and sectors. The advertising expenditures of companies and sectors will also be described, as well as how the respondents judged the companies and the sectors.

The results of the tests of the hypotheses will be presented in Chapter 5 for the respective companies and sectors. Some examples of what will be tested are whether the amount of media coverage influences reputation, whether different types of news influence reputation in different ways, and whether paid publicity influences reputation.

Chapter 6 will sum up the conclusions of this study and include a discussion. The limitations of this study will be addressed, and feedback will be given on the theories that were applied. This final chapter will be concluded with a reflection on a media monitor based on this study's starting points.

2

EFFECTS OF NEWS AND ADVERTISING ON REPUTATION

Investigations on the impact of media coverage on corporate reputation have gotten onto the corporate reputation research agenda. Three empirical studies conducted in this field will be described in section 2.1. Subsequently, in section 2.2 theories about the effects of the amount of media coverage on corporate reputation will be presented. In section 2.3 the need to distinguish different types of news will be elaborated upon, and the types of news that are employed in this study will be discussed. The focus in section 2.4 will be on the effects of *success &
failure news* on corporate reputation. Section 2.5 will highlight the effects of *support & criticism news* on corporate reputation. Following this, in section 2.6 the influences of *issue news* on corporate reputation will be discussed by focusing on issue-ownership theory, agenda-setting, and priming.

In addition to media coverage, paid publicity is also assumed to influence corporate and sector reputation. The effects of advertising intensity on corporate and sector reputation are described in section 2.7. An overview of the hypotheses is given together with the conceptual model in the last section of this chapter, section 2.8.

2.1 EMPIRICAL RESEARCH ON THE EFFECTS OF MEDIA NEWS ON CORPORATE REPUTATION[1]

Already in the 1960s, Stone et al. (1966) stressed the need for empirical research on the effects of media coverage on reputation:[2]

> Traditional research into product image has concentrated mainly on the image of the product as conveyed through mass media. Relatively little attention has been given to changes in that image as it is received and discussed by the audience.... In a large-scale survey, different aspects of an image can be picked up and monitored, both geographically and within various strata of consumers. The relative salience of certain aspects of the image in the mass media can be compared with the relative salience of the same aspects in the minds of respondents. (Stone et al., 1966, p. 620, p. 624)

Nearly 25 years later, Fombrun and Shanley (1990) performed an empirical study in this field. This study will be focused upon first. Following this, two other empirical studies on the effects of media coverage on reputation will be elaborated upon. Recently another empirical study on the effects of media coverage on reputation appeared. This empirical study of Carroll (2004) will be elaborated upon in the subsection about agenda-setting (2.6.1). Since empirical studies on the effects of media coverage on reputation are rare, they provided valuable input for the present study. The studies will be described by focusing on the two hypotheses that were tested by Fombrun and Shanley (1990) and that return in the two other studies. The first hypothesis assumes that the more visible an organization is in the media – operationalized as the total number of articles about an organization – the *better* its corporate reputation will be. Fombrun and Shanley's second hypothesis is that the more positive the tone of a firm's media coverage, the better its reputation will be.

Fombrun and Shanley (1990) measured media visibility as the total number of articles written about the company in 1985. A coder content-analyzed the articles and classified them as positive or negative news about a firm.[3] The survey by *Fortune* magazine, which is known to reputation scholars as "the *Fortune* data," served as the reputation data (see endnote 9 Chapter 1). The eight items of the Fortune data were used to construct an overall reputation index.[4] The study set consisted of 148 firms. Contrary to the hypothesis, it appeared that the greater the volume of media coverage about an organization, the *worse* its corporate reputation was. The hypothesis was not confirmed that the more positive the tone of a firm's media coverage, the better its reputation. It could also not be proved that media coverage or the tone of news has an interactive effect on reputation.

Wartick (1992) studied the effects of media coverage on corporate reputation as well. The sample he used consisted of 29 companies. The media coverage was measured by data compiled and reported by the Conference on Issues and the Media (CIM). The CIM data focused on 13 media outlets and included television and print news. A limitation of the dataset was that the data were not reported by the individual media outlets, but as an aggregate for all 13 outlets (Wartick 1992). The CIM reported the ad value (used to operationalize media exposure) and the tone of the total coverage in the two weeks, i.e., positive, negative, or mixed. The overall ratings for firms in the 1987 and the 1988 *Fortune* survey comprised the data for change in corporate reputation.

Wartick's study showed that the amount of media coverage does not correlate significantly with changes in corporate reputation on the level of the overall sample of 29 companies.[5] If the sample is divided into reputational subgroups, however, it appears that the amount of media coverage does correlate with corporate reputation. Wartick's study showed that for companies with "good" and "average" reputations, more media coverage is related to larger, positive changes in corporate reputation.

There is a positive correlation between the tone of media coverage and the direction (and magnitude) of change in corporate reputation on the level of the overall sample. In other words: positive media coverage of an organization is connected with a positive change of the corporate reputation of the organization concerned and vice versa. Within the subgroups of companies with good starting reputations and companies with average reputations, the tone of media coverage is not significantly correlated, however. For the companies with poor reputations that received *negative* media coverage, corporate reputation diminished.

In order to test cause-and-effect relationships, Verčič (2000) stressed the importance of time series analysis to study the effects of media coverage on corporate reputation.[6] Two of the hypotheses from Verčič' study will be discussed in this subsection: the larger the amount of media coverage and the more positive a firm's current media coverage, the better its reputation. Verčič studied three organizations: an organization in which people's trust increased (the British Post Office), an organization with no apparent trend in trust (British Airways), and an organization with a decline in trust (Shell). Media exposure was measured as the sum of stories per year per company over a period of ten years (between 1988 and 1997). The Reuters Business Briefing database was used to gather the media data; only print news was included. The tenor of media coverage was coded with two values, positive and negative. The analysis was conducted by observing the negative rather than the positive stories, since it was assumed they were a more powerful antecedent of changes in the levels of trust.[7]

Verčič used the annual corporate image public opinion poll conducted by Market and Opinion Research International (MORI) to measure reputation. In this survey, the respondents were asked to indicate their overall impression of any of the three focal companies. In order to match the media data, the opinion polls from 1988 to 1997 were used. Verčič was the first to use the MORI opinion polls as time series data. The hypotheses were tested by means of a two-step approach. In the first step, trends in the amount of media coverage, media favorability, and trust were detected. In the second step of the tests of the hypotheses, the variables within a trend were differenced to extract the trend. However, the two hypotheses could not be confirmed.

In sum, these three pioneering studies show mixed results. Fombrun and Shanley (1990) found a negative impact on reputation as a result of the amount of media coverage, while Wartick (1992) found a positive influence, and Verčič (2000) did not find a significant effect on reputation as a result of the amount of media coverage. The tone of media coverage did not influence reputation in either Fombrun and Shanley's study or in that of Verčič. Wartick found a positive impact of the tone of news on reputation.

How can these differences be explained? One possible explanation for the different results in the effects of the amount of media coverage may be that the effects of the amount of news differ per medium. In other words, the effects of the amount of television news may differ from the effects of the amount of print news. Moreover, none of the three studies took into account the media use of the respondents. Someone who did not use the media at all was treated the same as someone who watched the news and read three different newspapers a day. Another aspect is that the news was studied from a global perspective. Whereas political communication studies distinguish different types of news or different types of frames, the three studies discussed earlier did not make such a distinction. Each of these types of news may have a different impact on reputation. Another important point is that in all three studies, the coders were asked to determine – from the reader's perspective – whether an article was positive or negative for the organization concerned. As was indicated by Verčič (2000), however, an article can have different impacts on its readers. To give an example, environmental activists may interpret an increase in oil prices as a positive message, hoping that people will use their bikes instead of their cars. However, a fanatic driver will probably consider this to be bad news.

Verčič' study is described in detail, which makes it possible to see why he found reputation was not affected by either the amount of media coverage or by the tone of news. Verčič conducted his analyses by using Ordinary Least Squares (OLS) regression. The question can be raised, however, whether the number of cases was too low.[8] As becomes clear from the appendices in Verčič's study, the ten points in time per company in which media coverage and reputation were measured form the number of cases per company. Another aspect is that Verčič differenced the news and the reputation of the companies. With regard to the media variables, it can be wondered if this was necessary, since news is only news if an event is newsworthy. From a theoretical point of view it can be argued that for most companies, the trend in the amount of media coverage will be that there is "no trend." If nothing special happens, there will be no news, so therefore the news variables do not have to be differenced.

It can be concluded from the foregoing that a study on the effects of media coverage on reputation needs to: a) distinguish between the effects of the amount of television news and the amount of print news; b) take into account the media use of respondents; c) distinguish between different types of news; d) focus on the manifest content of the media coverage. These aspects will be elements of the present study.

The two hypotheses tested in the studies described earlier will also be tested in this study. The first hypothesis – about the effects of the amount of media coverage on corporate reputation – will be discussed in the next section.

2.2 THE EFFECTS OF THE AMOUNT OF MEDIA COVERAGE ON CORPORATE REPUTATION

Starting points for theorizing on the effects on reputation of the amount of media coverage are provided by studies like Zajonc's (1968) on the *mere exposure effect*. Zajonc demonstrated that subjects judge stimuli more positively the more they are exposed to them. One of his experiments was an experiment with fictitious "Chinese-like" characters. The subjects were told that the experiment dealt with the learning of a foreign language. They were asked only to pay close attention to the characters whenever they were exposed to them. The results showed that the subjects assigned a more positive meaning to the fake Chinese characters the more they were exposed to them. In other words, Zajonc's study revealed that mere repeated exposure to a stimulus object causes increased liking of that object.

On basis of Zajonc's study and the other studies to be discussed in the next subsection, the present study argues that media visibility will improve the company's reputation, since it evokes feelings of familiarity and stresses the power of the company. At very high levels of media exposure, however, the audience may start to wonder if there is something wrong with the company concerned. This will lead to an inverted U-shape relationship of the effects of the amount of media coverage on reputation.

Miller (1976) found an inverted U-shaped function in his mere exposure study. He conducted an experiment with a poster in the common areas of the dormitory that read "Reduce foreign aid". There were three experimental conditions: no exposure, moderate exposure (30 posters left up for three days), and overexposure (170 posters). Miller's results showed that respondents were in favor of decreasing foreign aid in the moderate exposure condition. Overexposure significantly lowered positive attitude towards reduction of foreign aid, however. In other words, Miller found an inverted-U function between the number of exposures to the poster and evaluative ratings.

Cacioppo and Petty (1979) studied the attitudinal effects of mere exposure in a communication context as well.[9] They conducted an experiment with a 2 × 4 factorial design in which the position advocated (proattitudinal versus counterattitudinal) and the number of presentations (0, 1, 3, or 5) served as between factors. The subjects were asked to indicate their agreement with a recommendation that university expenditures be increased. The results showed that the relationship between the repetition of the message and agreement with the message followed an inverted-U shape (non-monotonic relationship) for *both* the group that listened to the proattitudinal message and the group that listened to the counterattitudinal message. This means that repetition leads to first increasing and then decreasing agreement with the message. In a second experiment, the same results were found. An explanation for the inverted U-shaped relationship between exposure and affect is the two-factor theory (Berlyne, 1970; Cacioppo & Petty, 1979). The two-factor theory proposes that two factors, positive habituation and tedium, mediate the relationship between mere exposure and affective response. Cacioppo and Petty (1979) used the concept of *two-stage attitude-modification process*. In the first stage, repeated exposure to persuasive message arguments should increase peoples' opportunities to cognitively elaborate the message's arguments and to realize their favorable implications. Hence, counterargumentation declines at the moderate exposure frequency. At high exposure levels, however, tedium and/or reactance may have motivated the individual to again attack the now offensive communication. Counterargumentation is renewed and agreement decreases at high exposure levels. Therefore, this process is also referred to as the two-stage cognitive elaboration-then-tedium process. In studies on ad exposure and ad effectiveness, inverted U-shape relationships were found as well (see subsection 2.7).

If the two-factor theory is applied in a business news setting, it can be argued that at low levels of media exposure, the audience will be more familiar with the company, which leads to a better reputation for the company. At very high levels of media exposure, the audience may start to wonder if there is something wrong with the company concerned, and the reputation of the company will worsen.

As will be elaborated upon in the next subsection, the present study postulates that the second stage of attitude-modification – the higher media exposure levels in which tedium occurs – is reached sooner by newspaper readers than by television viewers. It is assumed that newspaper readers have a higher chance of receiving a message than television viewers, since they are more involved with news about organizations.

This idea is inferred from one of the axioms from Zaller's (1992) Receive-Accept-Sample (RAS) model. The RAS model explains how people acquire political information from the mass media and convert it into political attitudes. According to Zaller (1992, p. 51), opinion statements are "the outcome of a process in which people *receive* new information, decide whether to *accept* it, and then *sample* at the moment of answering questions". The 'reception axiom' of the RAS model (Zaller, 1992) postulates that the greater a person's level of cognitive engagement with (or political awareness of) an issue, the more likely it will be that he or she will receive political messages concerning that issue. Or to put it in a business news context, the greater a person's level of "organizational awareness," the more likely it will be that he or she receives messages about organizations.

The application of the two-stage cognitive elaboration-then-tedium process of Cacioppo and Petty (1979) and the reception axiom from Zaller's model in a context of the effects of mere exposure to business news lead to a "two-stage familiarity-then-something wrong model". This model postulates that the greater a person's level of cognitive engagement with organizations – organizational awareness – the more likely it will be that the person receives business news. Moreover, it is assumed that the organizational level of awareness of television viewers is lower than the level of organizational awareness of newspaper readers. It is expected there will be an inverted U-shaped relationship between the amount of news and the reputation of an organization.

In the first stage of the model, more news about the organization increases familiarity with the organization. More familiar companies will be rated better (Fombrun & Van Riel, 2004). It is expected that television viewers will remain in the first stage of the model. This is due the lower level of organizational awareness of television viewers, which results in a lower exposure to the amount of television news. This leads to the following hypothesis:

Hypothesis 1a: The amount of television news about a certain organization is positively related to its reputation.

Newspaper readers, on the contrary, possess a higher level of organizational awareness. Therefore it is assumed that newspaper readers will follow the two stages of the model. If there is relatively little print news about an organization the reputation of the organization will be improved, since familiarity results in a better reputation. If the amount of print news reaches a certain level, the respondents will start to think there is something wrong with the organization concerned and the reputation of the organization will deteriorate. This leads to the following hypothesis:

Hypothesis 1b: If the amount of print news about an organization is low to medium, there will be a positive relationship between the amount of print news and the reputation of the organization. If the amount of print news about an organization goes beyond a certain point, the relationship between the amount of news and the reputation of the organization will be negative.

This model may explain the results found by Fombrun and Shanley (1990). The high amount of print news about the focal organizations resulted in a negative reaction by the newspaper readers. It is likely that the respondents (directors who rated only firms in their own industry) possessed a high level of organization awareness, and that they remembered there was something wrong with the organization in question. Since Wartick (1992) investigated both the amount of television news and the amount of print news without determining their separate effects, it is not clear if this model can also explain his results.

2.3   THREE TYPES OF NEWS

From the empirical studies discussed in subsection 2.1, it appeared that the valence of news does not seem to have a systematic effect on reputation. This fits
with findings in the field of political communication. Shah, Watts, Domke, and
Fan (2002) concluded that a valence model of media effects could not adequately
explain Clinton's job approval ratings. This study will be elaborated upon in subsection 2.5, about the effects of support & criticism news on corporate reputation.

A possible explanation for the non-systematic effects of the tone of news on
reputation could be that the coders were asked to determine – from the perspective of the reader – if an article would be positive or negative for the organization
in question. This can be exemplified best by Verčič's (2000) dissertation, as he
explained his coding instruction most explicitly. Verčič reported there were 36
stories about Shell and the increasing prices of oil and gas, articles originally classified as negative. However, Verčič argued that fluctuation in prices in oil industry can also be seen as "business as usual," although it will be perceived as "negative" news for the general audience. For this reason, Verčič conducted two types
of analyses, one that included a category on the increasing prices of oil and gas,
and one that did not. He also explicitly addressed this problem, stating that it is a
very strong assumption to assume the same valence of every report for every
member of a population. It is assumed in his study that the news a company is
laying off employees is "bad" news with a negative valence. According to Verčič,
this is a very strong assumption because financial markets often reward companies for lowering the number of employees, which shows that for some members
of a population the lay-offs are good news.

The present study does indeed postulate that an article can have a different effect on various stakeholders. It can be argued that environmental activists will interpret increasing prices of oil as a positive message: hopefully people will avoid
their cars and use their bikes instead. The fanatic driver will probably consider it
to be bad news. To prevent problems like this, the coders should be instructed to
code the *manifest* content of the text (oil prices are increasing) and not how they
think the public will perceive the news. Moreover, much information from the articles gets lost in the studies described earlier. If they investigate only whether the
amount of media coverage and the tone of news have an effect on reputation, it
remains unclear what other actors and issues the organization in question was
connected with in the news. Since a great deal of the social psychological theories
used in communication science are situated on this *relational* level – such as the
balance theory (Heider, 1958), cognitive dissonance theory (Festinger, 1957), or
the theory of reasoned action (Fishbein & Ajzen, 1980) – content analysis should
also be situated on a relational level and a distinction should be made between
various types of news.

The three types of news distinguished in this study are success & failure news, support & criticism news, and issue news. The various types of news are based on the assumption that news contains information about *actors* (such as organizations and the government) and/or *issues* (such as the environment and employment). If success is attributed to an actor (to a company, for example), this type of news is called success & failure news. If there are two actors in the news, this is called support & criticism news. If there is an issue in the news, this is called issue news.

The first sentence in Table 2.1 belongs to the success & failure news category, since failure ("took a beating") is attributed to an actor, in this case Shell. In the second sentence "BP and Amoco are joining forces," two actors (BP and Amoco) are supporting each other. Therefore, this type of news is called support & criticism news. The sentence "Mamma's boy leads BP Amoco" is a special type of support & criticism news. Something unkind is being said about John Browne, Chief Executive Officer of BP Amoco. Since the media do not state who made this statement, it is attributed to the media. In other words, the head of BP is being criticized by the media.[10]

Issue news can take different forms. An organization can express its attitude towards an issue, such as in the sentence "Shell to plant forests" (see Table 2.1).

TABLE 2.1.   **Three types of news**

Type of news	*Example of a headline*
Success & failure news	
Success or failure of an actor	Shell took a beating
Support & criticism news	
Two actors in the news	BP and Amoco are joining forces
Media evaluation of an actor	Mamma's boy leads BP Amoco
Issue news	
Attitude of organization towards issue	Shell to plant forests
Decrease or increase of an issue	Lower profits for Shell

In addition, factual information (or information that is presented as factual) about the decrease or increase of an issue can be given in a sentence like "Lower profits for Shell." Complex sentences may contain various types of news, as will be elaborated upon in Chapter 3. Additional background information about the three types of news will be given in the next subsection.

The study of different types of news is relatively new to the field of business communication. In the field of political communication, the emphasis has been on issue news during the past 20 years of empirical research on the agenda-setting process (McCombs, 1992). McCombs (1992) was surprised there was so little study of the effects of other types of news in the field of political communication.

McCombs and Shaw's (1972) widely cited article already mentioned that a considerable amount of campaign and other news was devoted *not* to discussion of the major political issues, but rather to analysis of the campaign itself. Although this study focuses on business news, it is likely that news about issues will not be the only type of news. Therefore, two other types of news – support & criticism news, and success & failure news – were studied as well.

The next subsections will further elaborate on the three different types of news. The concept success & failure news (or "horse race news") is derived from research in the field of political communication.[11] The term horse race news is used to indicate that during elections, the press is mostly concerned with who is ahead in the opinion polls ("Democratic party leads"). In the case of success & failure news there is always only one actor involved.

Kleinnijenhuis, Oegema, De Ridder, Van Hoof, and Vliegenthart (2003) found that during political campaigns in the Netherlands,[12] 19% of the news in the largest five newspapers and on three television channels was dominated by news about success & failure in 1994; this percentage was 16% in 1998, and 20% in 2002.

The concept support & criticism news (or *conflict frame*) is derived from studies in the field of political communication, like the concept of success & failure news. In political communication, the conflict frame refers to disagreement and conflict between different actors, including individuals, groups, institutions, and countries (Neuman, Just & Crigler, 1992; Price, Tewksbury & Powers, 1997; Semetko & Valkenburg, 2000; Shah et al., 2002; De Vreese, 2003). In the Dutch elections of 1994, 1998, and 2002, roughly half the campaign news was dominated by this type of news (Kleinnijenhuis et al., 2003). In the present study, the term support & criticism news is preferred over the term "conflict news," since in the case of business news, companies don't just compete with each other, they also cooperate.

In issue news, the organization is tied to a certain topic in the media coverage. The "issue" concept is used not only in the field of political communication (for example, in the studies of the agenda-setting hypothesis that will be discussed later), but also in the field of organizational and management theory. The term *issues management* received its name in 1976 from a veteran corporate public relations officer, W. Howard Chase (Heath & Nelson, 1986). As mentioned by Wartick and Heugens (2003), the term had already been widely adopted among academics by the following decade, for example, in books such as *Management Response to Public Issues* (Bucholz, Evans, & Wagley, 1985) and *Managing Public Relations* (Gruning & Hunt, 1984). Dutton and Dukerich (1991) described issues as events, developments, and trends that employees collectively recognize as being of some consequence for the organization. Since this book is about the effects of media coverage, news about issues is described as events, developments, and trends that the media (rather than the employees) recognize as being of some consequence to the organization.

This division into three types of news is often employed by Kleinnijenhuis and his colleagues to predict the effects of media coverage about politicians and political parties on voting behavior in the Netherlands (Kleinnijenhuis, Oegema, De Ridder, Van Hoof, & Vliegenthart, 2003; Kleinnijenhuis, Oegema, De Ridder, & Ruigrok, 1998; Kleinnijenhuis, De Ridder, Oegema, & Bos, 1995).

In the next subsection, the effects of these different types of news on reputation will be discussed from different theoretical perspectives. Since theories about the effects of business news on reputation are rare, theories are used that originate in the field of political science. The field of political communication has a long tradition of investigating the content of media coverage and of predicting its effects on public opinion. When applicable, it will be mentioned whether a particular theory is also used in marketing research, which is one of the academic disciplines that has contributed to theory-building in the field of corporate reputation (Fombrun & Van Riel, 1997).

2.4 THE EFFECTS OF SUCCESS & FAILURE NEWS ON CORPORATE REPUTATION

As hypothesized by several authors (Fombrun & Shanley, 1990; Wartick, 1992; Carroll & McCombs, 2003), it is likely that media coverage with a positive tone will improve corporate reputation. This means that in the case of success & failure news, the reputation of an organization will improve if success is attributed to the organization in the media (people want to support the winner). If the organization is in the news with a failure, its reputation will deteriorate. Lazarsfeld et al. (1944) call this the *bandwagon effect*.

To put it in other words, the bandwagon effect can also be described as "everybody loves a winner." In his introduction, Rohlfs (2001) writes, "A bandwagon seems to mean what sports commentators call "momentum" – that is, a general tendency of success to breed further success." Lazarsfeld et al. (1944) argued that campaign managers of political candidates take full advantage of the bandwagon effect by suggesting in the media that everyone is supporting one candidate. For example, in sentences like, "I have heard *a number of Democrats* on radio voting for Willkie – Irvin Cobb, John L. Lewis. I therefore feel that Willkie will get it, with such backing" and "I heard over the radio the other night that one of the *largest betting houses* in New York was betting 9 to 5 on Roosevelt and when they give those odds, there is really something to it" (Lazarsfeld et al., 1944, p. 109). Moreover, Lazarsfeld, Berelson, and Gaudet remarked that public opinion polls are mentioned by respondents as a source of change in expectation. Since the focus of the news is on who wins or loses, news about success & failure is also called horse race news.

Marsh and O'Brien (1989) conducted an experiment in which they studied opinion bandwagons in attitudes on the Common Market. Respondents who were told that public opinion was moving towards staying in the Common Market were significantly more in favor of staying in the market than respondents who were told that public opinion was moving towards getting out of the Common Market. Patterson (1993) demonstrated that when a political candidate's support in the polls increases sharply, the news of his candidacy becomes more favorable. This means that in his use of the bandwagon effect, not only "the public" was influenced by the opinion polls, but the media as well.

Leibenstein (1950) put the bandwagon effect into an economic perspective. He described the bandwagon effect as the case where an individual will demand more (less) of a product at a given price because some or all other individuals in the market are also demanding more (less) of the product.[13] Biddle (1991) modeled a product demand curve with a bandwagon effect using data on sales of personalized license plates. Although he found some positive evidence of a bandwagon effect, he could not rule out that this effect could also be explained by another model, the product diffusion model. In the diffusion model, the demand for a new product grows as information about the new product spreads (Biddle, 1991).

Fombrun (1996) described several horse races and explained how they work in business. When products or companies get top ratings in contests, they are generally awarded a prize or other award that confirms the company's reputation. Winners get more favorable media coverage, increasing their visibility and, indirectly, their ratings by stakeholders. Fombrun mentioned as examples the annual rankings of admired companies by *Fortune* magazine, and the rankings of the best business schools by the magazine *Business Week*. The *Fortune* rankings of admired companies are widely used and described in corporate reputation research (Deephouse, 2000; Fombrun, Gardberg, & Sever, 2000; Fombrun & Shanley, 1990; Fryxell & Wang, 1994; Jones et al., 2000; Van Riel, 1997; Wartick, 1992).

Recently, Rohlfs (2001) focused on bandwagon effects in high-technology industries. Rohlfs illustrated the bandwagon effect with the following case of videocassette recorders (VCRs). In 1975, Sony released its VCR with a technical standard (Beta). Masushita, its competitor, brought out its own VCR with an incompatible VHS format the following year. The Beta format had superior recording quality, while the VHS format had longer playing times. It should be noted that the original intended use of the VCR was solely to record television programming and watch it at a different time ("time shifting"). The sales lead achieved by VHS in the second and third years would not have been enough to beat Beta if time shifting had remained the main reason to buy a VCR. At that time, however, an independent supply of prerecorded videocassettes appeared spontaneously. The sales lead of VHS was just enough to lead video stores – who wanted to offer their clients the largest possible selection of movies – to carry more VHS than Beta titles. This in turn led to more people buying VHS machines. By the late 1980s, Beta had disappeared entirely from the home market.

Negative success & failure news about a company (when failure is attributed to the company) can also improve the reputation of the company concerned. This is what is called the *underdog effect* in political party preference studies: if voters see that a party is losing in the opinion polls, they will vote for that party out of compassion. Marsh (1984) described that the thesis in bandwagon literature is that people decide how to vote by conforming (bandwagon) or going against (underdog) the perceived majority opinion. She remarked that literature has concentrated on bandwagon effects, while underdog effects were often neglected. Some literature does study the underdog effect. In two rather similar experiments, artificial elections were constructed in which a series of votes were taken (Fleitas, 1971; Laponce, 1966). Both these experiments suggest support for the underdog effect. However, as suggested by Marsh (1984), the experiments were unlike real-life situations because there was only minimal information available – the voter knew nothing about the candidates except for some brief background commentary. Based on figures on the wins and losses of political parties, Irwin and Van Holsteyn (1988) argued that an underdog effect could have been present in the 1986 elections in the Netherlands. In two studies that will be discussed underneath, both bandwagon and underdog effects showed up.

Mutz (1995a) studied the effects of media coverage on individual donations received by Democratic presidential candidates. Her time series analysis of contributor behavior suggests that some contributors are motivated to donate by coverage suggesting their strongly favored candidate is losing ground, while other candidacies benefit from coverage suggesting increased viability. The bandwagon effect was stronger than the underdog effect. The Democratic presidential candidates portrayed as losing ground gained around 20 or 30 additional donations, while the candidates portrayed as successful and who received an equal amount of news gained one hundred or more donations. For a more or less similar study with Republican presidential candidates, see Mutz (1995b).

Kleinnijenhuis and De Ridder (1997) found an underdog and a bandwagon effect in their analysis of the 1994 election campaign for the Dutch parliament (see also Kleinnijenhuis 1998). The news that the Christian Democrats (CDA) would lose the elections dramatically made politically sophisticated voters decide to vote for the CDA. This underdog effect did not counterbalance the bandwagon effect, however (Kleinnijenhuis, 1998). The more the party of the economically conservative Liberals (VVD) was in the news with its success in the opinion polls, the more votes it gained (bandwagon effect). A bandwagon effect was found as well in the analysis of the 2002 election campaign for the Dutch parliament (Kleinnijenhuis et al., 2003). The present study expects that success & failure news about corporations and sectors will lead mainly to a bandwagon effect, like success & failure news in the election campaigns.

Hypothesis 2a: The tone of success & failure news is positively related to reputation.
Hypothesis 2b: The tone of success & failure news is positively related to the perception of
the success of the policy of the organization.

2.5 THE EFFECTS OF SUPPORT & CRITICISM NEWS ON CORPORATE REPUTATION

As indicated in the previous subsection, several authors (Fombrun & Shanley,
1990; Wartick, 1992; Carroll & McCombs, 2003) hypothesized that the more positive the tone of the news, the better for corporate reputation. If this hypothesis is
applied to support & criticism news, it would mean that the reputation of an organization will get better the more it is supported in the news by another actor, and
get worse the more it is criticized by another actor (effect regardless of the source).

However, the congruity principle of Osgood and Tannenbaum (1955) makes
clear that taking the source of criticism into account deserves attention. The congruity principle of Osgood and Tannenbaum (1955) is a special case of Heider's
balance theory (1946; 1958). The balance theory is a theory of cognitive consistency developed around the idea that people strive for consistency between their
attitudes, beliefs, values, and behaviors. This striving for consistency has a social
aspect, because it is often others' perceptions that puts pressure on people to be
consistent. Inconsistency (an unbalanced state or incongruity) among attitudinal
elements creates "tension," which forms a motivational force for cognitive change
towards a balanced state (congruity).

The congruity theory of Osgood and Tannenbaum (1955) deals specifically with
the attitude of a person towards a source of information and the object of the
source's assertions. The point is how relationships among these three entities are
organized in the mind of an individual (p). In the congruity paradigm, a person
(p) is confronted with an assertion from a source (s), towards which he or she has
an attitude about an object (o), towards which he or she has an attitude. In contrast to the balance theory, the congruity theory is able to predict changes in p's attitude towards o and s.

One example of congruity and two examples of incongruity are shown in Figure 2.1. In the first triangle, the newspaper reader evaluates Greenpeace positively and BP negatively. If Greenpeace criticizes BP (–) in the news for its environmental policy, a stable state of congruity is said to exist (see Triangle A of Figure
2.1). If the newspaper reader evaluates Greenpeace positively and BP negatively,
and given that Greenpeace (+) supports BP (–) in the news for its environmental
policy, incongruity is said to exist (see Triangle B). Given that Greenpeace and BP
are both evaluated negatively by the newspaper reader (see Triangle C), and given
that Greenpeace (–) criticizes BP (–), incongruity is said to exist as well.

FIGURE 2.1. **An example of congruity (A) and two examples of incongruity (B, C)**

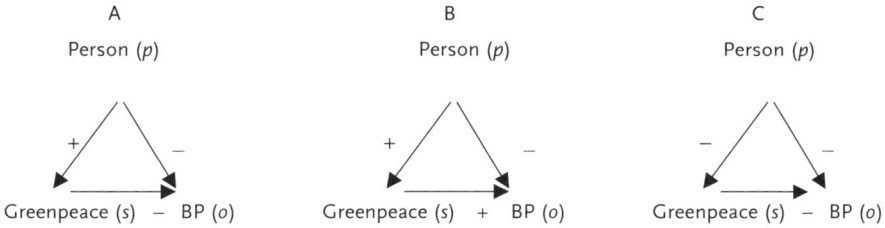

Incongruity is present when the attitudes towards the source and the object are similar and the assertion is negative, or when they are dissimilar and the assertion is positive. It is assumed that in the case of incongruity, both the attitude towards the object and the attitude towards the source change. The direction and magnitude of attitude change can be illustrated with the second image of Figure 2, the criticism of Greenpeace on BP.

FIGURE 2.2. **An incongruity model (Osgood and Tannenbaum, 1955; Zajonc, 1960)**

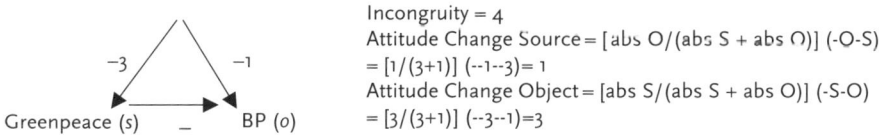

Incongruity = 4
Attitude Change Source = [abs O/(abs S + abs O)] (-O-S)
= [1/(3+1)] (--1--3) = 1
Attitude Change Object = [abs S/(abs S + abs O)] (-S-O)
= [3/(3+1)] (--3--1) = 3

A scale running from −3 to 0 to +3 is used to operationalize the intensity (as well as sign) of the person's attitudes. The attitudes of the person towards Greenpeace and towards BP are not equally polarized. The person holds a more extreme attitude towards Greenpeace (−3) than towards BP (−1). The criticism by Greenpeace of BP for its environmental policy creates an incongruent situation. In this example, incongruity equals 4. The person's attitude towards Greenpeace would have to shift 1 unit (from −3 to −2), and the person's attitude towards BP would have to shift 3 units (from −1 to +2) for congruity to be restored. This means that the person's attitude towards BP is improved by the criticism of Greenpeace. Close examination of the model's change equations reveals Osgood and Tannenbaum's assumption that highly polarized judgments are more resistant to change. The person's attitude towards Greenpeace will shift only 1 unit, while the person's attitude towards BP will shift 3 units.

The congruity principle has not fared particularly well in terms of the accuracy of its attitude change predictions (see Kerrick, 1958; Rokeach & Rothman, 1965; Tannenbaum, 1968). Eagly and Chaiken (1976) found that most subjects' common reaction to experimental materials suggesting that a well-liked source supported a disliked attitude object is no attitude change at all: the subjects did not believe that the source really made the statement, or thought that the source was misquoted. Osgood and Tannenbaum (1955) called this denial of an assertion the "incredulity reaction." Since the present study uses a real-life situation (with "real" news), it is unlikely that the incredulity reaction will occur as much as in experimental situations where highly incongruent situations were created (such as "Eisenhower praises communism").

Nevertheless, incongruity theory demonstrates that it is not only important to know if the organization was supported or criticized in the media, but also to take into account the reputation of the actor which supported or criticized the organization.

In line with the incongruity theory, several recent political communication studies (Kleinnijenhuis & De Ridder, 1997; Kleinnijenhuis et al., 1998; Shah et al., 2002; Kleinnijenhuis et al., 2003) found a *negative* effect from support & criticism news (or conflict frames) on the popularity of the political party or president at whom the criticism was being directed. In other words, *criticism* in the news *improved* the position of the besieged parties.

Kleinnijenhuis & De Ridder (1997) examined the effects of strategy-framed news (success & failure news and support & criticism news) on political party preferences during the Dutch election campaign of 1994, and in agreement with their hypothesis, they found a *boomerang effect*. The more the liberal political party VVD was *criticized* by other political parties, the more public support it received from newspaper readers. Boomerang effects were also found in the next studies of the Dutch general elections. Kleinnijenhuis et al. (1998) examined the effects of success & failure news, support & criticism news, and issue news on political party preferences during the election campaign of 1998. The results showed that the more the liberal political party VVD and the left-wing party GroenLinks were *criticized* by other political parties, the better they both did. Criticism of opponents benefits the besieged party because it contributes to its profile (Kleinnijenhuis et al., 1998). In the analysis of the 2002 election campaign for the Dutch parliament, a boomerang effect was present as well. The more the controversial populist politician Pim Fortuyn was criticized in the media by the other political parties, the larger his following became (Kleinnijenhuis et al., 2003).

Shah et al. (2002) studied the effects of different types of frames on president Bill Clinton's job approval ratings during the period of the Monica Lewinsky affair. The results of their study showed that citizens were more *positive* about Clinton the more he was *attacked* by the conservative elites ("the conservative attack frame"), even though the media content was overwhelmingly negative about Clinton. Moreover, news coverage that questioned the motives of the Republicans for attacking Clinton ("the liberal response frame") had a positive influence on Clinton's job approval ratings as well. Shah et al. (2002) concluded that these two frames reinforce each other.

In agreement with these results, the present study expects that newspaper readers and television viewers will react suspiciously if an organization is criticized by its competitors or supported by its friends. Moreover, the present study postulates that the reputation of an organization will worsen the more it is criticized by a credible actor.

Hypothesis 3a: The tone of support & criticism news will be negatively related to reputation if its competitors criticize the organization. The tone of support & criticism news will be positively related to reputation if impartial actors criticize the organization.

Hypothesis 3b: The tone of support & criticism news will be negatively related to the agreement with the policy of the organization or sector if its competitors criticize the organization. The tone of support & criticism news will be positively related to the agreement with the policy of the organization if impartial actors criticize the organization.

2.6 THE EFFECTS OF ISSUE NEWS ON CORPORATE REPUTATION

The effects of issue news on corporate reputation will be viewed from three theoretical perspectives: *agenda-setting, priming,* and *issue ownership.* These three perspectives will be used and can be distinguished as follows. In the agenda-setting hypothesis, the relationship between issue news and the *salience* of an issue or the salience of an aspect of an issue is studied. Or to put it in a business news context, in the case of agenda-setting, what is studied is whether people think more about an organization the more that organization is in the news. Agenda-setting does *not* predict how people will *evaluate* the organization.

Priming and issue ownership are consequences of agenda-setting. With both theories, the focus is on the effects of issue news on the *overall evaluation* (the reputation) of the organization. The priming hypothesis postulates that a certain issue will be *more important* in the formation of the overall evaluation of the organization the more that issue is in the news. It does not predict whether this will result in a more negative or more positive evaluation of the organization that depends on the issue concerned. The issue-ownership hypothesis, on the other hand, postulates that the reputation of an organization will be *improved* if the organization is in the news with issues on which they are advantaged and their competitors are less well regarded. In other words, in the case of agenda-setting, the dependent variable is *corporate image* (the aspects people think of), while in the case of priming and issue ownership, *corporate reputation* (the evaluation of the organization) is the dependent variable (see the conceptual model, Figure 2.4). In the next subsections, agenda-setting, priming and issue-ownership theory will be elaborated upon.

2.6.1 Agenda-setting

As described earlier in Chapter 1, the oil spill caused by the supertanker Exxon Valdez was used by Dearing and Rogers (1996) to illustrate agenda-setting. Dearing and Rogers stated that prior to the 1989 Exxon Valdez incident, several indicators suggested that a gradual worsening of the global environment was taking place. Although scientific conferences were held and the head of state discussed national environmental problems, the environment was not on the national agenda in the United States. It was the massive media attention given to the Exxon Valdez incident that convinced the U.S. public that the environment was an important issue facing the nation.

Agenda-setting research has shown that the media determine the importance of issues, or to use Cohen's famous statement (1963, p. 13), "...the press is significantly more than a purveyor of information and opinion. It may not be successful much of the time in telling people *what* [italics added] to think, but it is stunningly successful in telling its readers what to think *about*." In their pioneering work "The agenda-setting function of mass media," McCombs and Shaw (1972) matched what Chapel Hill voters said were key issues of a political campaign with the actual content of the mass media used by them during the campaign. They found a high correlation between the campaign issues carried by the media and the voters' independent judgments of what were the important issues. After McCombs and Shaw's publication, the agenda-setting hypothesis was investigated widely, and more than 300 empirical studies focused on the public agenda composed of issues in the media (Dearing & Rogers, 1996). Iyengar and Kinder (1987) tested the agenda-setting hypothesis with six experiments, and concluded that their evidence decisively sustains the agenda-setting hypothesis. Dearing and Rogers (1996) concluded on the basis of their review of 112 empirical studies that 60% of the studies support the hypothesis that the position of an issue on the media agenda is important in determining that issue's salience on the public agenda. As mentioned by Dearing and Rogers, most of the studies were cross-sectional. McQuail (1994; 2000) stressed the importance of longitudinal research. He argued that in order to prove there is a causal connection between the agenda of the media and the public agenda, content analysis is needed plus evidence of opinion change over time in a given section of the public (preferably with panel data). In addition, an indication is needed of relevant media use by the public concerned. In Trumbo's longitudinal investigation (1995), the agenda-setting process was applied to the issue of global warming. The results showed there is a simultaneous relationship between television attention and concern about the environment.

McCombs, Llamas, Lopez-Escobar, and Rey (1997) distinguished between the *first level* and the *second level* of agenda-setting. The first level of agenda-setting deals with the salience of an object or an actor, while the second level of agenda-setting deals with the *attributes* of the object or the actor concerned (Carroll &

McCombs, 2003). The first and second levels of agenda-setting can be illustrated as follows. If there is a lot of news about the environment and respondents mention the environment as one of the most important problems facing the world today, this indicates a first level agenda-setting effect. If Shell is repeatedly in the news with its pollution of the environment and respondents associate the actor "Shell" with the attribute "environment," a second level agenda-setting effect is said to be present. Or to put it simply: in the case of the first level of agenda-setting, only one actor or issue is involved, whereas in the case of the second level of agenda-setting, *the attributes* of that actor or issue are involved as well.

McCombs et al. (1997; 2000) tested the second level of agenda-setting in a Spanish General Elections study. McCombs et al. (1997; 2000) studied the images of three political candidates (the objects) and their various traits (the attributes) that defined their images. They distinguished two dimensions of the attribute agenda: a substantive (or cognitive) and an evaluative (or affective) dimension. Some examples of the substantive aspects of the candidates' images were: issue positions and ideology, perceived qualifications, and integrity. The affective aspects of the attribute agenda captured the tone (positive, negative, or neutral) of the substantive statements. Content analysis of newspapers, television news, and political advertising slots was used to determine the candidate images in the mass media. Five days after the election, a survey was conducted among 363 citizens of Pamplona. In 15 of the 21 tests (7 media x 3 candidates), there appeared to be a significant positive correlation between the media agenda and the public agenda. In other words, the hypothesis that the salience of attributes on the media agenda influences the salience of attributes on the public agenda was confirmed.

The study of Hertog and Fan (1995) can also be classified as a second level agenda study. Hertog and Fan showed that newspaper coverage and news magazine coverage influenced public beliefs concerning the likelihood of HIV transmission through toilets, sneezing, and insects. They analyzed media coverage from 1987 to 1991; results from 23 surveys were available for modeling the time series of each of the three beliefs.

Carroll and McCombs (2003) applied the theoretical framework of agenda-setting in the field of business communication. They stated that although agenda-setting effects were studied in political communication settings, it is rational to assume that the central theoretical idea – the transfer of salient issues from the media agenda to the public agenda – fits equally well in the field of business communication. Carroll (2004) investigated this idea empirically. He tested several hypotheses, including the first and the second levels of agenda-setting. The hypothesis that was designed to test the first level of agenda-setting predicted that more media coverage about a firm would result in a higher degree of public awareness of that firm. The second level of the agenda-setting hypothesis proposed a correlation between the amount of news coverage devoted to particular attributes of the firms and the proportion of the public describing the firm in terms of these attributes. Carroll analyzed the business news in the New York Times, which served as a proxy for a national newspaper. Corporate reputation data were gathered from the poll results from the Annual Reputation Quotient^{SM} study. The Reputation Quotient^{SM} is a well-known measure of reputation (see Fombrun, Gardberg, and Sever, 2000 for a description of the development of this measure). Carroll's study included the U.S. firms that were part of the 1999 and 2000 surveys.

Carroll (2004) found support for the first level of the agenda-setting hypothesis. After controlling for the previous year's public agenda, it emerged that the media influences which firms are thought about in the first place. The correspondence between the amount of media coverage devoted to executive performance and workplace environment and the use of these attributes by the respondents provided support for the second level of the agenda-setting hypothesis. No relationship was found between media coverage and reputation descriptors for the attributes of financial performance, products and services, or social responsibility.

The second level agenda-setting hypothesis will also be tested in the present study.

Hypothesis 4: The amount of media coverage devoted to particular issues is positively related to the proportion of the public that defines the organization by those issues.

On the basis of this hypothesis, for example, it is expected that if there is more media coverage about the Dutch Railways (NS) and the environment, the respondents will associate the NS more with "the environment." The second level agenda-setting hypothesis does not predict whether corporate reputation will improve or deteriorate if the organization is associated more with a certain issue. It should be noted that this hypothesis relates to the substantive dimension of the attribute agenda only, and the evaluation of the issues was not taken into account.

2.6.2 Priming

As was outlined in the previous subsection, the second level of agenda-setting hypothesis predicts that the more the attributes of a firm are in the news, the more people will associate an organization with a certain attribute (such as the environment or safety). However, the second level of agenda-setting hypothesis does not predict how this affects the overall evaluation (in terms of good or bad) – in other words, the reputation – of the company. The priming hypothesis, on the contrary, does focus on the effects of issue news on its importance in the formation of the *reputation* of the company.

Iyengar and Kinder (1987) have extended the agenda-setting theory with a theory of priming. It is a matter of priming if the public judges political or other actors by using criteria based on matters in the news. According to the priming hypothesis, if television news becomes preoccupied with the environment, then citizens would *evaluate* Shell primarily by its environmental policy. Iyengar and Kinder (1987) drew upon ideas developed within the information-processing perspective in psychology. Their point of departure was Simon's (1979) observation that people have a limited cognitive capacity and do not pay attention to everything. "Satisficing" – aiming at the good when the best is incalculable – is the solution. A second point of departure by Iyengar and Kinder was that rather than undertaking exhaustive analysis, people ordinarily prefer heuristics – intuitive shortcuts and simple rules of thumb. One such rule of thumb is reliance upon information that is most accessible.

Iyengar and Kinder conducted three experiments. In the first experiment, participants viewed newscasts that either emphasized inadequacies in U.S. defense preparedness or did not. In the second experiment, one group of participants watched newscasts emphasizing defense while another group watched newscasts dotted with stories about inflation. In the third experiment, participants watched newscasts that either paid special attention to unemployment or did not. On the final day of each experiment, respondents were asked to rate the president's performance with regard to various problems, including "maintaining a strong defense" (experiments 1 and 2), "reducing inflation" (experiment 2), and "keeping employment down" (experiment 9). Participants were also requested to evaluate the president's general performance. The results were in line with the priming hypotheses, and viewers who were shown stories about a particular problem gave more weight to the president's performance on that problem when evaluating the president's overall performance.[14] Furthermore, Iyengar and Kinder investigated who was more or less vulnerable to priming. Their findings made it clear that priming was not related in any consistent way to education: sometimes viewers with little formal education were primed a bit more by television news coverage, and sometimes they were primed a bit less.

More recently, Domke (2001) conducted an experiment in which the news discourse about crime was systematically altered – either including or excluding racial cues – to examine how individuals use issue information to form political judgments. One of his findings was that participants presented with news coverage of crime imbued with racial cues drew on such considerations in their thoughts about crime. Other recent priming research also indicates that mass media emphasis on particular political issues increases the salience of issues (Domke, Shah, & Wackman, 1998), which then shape the criteria that are applied while forming judgments about political actors (Goidel, Shields, & Peffley, 1997; Krosnick & Kinder, 1990; Pan & Kosicki, 1997). Furthermore, priming is used as a theoretical perspective to study the effects of depictions of aggression in the mass media on the actions of people. For an overview of this type of research, see also Jo and Berkowitz (1994).

Priming is used not only in political research on the effects of mass media on the judgment of political actors, but also in consumer research. This is because of the theoretical foundation of priming, which can be traced back to psychological concepts of priming in work on cognitive processing of semantic information. In semantic priming studies (Collins & Loftus, 1975; McKoon & Ratcliff, 1995; McNamara, 1992), subjects decided whether an item such as "pear" was a word or non-word and responded more quickly and accurately when the item was preceded by an associated word such as "apple." Two more kinds of priming studies can be distinguished (Mandel & Johnson, 2002): categorical priming[15] and feature priming. Feature priming comes closest to the priming concept as used by Iyengar and Kinder. In feature priming, a subject is exposed to a prime that is associated with a particular feature, and this feature is then weighted more heavily in evaluation. Mandel and Johnson (2002), for example, manipulated the background pictures of a web page in a way that affected consumer choice. If the background of a car website was green with dollars (to prime price), subjects gave on average more points to a car that was cheaper and less safe than did subjects who were primed on safety.

In this study on the effects of media coverage on reputation, the priming hypothesis is formulated as follows:

Hypothesis 5: If respondents are asked to give an overall evaluation of the organization, they will weigh aspects/beliefs more heavily in their overall evaluation the more these aspects have been in the news.

Lets take the example of the NS and the environment again in order to illustrate the difference with the previous two hypotheses. It follows on from the priming hypothesis that the more the NS is in the news along with the issue of the environment, the more important this issue will become in the *overall evaluation* of the NS. The priming hypothesis does not indicate whether the reputation of the organization will improve or deteriorate.

2.6.3 Issue ownership

As was indicated in the previous subsection, the priming hypothesis postulates that a certain issue will become more important to the overall evaluation of a company the more that issue is in the news. It does not focus on the direction of change in reputation. Issue-ownership theory predicts whether the issue news about an organization will improve or worsen the reputation of the organization concerned.

In issue-ownership theory[16] (Budge & Farlie, 1983; Petrocik, 1996), reputations are crucial. The theory is used in political communication research, and views political party competition in terms of varying emphases on policy areas by the various political parties. In their public pronouncements, parties do not discuss each topic of public interest, contrasting their own detailed policies with those of their competitors. Instead, they devote most attention to the types of issues which favor themselves, and give correspondingly less attention to issues which favor their opponents (Budge & Farlie, 1983). If the budget deficit is a prominent issue in the news, voters will probably choose to vote for a right-wing party, because right-wing parties have a better reputation for controlling government expenditures than left-wing parties. Right-wing parties are considered to be the issue owner of cutbacks in expenditures. Budge and Farlie (1983) tried to predict the elections in ten countries using the issue-ownership model. They distinguished 14 broad types of issues (such as civil order, foreign relationships, and defense). For each issue, it was indicated whether the political parties were expected to gain or lose by its emergence. Then the elections in ten countries were predicted using issue-ownership theory and by using naïve methods of prediction, like the mean of past outcomes or the last result. A better forecast was made by using issue-ownership theory than by using a naïve method.

That issue ownership also seems to occur with regard to paid publicity can be illustrated by using the clothing firm Hennes & Mauritz (H&M) as an example. In the Dutch newspaper *de Volkskrant*, the advertising director of H&M stated that because of the limited size of the budget, the whole advertising budget is spent on billboards. That H&M succeeded in becoming issue owner of famous models in cheap women's underwear was made clear by the following quote from the advertising director: "Three years ago our competitor C&A started a campaign for women's underwear. It could have been one of our billboards. There was a great demand for bras that week – everybody thought it was our campaign" (Faber, 1996).

If issue-ownership theory is applied in a business news context, it is assumed that with regard to the issues in the news, the public will ask itself what organizations have the best reputations on these issues. If there is news about an issue the public perceives the organization to be handling successfully – an *owned issue* – the reputation of the organization will improve. However, the opposite is the case as well. If there is news about an issue the public regards the organization as being incapable of handling, a *poorly owned issue*, the reputation of the organization will worsen if there is news about that issue. The issue-ownership hypothesis can be formulated as follows:

Hypothesis 6: The more a company or sector is in the news with owned issues, the better its reputation will be.

Assume that NS has a positive reputation with regard to the issue of "the environment," since trains are less harmful than cars. In that case, the issue-ownership hypothesis would predict that the reputation of the NS will improve the more the environment is in the news. Note, however, that Shell may have a bad reputation with regard to the environment, which means Shell's reputation will get worse if there is more news about the environment.

2.7 THE EFFECTS OF PAID PUBLICITY ON CORPORATE REPUTATION

In addition to the effects of media coverage, it is assumed that paid publicity has an impact on corporate reputation. The effects of advertising intensity on corporate reputation were also examined in the study by Fombrun and Shanley (1990) described in section 2.1. The effects were controlled for the influences of media coverage and other antecedents of reputation. Fombrun and Shanley found that the higher a firm's advertising intensity, the better its reputation. This finding is in agreement with their expectation, which is based on the notion that firms develop strategic postures from advertising allocations (Fombrun & Ginsberg, 1990). In his summary of research on corporate reputation, Brown (1998) mentioned the studies by Winters (1986; 1988) as evidence that both product advertising and corporate advertising exert a positive influence on the reputation of oil company Chevron (Winters, 1986; 1988).

In the marketing field, the effects of advertising repetition are widely studied. Several authors (Berger & Mitchell, 1989; Nordhelm, 2002; Sawyer, 1981; Vakratsas & Ambler, 1999) mentioned mere exposure theories as one of the perspectives from which the effects of advertising repetition are examined. The findings of Fombrun and Shanley (1990) and Winters (1986; 1988) can also be explained from this perspective, which postulates that the mere repeated exposure of the individual to a stimulus object enhances his attitude towards it.

Many experimental studies reported or focused on an inverted U-shaped (nonmonotonically) relationship between ad exposure and ad effectiveness (Anand & Sternthal, 1990; Calder & Sternthal, 1980; Craig, Sternthal, & Leavitt, 1976; Nordhelm, 2002; Schumann, Petty, & Scott Clemons, 1990). Based on his meta-analysis of 208 experiments, Bornstein (1989) concluded that the frequency affect curve generally levels off after 10-20 stimulus presentations. This means that repeated exposure leads initially to liking, but ultimately leads to disliking or in other words to "advertising wearout."[17] As was indicated earlier in section 2.2, an explanation for this inverted U-shaped relationship between exposure and affect is the two-factor theory (Berlyne, 1970; Cacioppo & Petty, 1979).

According to Nordhelm (2002), the different relationships between the repeated exposures and affective response, the inverted U-shaped pattern and the monotonically increasing pattern, may reflect a calibration issue: the point at which affective response begins to decline has not yet been reached in the monotonically increasing pattern.

With reference to this point, it is important to note that many of the studies reporting an inverted U-shaped pattern (Anand & Sternthal, 1990; Craig et al., 1976; Nordhelm, 2002) are based on experiments. The question can be raised as to whether the inverted U-shaped relationship between ad exposure and ad effectiveness will also occur in a real-life situation. People have more opportunities of escaping from the ad in a real-life situation (i.e., by getting something to drink during a commercial break) than in a laboratory. It is highly unlikely in daily life that people will see an ad 25 times in 20 minutes,[18] as was the case in Nordhelm's (2002) experiment. Or to translate this into the concepts of the RAS-model: contrary to print news, the chance of reception of the ad may be relatively low, due to a generally low level of "ad awareness."

Classical conditioning is another approach frequently used to study the effects of advertising. Nord and Peter (1980) suggested that classical conditioning (or "respondent conditioning," as they call it) might alter consumer preferences. Classical conditioning occurs when a previously neutral stimulus, by being paired with an unconditioned stimulus, comes to elicit a response very similar to the response that was evoked by the unconditioned stimulus. For example, when a new product (the conditioned stimulus) of which people have no opinion is repeatedly advertised with exciting parties (the unconditioned stimulus), it is possible for the new product to eventually generate excitement on its own (the conditioned response) solely through the repeated pairing with exciting events. In the case of classical conditioning, the repeated pairing of the product with the pleasant stimulus leads to an increase in consumer preference for the product (Allen & Madden, 1985; Bierley, McSweeney & Vannieuwkerk, 1985; Kim, Allen & Kardes, 1996; Stuart, Shimp & Engle, 1987). In order to apply classical conditioning, the content of the ad or the commercial should be taken into account.

Since the present study has most in common with the study by Fombrun and Shanley (1990), mere exposure theory is used and a positive linear relationship between advertising and reputation is expected. The hypothesis is formulated as follows:

Hypothesis 7: Advertising intensity is positively related to reputation.

2.8 Overview of the hypotheses and conceptual model

The field in which the effects of media coverage on corporate reputation are studied is relatively young. Empirical studies are rare, and show mixed results. For this reason, political communication theories (such as agenda-setting, priming, and issue ownership) are used in the present study. The field of political communication has a long tradition of investigating the content of media coverage and of predicting its effects on public opinion.

The following hypotheses were formulated:

H. 1a *The amount of television news about a certain organization is positively related to its reputation.*

H. 1b *If the amount of print news about an organization is low to medium, there will be a positive relationship between the amount of print news and the reputation of the organization. If the amount of print news about an organization goes beyond a certain point, the relationship between the amount of news and the reputation of the organization will be negative.*

H. 2a *The tone of success & failure news is positively related to reputation.*

H. 2b *The tone of success & failure news is positively related to the perception of the success of the policy of the organization.*

H. 3a *The tone of support & criticism news will be negatively related to reputation if its competitors criticize the organization. The tone of support & criticism news will be positively related to reputation if impartial actors criticize the organization.*

H. 3b *The tone of support & criticism news will be negatively related to the agreement with the policy of the organization or sector if its competitors criticize the organization. The tone of support & criticism news will be positively related to the agreement with the policy of the organization if impartial actors criticize the organization.*

H. 4 *The amount of media coverage devoted to particular issues is positively related to the proportion of the public that defines the organization by those issues.*

H. 5 *If respondents are asked to give an overall evaluation of the organization, they will weigh aspects/beliefs more heavily in their overall evaluation the more the aspects have been in the news.*

H. 6 *The more a company or sector is in the news with owned issues, the better its reputation will be.*

H. 7 *Advertising intensity is positively related to reputation.*

Figure 2.3 shows the schematic reproduction of the hypotheses in a conceptual model. The media variables (the independent variables) are presented on the left side of the model. The public opinion variables (the dependent variables) are presented on the right side of the model. In the bottom section of the conceptual model, attention is given to the influence of the educational level of respondents.

FIGURE 2.3. **Conceptual model with hypotheses**

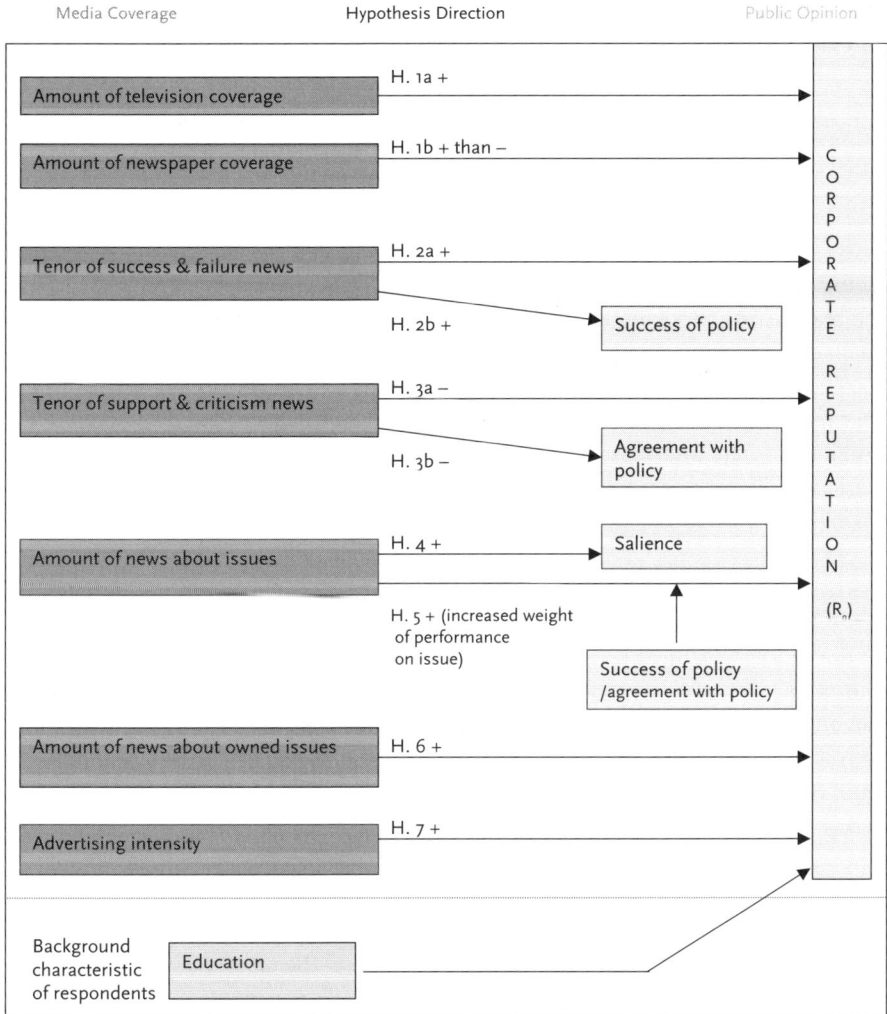

3
METHOD

In this chapter the research design will first be outlined. In the same subsection, the reasons for choosing the concerned companies and sectors for this study will be elaborated upon.

The type of content analysis used in this study will be addressed in section 3.2. The measurement of corporate reputation will be focused upon in section 3.3. Finally, the data analyses techniques that were either used in this study or that were considered for use will be described in section 3.4.

3.1 RESEARCH DESIGN

In the present study, three types of data were used: media data, advertising data, and public opinion (reputation) data. In order to gather the public opinion data, TNS NIPO (Dutch Institute for Public Opinion and Market Research), a large market research agency in the Netherlands, kindly conducted our survey among their *panel* of respondents. Since the media use of each of the respondents was known, it was possible to take the individual media use of the respondents into account when assigning the news and advertising data to the respondents. As far as we know, this has never been done in any other study on "media effect on reputation."

The Netherlands is a suitable country for testing the effects of media coverage in a real-life situation because subscriptions to daily newspapers are responsible for 85% of the total circulation here (Bakker, 2003). This means that a respondent's subscription(s) to a certain newspaper is a good predictor of the print news consumption of the respondent, since the impact of single issues is relatively small.

Media coverage about the focal companies in newspapers and on television was analyzed for the period from July 24, 1997 to July 22, 2000. This period was selected in agreement with the public opinion polling data: all the media coverage from the year preceding the poll was analyzed. The newspaper articles were selected from the five largest Dutch daily newspapers: *De Telegraaf, Algemeen Dagblad, NRC Handelsblad, Trouw,* and *de Volkskrant. De Telegraaf* is the biggest newspaper in the Netherlands. *NRC Handelsblad* and *de Volkskrant* are "quality" newspapers. *Algemeen Dagblad* and *Trouw* can be placed in the middle, between the quality newspapers and *De Telegraaf*. National newspapers were analyzed because it was assumed that national newspapers have the most impact on corporate reputation. A major part of the respondents in this study also read a regional newspaper (59%). It should be realized, however, that regional newspapers reach only part of the public (Kleinnijenhuis et al., 1995). Of the television news, the news broadcasts of the public broadcasting company NOS and the commercial channel RTL 4 were analyzed. In total, the media coverage on the focal organizations and sectors consisted of 9,285 newspaper articles and 2,225 news items, which together resulted in 15,883 assertions.

The advertising data were measured in advertising expenditures per focal company per medium. The data were obtained from the Dutch agency BBC. The media and the period in which advertising expenditure data were gathered were selected to match the media and the period that were selected for the content analysis. This means that the five largest Dutch daily newspapers were selected together with the public broadcasting company NOS and the commercial channel RTL 4. The period in which the advertising expenditures were measured was from July 24, 1997 to July 22, 2000.

The panel survey was conducted at three periods in time, during the summers of 1998, 1999, and 2000. For every period, the questionnaire consisted of the same questions except for the overall reputation measure, which was only asked in 1999 and 2000. The survey was designed by the Department of Communication Science at the Vrije Universiteit as part of the Media Monitor project. During each of the three years, 446 respondents[1] filled in the survey; 606 respondents took part once, and 306 respondents took part twice. TNS NIPO gathered the data by using the "NIPO Telepanel."[2] The NIPO Telepanel is a national representative panel made up of approximately one thousand households, which are provided with computers. The respondents received questionnaires via a modem. They could fill in the survey at a suitable moment, after which the answers were returned automatically to TNS NIPO. To prevent the respondents from getting tired of the questions, they only had to fill in the questions for six organizations (and not for all ten). Since the respondents formed a national representative panel, a weight factor was not applied. See subsection 3.3.5 for a description of the background characteristics of the respondents.

Selection of the focal companies.
Bearing in mind the generalizability of the results, we chose to examine a wide variety of companies from different industries. In addition, two sectors were taken into account. The industries focused upon were the oil industry, the banking industry, the retail trade food industry (supermarkets), the transport industry, and two professional sectors (see Table 3.1).

TABLE 3.1.   **The focal organizations per industry or per sector**

Industry or sector	Names and characteristics of the companies	
Oil industry	Shell: Dutch-British company, visible	BP: British company
Banking industry	ABN AMRO: visible, stock quoted	Rabobank: visible, not stock quoted
Supermarkets	Albert Heijn: market leader	Super de Boer: large supermarket chain
Transport	Dutch Railways (NS): monopolist	Dutch airport Schiphol: main airport
Sectors	Dutch police	Dutch agricultural sector

Two companies from the same industry were selected in order to make sure that, in general, they were coping with more or less similar economic developments, environmental problems, consumers, stakeholders, etc. In other words, the background characteristics of two companies from the same industry were roughly the same. Large organizations and professional sectors were focused upon in order to make sure there was enough media coverage about the companies to study its effects. In 2001, Shell, Rabobank, and ABN AMRO belonged to the 30 most visible companies in the Netherlands. Ahold, parent company of Albert Heijn, was the most visible company in the Netherlands in 2001 (see Fombrun & Van Riel, 2004). The police and the agricultural sector were selected because of the link with current research in our department. See for example the following studies about the police and police science: Huberts, 1998; Huberts, 2001; Lasthuizen, Huberts, and Kaptein, 2002. See Berendse (2003) for a study on the agricultural sector. In addition to the link to the current research in our department, it was decided to take sectors into account because it would be interesting to examine whether media coverage has the same impact on "sector reputation" as on "corporate reputation."

In addition to the similarities, there were also some differences between several pairs of companies that may have influenced the amount of media attention. Shell is a Dutch-British company, while BP is a British oil company. It is likely that Shell received more media attention than BP, because "distance" is a news selection factor (Van Cuilenburg, Scholten, & Noomen 1992). As was addressed earlier, it is rational to assume that the power of an organization is expressed in the size of the organization. Organizations with many employees and a quotation on the stock exchange will, in general, receive more media attention than smaller organizations. Therefore, it can be expected that more media attention will be paid to stock exchange-listed ABN AMRO than to Rabobank, which is not listed on the stock exchange. Moreover, it can be expected that Albert Heijn will receive more media coverage than Super de Boer, since Albert Heijn is market leader and reaches more customers (75%) than does Super de Boer (39%) (GFK, 2002). See Appendix A for additional background information for each of the focal organizations/sectors.

3.2 NEWS

In the previous chapter, it was outlined that the effects of media coverage on corporate reputation should be studied at a relational level. It is important to know how an organization is portrayed in the media in relation to its competitors, and with which issues the company profiles itself or with which issues the company is connected in the news. If the source of criticism is known, it becomes clear why the reputation of a company can be *improved* if the organization is *criticized* in the news. In order to study the news at a relational level and to make an elaborated media profile of the companies, a network approach to text analysis was used. The present study used the Network analysis of Evaluative Texts (NET) method (Van Cuilenburg, Kleinnijenhuis, & De Ridder, 1988; Van Cuilenburg et al., 1992; Kleinnijenhuis, De Ridder, & Rietberg, 1997; De Ridder, 1994a).

The NET-method will be described in subsection 3.2.1 Subsequently, the selection and weighting of the articles will be presented in subsection 3.2.2. In the next subsection, the coding procedure (3.2.3) will be described together with the reliability of the present study. Finally, the measures of the media variables and advertising intensity will be discussed (3.2.4).

3.2.1   The Network analysis of Evaluative Text (NET)

The NET-method was developed in the 1980s at the Vrije Universiteit in Amsterdam (for the main publications about the NET-method, see Van Cuilenburg et al., 1988; Kleinnijenhuis et al., 1997; De Ridder, 1994a). The method is based on Osgood, Saporta, and Nunnally's (1954) Evaluative Assertion Analysis. Osgood et al. (1954) assumed that the English language consists of two types of words: *common meanings* and *attitude objects*. Common meanings are signs whose meaning is the same for all users of English, e.g., all users of English will agree that "peace" is something favorable and that "atrocity" and "thief" have negative meanings. Most adjectives are common-meaning terms (e.g., good, bad, beautiful, ugly). Without common meanings, communication is impossible.

Attitude objects, which are referred to in the NET-method as *meaning objects,*[3] can have a different evaluative meaning to its users. Evaluation of the words "socialism" and "communism" depends upon the past experiences (sociological, educational, etc) of the individual encoding or decoding the term (see Osgood et al. 1954). Or, to put it in the context of business news, the evaluative meaning of the "lay-off of staff" can be different for several groups of receivers, as was remarked by Verčič (see Chapter 2). It is very bad news for the employees concerned, but the financial markets often reward companies for lowering the number of employees. The NET-method distinguishes various types of meaning objects such as actors and issues. Examples of actors are individuals, organizations, institutions, and NGOs. Issues refer to abstractions like "employment" and "environment."

Predicates

Another element of evaluative assertion analysis used in the NET-method is based on the notion that language contains "verbal connectors" – words that express the relationship between two meaning objects. Verbal connectors are called *predicates* in the NET-method. A predicate contains at least one common meaning term.

For example, in the sentence "People of good will denounce these aggressors" the predicate is "will denounce," since that part of the sentence connects the two meaning objects (the people of good will and the aggressors) together. It is assumed that the coders agree that the predicate "will denounce" is a negative or dissociative assertion (Osgood et al., 1954).

Box 3.1.   Several types of predicates

- *WIL* predicates are used to express the attitude of an actor towards an issue, as in the sentence "Shell to plant forests."
- *REA* predicates relate an issue or an actor to the author's reality. Examples of "reality predicates" are "Fifth runway close at hand," "Shell took a beating," and "Shell: bad climate for the oil industry."
- *AFF* predicates are used to express an affective relationship between two actors (e.g., "Cooperation between Schuitema and BP"). Note that the two meaning objects are always actors.
- *EVA* predicates are used to evaluate an actor or an issue in terms of good or bad, as in the sentence "Mamma's boy leads BP Amoco." The head of BP is referred to as a mamma's boy, which for most readers will mean that the chief executive is being evaluated negatively.
- *CAU* predicates express a causal relationship between two issues, as in the sentence
-   "High price of gas increases Shell's profits." This sentence is coded as: "Gas price / increases (+) / profit Shell."
- *TCH* predicates express a relationship of "touch," which implies a relationship in which an issue hits an actor, like in the sentence: "Crisis in southern Asia hits Shell." This sentence is coded as: "World economy/ hits (–) / Shell."

To serve the objective of this study, two extra types of predicates were formulated:
- *FUS* predicates were used to express mergers between two companies, such as in the headlines "Merger BP and Arco" and "BP no longer wants Rosneft." The FUS predicate was introduced in order to distinguish news about mergers from other news about support & criticism, and success & failure. This distinction was important since "mergers" was one of the associations mentioned by the respondents.
- *RES* predicates were used when coding media coverage about the police. RES was used to express that not only did the police have an attitude towards an issue or an actor, they had also succeeded. Without the type RES, two sentences with a different meaning would have rendered the same assertion. For example, the sentence "Police find 85 kilos of cocaine" would have been coded as "Police/ finds/ WIL/-1/cocaine," but the sentence "Police want to take action against cocaine" would have resulted in the same assertion, namely "Police/ want to take action against / WIL/-1/cocaine." This would have ignored the fact that the police had success in the first sentence, and in the second sentence only intentions.

"Quality" refers to the value of a predicate
If the predicate is used to dissociate ("conflict") the meaning objects of each other, the value of the predicate is –1. If the predicate is used to associate ("cooperation") the meaning objects of each other, the value of the predicate is +1. The value of a predicate is termed *quality*. The values "+0.5" and "-0.5" are used when a refinement is made in the text (such as "maybe").

Several types of predicates.
In the NET-method, a number of predicates are distinguished. Six types of predicates described by De Ridder (1994a) will also be used in this study.[4] In addition, two different types of predicates were formulated to serve the purpose of this study. The different types of predicates will be illustrated in Box 3.1.

As shown in Box 3.1, the type of predicate is determined by the types of meaning objects. If a predicate expresses a relationship between two actors, the type of predicate is always AFF or FUS, the latter in the case of a merger between two companies. If a predicate expresses a relationship between two issues, the type of predicate is always CAU. The type of predicate is also determined by the position of the meaning object. This will be further explained in the subsection on the combination of the different types of predicates with the concepts "subject" and "object".

Assertion, subject, object, and source
In order to code sentences, they are parsed into *assertions*[5] that connect one meaning object to another. Each assertion contains a subject, a predicate, and an object. A *subject* is the "active" meaning object (De Ridder, 1994a). The *object* is the meaning object that is influenced or implied by the subject (Kleinnijenhuis et al., 1997; De Ridder, 1994a). In the sentence "John hits Peter," John is the subject while he is the active meaning object, and Peter is the object.

Note that the terms "subject" and "object" as used in the NET-method do not necessarily correspond with the grammatical concepts of subject and object (Van Cuilenburg et al., 1988). For example, in the sentence "Peter gets a cookie from John," Peter is the grammatical subject and John is the grammatical object. In the NET-method, however, John is the subject and Peter is the object, since John fulfils the active role (he gives a cookie) while Peter fulfils a passive role (he receives a cookie).[6]

The concept *source* means the source of the text (De Ridder, 1994a). In the sentence "Bush: Bin Laden behind U.S. terror threat," Bush is the source, while Bin Laden is the subject and the U.S. is the object.

Combining and elaborating upon the concepts discussed earlier
The concepts source, subject, predicate, type of predicate, quality of the predicate, and object are amplified and applied to example sentences of business news in Table 3.2.

TABLE 3.2.   **Different assertions**

Sentence	Source	Subject	Predicate	Type	Quality	Object
Success & failure news						*Actor*
1. Shell took a beating		Reality	took beating	REA	-1	Shell
2. Shell: bad climate for oil industry	Shell	Reality	bad climate	REA	-1	oil industry
Support & criticism news		*Actor*				*Actor*
3. Cooperation between Schuitema and BP		Schuitema BP	cooperates with cooperates with	AFF AFF	+1 +1	BP Schuitema
4. Shell is on the right track		Shell	right track	EVA	+1	ideal
Issue news		*Actor*				*Issue*
5. Shell to plant forests		Shell	is going to plant	WIL	+1	environment
6. Fifth Schiphol runway close at hand		Reality	close at hand	REA	+1	fifth runway

The first two sentences are examples of success & failure news, which were also
indicated in Table 2.1. In the sentence "Shell took a beating" the author of the text
presents "factual information" ("failure," in this case "took a beating") about Shell.
This sentence does not contain a subject. It is left unclear who or what is responsible for Shell's defeat. If the subject is lacking and an actor or issues proclaim
success (+1) or failure (–), or variables increase (+) or decrease (–), the *reality* is
coded as the subject (Kleinnijenhuis et al., 1997). Note that reality is never coded
as the object. In the second sentence, Shell fulfils the role of the *source*, a cited (or
quoted) actor who in this case states that the climate in the oil industry is bad.

The third and fourth sentences belong to support & criticism news. In the
third sentence, an affective relationship between two actors is expressed. In this
case, the two actors support each other. Since "cooperation" expresses a mutual
relationship, two assertions are coded. Table 2.1 showed that this sentence is an
example of support & criticism news, while an affective relationship between two
actors is expressed. The fourth assertion is a special type of support & criticism
news. Shell is related to a common meaning term, namely the words "on the
right track." In evaluative statements the object is unspecified. In other words, in
this example it remains unclear who or what is influenced by the assertion that
Shell is on the right track. Therefore the object *ideal* (i.e., "good") is coded. Since
the expression "on the right track" is usually seen as an association with the good,
the quality of the predicate is positive: +1. In the case of evaluative statements,
judgment extrapolation (Van Cuilenburg et al., 1988) can be applied. This means
that evaluative statements are expected to indicate the relationship between the
source and the subject. In this example, the assertion "Shell is on the right track"
is cited by the newspaper without indicating the source of the assertion. If the
source of an evaluative assertion is unknown, the media are treated as the source.

In the case of judgment extrapolation of this evaluative statement, the media are transferred to the subject position and Shell is transferred to the object position. Since "on the right track" is an association with good, the quality of the predicate is positive, like before the judgment extrapolation. After judgment extrapolation, the assertion is coded as: media (subject)/ on the right track (predicate)/+1/Shell (object). Note that after judgment extrapolation, there is both an actor in the subject position and in the object position. Therefore, after judgment extrapolation, evaluative statements belong to support & criticism news.

In the fifth sentence "Shell to plant forests," "Shell" is the *subject* because it is the active meaning object. The "environment" (the forests) is the object because it is influenced by the subject, Shell. The predicate "to plant" connects the two meaning objects with each other. The type of the predicate is WIL, because there is an actor in the subject position and an issue in the object position. The value of the predicate is positive (+1), since planting forests is seen as a positive attitude by Shell towards the environment. In the sixth sentence "Fifth Schiphol runway close at hand," success is proclaimed for the fifth runway at Schiphol. This sentence does not contain a subject. It is left unclear who or what is responsible for the success of the Schiphol runway. Therefore, the fictive meaning object *reality* is coded as the subject. Table 2.1 shows that the last two sentences belong to issue news.

Different types of news in one sentence
Complex sentences may contain various assertions of different news types. Success and support may appear together (see Box 3.2, example sentence 1). Reaching a goal concerning an issue may be labeled as a success (example sentence 2). However, this does not have to be the case. There are lots of "dissonant" sentences where someone gets support even though he is losing (example sentence 3), or where the initiator of a successful policy has to leave (example sentence 4).

Box 3.2.   Examples of sentences with several types of news (Meijer, Oegema, & Kleinnijenhuis, 2000)

1. Fine results (success) for joint venture by BP and Amoco (mutual support).
2. HarvardNet recognized as top-500 firm (success) because the firm manages stay ahead with technological developments (issue).
3. Cenexto bankrupt (failure) in spite of government grants (support).
4. Now that the takeover of KLM has succeeded (success), the chief executive officer can go (failure/loss).

Combining the different types of news with the various types of predicates
In table 3.3, the different types of news are combined with the different types of predicates. Each type of news is illustrated with an example of an assertion.

In the case of success & failure news, the fictive meaning object "reality" is always in the subject position and there is an actor in the object position of the assertion. The type of predicate is REA. In addition, assertions with RES predicates belong to success & failure news about actors. As is explained in Box 3.1, RES predicates were used to express that the police did not only have an attitude towards an issue or an actor, they also succeeded in catching the thieves or in discovering the drugs. This means that in the case of a RES predicate, an extra assertion is always coded that encompasses the success of the police.

TABLE 3.3.   **The different types of news combined with the different types of predicates**

Type of news	Type of predicate	Subject	Object
Success & failure news about actors "Shell took a beating"	REA	Reality	Actor
Success & failure news about actors "Police catches thieves"	RES	Reality	Actor
Success & failure news about actors "Police find 85 kilos of cocaine"	RES	Reality	Actor
Support & criticism news "Cooperation between Schuitema and BP"	AFF	Actor	Actor
Support & criticism news "Merger between BP and Arco"	FUS	Company	Company
Support & criticism news, media evaluation of an actor "Shell is on the right track"	EVA	Actor	Ideal
Support & criticism news "Police catch thieves"	RES	Actor	Actor
Issue news "Shell to plant forests"	WIL	Actor	Issue
Issue news, the success (increase) and failure (decrease) of an issue: "Fifth runway close at hand"	REA	Reality	Issue
Issue news, media evaluation of an issue "Code of conduct hypocritical"	EVA	Issue	Ideal
Issue news "Gas price increases Shell's profits"	CAU	Issue	Issue
Issue news "Crisis in southern Asia hits Shell"	TCH	Issue	Actor
Issue news "Police finds 85 kilos of cocaine"	RES	Actor	Issue

There are several types of support & criticism news. The most common form is one in which there is one actor in the subject position and one actor in the object position. The FUS predicates were used to express mergers between two companies. This type of predicate was used for the coding of news about Shell and BP, and is a further specification of support & criticism news. As is explained in above subsection "Combining and elaborating on the concepts discussed earlier," news in which an evaluation is given about an actor ("Shell is on the right track") belongs to support & criticism news. After judgment extrapolation, this results in an assertion in which the media criticize or support the concerning actor. RES predicates also belong to support & criticism news if there is an actor in the subject position and in the object position. In the assertion "Police catches thieves," a negative affective relationship is expressed between the police and the thieves.

There are six types of issue news. The most common type of issue news is news in which an actor expresses an opinion on an issue; WIL predicates are used in this case. Second, in the case of success & failure news about an issue, the fictive meaning object "reality" is always in the subject position, and there is an issue in the object position of the assertion. The type of predicate is REA. A third type of issue news is news in which an evaluation is given about an issue. After judgment extrapolation, this results in an assertion in which the media express their attitude towards an issue (see the subsection "Combining and elaborating on the concepts discussed earlier" for more information about judgment extrapolation). Fourth, assertions containing a CAU or a TCH predicate belong to issue news. In the case of a CAU predicate, there is one issue in the subject and one issue in the object position. Fifth, in TCH predicates, there is an issue in the subject and an actor in the object position of the assertion. Finally, RES predicates also belong to issue news if there is an actor in the subject position and an issue in the object position. In the assertion "Police find 85 kilos of cocaine," the police decrease the issue "crime."

Combining the different types of predicates with the concepts of "subject" and "object"
As mentioned earlier, the type of predicate is determined by the types of meaning objects (an issue or actor meaning object). If a predicate expresses a relationship between two issues, the type of predicate is always CAU. As will be further explained in this subsection, the type of predicate is also determined by the position of the meaning object in the assertion.

In the case of both WIL and TCH predicates, the assertion consists of an actor and an issue. Remember that the sentence "Shell to plant forests" contains a WIL predicate, while the sentence "Crisis in southern Asia hits Shell" contains a TCH predicate. In other words, in the case of a WIL predicate, the actor is in the subject position and the issue is in the object position, while this is reversed in the case of a TCH predicate. This is illustrated in Figure 3.1. In the case of a TCH predicate, the issue is in the subject position (the starting point of the arrow), and the actor occupies the object position (which is where the arrow ends).

FIGURE 3.1   **Schematic overview of the different types of predicates**

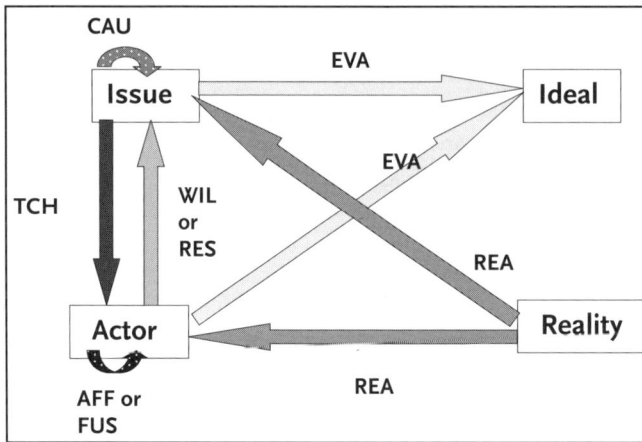

REA predicates can express a relationship between the author's reality and an issue ("Fifth runway close at hand") and a relation between the author's reality and
an actor ("Shell took a beating"). Note that the meaning object "reality" is always
coded in the subject position and never in the object position. Therefore, although the arrows are departing from "reality," they don't end there. This is the
opposite in the case of the meaning object "ideal," which is always coded in the
object position and never in the subject position of an assertion. In other words,
in REA predicates, an object is related to the *reality* of the author of the text, like in
the sentence "Shell took a beating." Since the subject is unknown, the fictive
meaning object "reality" is coded as the subject. In EVA predicates, the author or
the source cited by the author *evaluates* a meaning object. Since it is not clear who
or what is influenced by the predicate, the fictive meaning object "ideal" is coded
as the object.

Mapping the assertions into a network
Network text analysis originated with the idea that after semantic links among
concepts are coded, one can proceed to construct *networks* of semantically linked
concepts (Van Cuilenburg et al., 1988; Kleinnijenhuis et al., 1997; Popping, 2000;
De Ridder, 1994a). This can be illustrated by using the following four assertions:
1. "Shell / wants (+) / code of conduct."
2. "Code of conduct/ is hypocritical (–) / ideal."
3. "Shell/ wants a better (+)/ environment."
4. "Environment/ good for all (+) /ideal."

These assertions can be presented in the following networks. Meaning objects
can be presented as "points" and the connections as "arrows" (Van Cuilenburg et
al., 1988). The four assertions are represented by the four arrows in diagram A of

Figure 3.2. Every arrow represents the relationship between two meaning objects. Assertion 1 connects the subject "Shell" with the object "code of conduct." If there had been more assertions containing the subject "Shell" and the object "code of conduct," the quality of these assertions would have been averaged. Assertion 2 connects the subject "Code of conduct" with the object "ideal" and so on.

FIGURE 3.2. **Assertions represented as networks[7] (based on De Ridder, 1994a)**

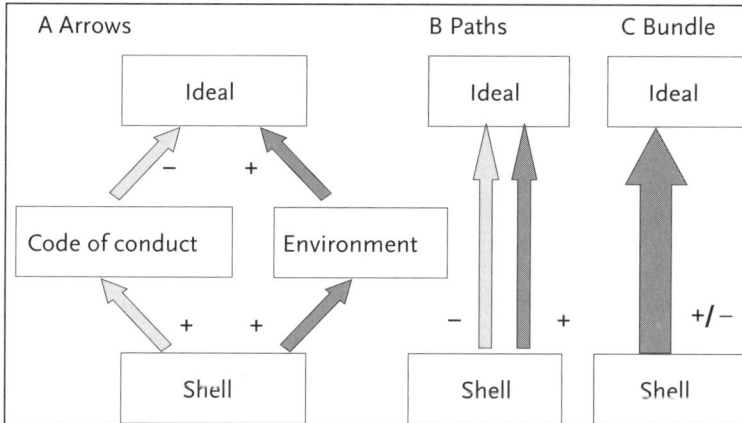

The elements of a network are not only connected directly by the arrows, but also indirectly by paths. In other words, it can be inferred from the text how the writer of the text judges[8] Shell, although this is an indirect judgment. In diagram B of Figure 3.2, the arrows of diagram A are merged into paths. In this example, there are two paths from Shell to "ideal": one via the "code of conduct" and one via the "environment." These two assertions can be combined in a path by applying the *transitivity axiom* (Van Cuilenburg et al., 1988).

The transitivity axiom postulates that calculation rules ("$+ \times + = +$"; "$+ \times - = -$"; "$- \times + = -$"; en "$- \times - = +$") can be applied to combine assertions. In other words, the friends of our friends are our friends ("$+ \times + = +$"), and the enemies of our friends are our enemies ("$- \times + = -$"). If the transitivity axiom[9] is applied, it follows that the first path indirectly expresses a negative judgment of Shell, because Shell is in favor of the code of conduct that is evaluated negatively, $+ \times - = -$. The second path expresses indirect a positive judgment of Shell, because Shell wants a better environment and the environment is evaluated positively, $+ \times + = +$. The paths can be merged into a bundle, which give an overall impression of the judgment about Shell (diagram C). The calculation of the bundle is based on the average direction of the paths. The transitivity axiom does not play a major role in the present study. It is not necessary to construct paths (diagram B) or bundles (diagram C) to test the hypotheses.

3.2.2   The selection and weighting of newspaper articles and television news

The FactLANE database (known as Nederlandse PersDatabank at the time of data gathering[10]) was used to collect all the newspaper articles with the exception of *De Telegraaf.* The articles from FactLANE contained not only the full text of the articles, but also the page numbers and the numbers of words. The key terms entered at the website used to retrieve the articles from the FactLANE database contained the name or, if available, the abbreviation of the focal company (see Appendix B). For the agricultural sector, various professions were entered such as farmer, market gardener, and cattle breeder.

The articles from *De Telegraaf* were retrieved from the LexisNexis database.[11] This database includes articles from January 1, 1999 onwards; the articles before this date were selected manually[12, 13] from the library of the University of Amsterdam.

This resulted in 9,285 articles. With regard to the number of articles and our capacity in terms of time and number of coders, a stratified sample was considered. Because this would not have provided enough articles per stratum to draw conclusions, it was decided to code only the *headlines* of all the articles and not the full text. Pictures were not taken into account. A "weight factor" was used to take the position of the article (the page number) and the magnitude of the article into account (see Box 3.3).

Data about news broadcasts by the public broadcasting company NOS and the commercial channel RTL 4 were gathered in cooperation with NOS and RTL. Both organizations have an archive with items of spoken text (the transcriptions). A total of 2,225 news items were analyzed. In addition to the transcriptions, the documents contain information about the broadcasting time of the news, images (in the case of RTL 4) and quotes (in the case of NOS). The period in which the transcriptions were gathered is the same as that for the newspaper articles, from July 24, 1997 to July 22, 2000. A weight factor was used to take the viewer ratings into account (see Box 3.3).

In total, the media coverage about the focal organizations and sectors consisted of 15,883 assertions. The weight factor was applied in such a way (see Box 3.3) that the dataset consisted of approximately the same amount of assertions (15,884 assertions) before and after weighting. There was a change in the relationship between print media and television news, however. Before weighting the news, 66% came from the print media while 34% came from television news. After a weight factor was applied, nearly half of the news (47%) came from the print media while the other half (53%) came from news on television channels. See Appendix C for the difference in unweighted and weighted assertions per medium.

Because television news and print news are treated as separate variables, this difference in the ratio of television news to print news did not influence the testing of the hypotheses. The same holds true for the different formats of the me-

---

Box 3.3.   The weight factor to weight the news

The weight factor for newspaper articles is composed of two weight factors: one weight factor that takes into account the page number of the article, and another weight factor that takes into account the total number of words in the article. An article on the front page is assigned the value "2." The articles on pages 2 and 3 are assigned the same weight factor, namely 0.91 (which is 1/ln (page number) in the case of even page numbers, and 1/ln(page number +1) in the case of uneven page numbers. The same formula is applied for page numbers 4 and 5, which have the weight factor 0.62; page numbers 6 and 7 have the weight factor 0.51. It was assumed that the economic quire is important for people with an interest in news about companies. While the economic quire usually starts from page 11, page 11 is assigned the same value as an article on page 3 in the case of news about companies (Shell, BP, AH, Super de Boer, ABN AMRO, and Rabobank). Pages 12 and 13 are assigned the same value as articles on pages 4 and 5, and so on. In the weight factor of the page numbers concerning the police, the agricultural sector, the NS, and the Dutch airport Schiphol, lower values were assigned to page number 11 than to page number 3. The weight factor, which is composed of the number of words, is a linear weight factor that is topped off at the top and at the bottom. It assigns the value "2" if an article contains more than 1,600 words, the value "1.9" if an article contains between 1,500 and 1,600 words, the value "1.8" if an article contains between 1,400 and 1,300 words, and so on. Articles containing 100 words or less are assigned the value 0.4. The two weight factors are multiplied and divided by 2 (the weight factor of the page number and the weight factor of the number of words have the same weight in the total weight factor).

The weight factor of television news is based on the viewer ratings of a random week in the middle of the research period (from March 15 to March 21, 1999). The viewer ratings were provided by the market research company Intomart, which provides ratings from before the year 2000. The 8:00 o'clock evening news, which is watched more frequently than any other newscast in the Netherlands, was assigned the maximum value of "2." The newscast watched the least was assigned the minimum value of "0.5."

The "total" weight factor of newspaper and television news is applied in such a way that before and after weighting, the dataset consisted of approximately the same number of assertions. This is done by first calculating the sum of the weight factor. Subsequently, the sum of the weight factor is divided by the original number of assertions. That outcome is multiplied by the weight factor. After multiplication, the sum of the new weight factor equals the original number of assertions.

---

dium types. In the case of television news, the transcriptions were analyzed, whereas in the case of print news, only headlines were coded. This does not influence the testing of the hypotheses, since separate variables were made for television and for print news.

3.2.3   The coding procedure

Although the NET-method provides numerous guidelines for the analysis of texts, a tailor- made coding instruction was needed to take the aim of the present study into account. The operationalization of theoretical concepts coincides for an important part with writing the coding instruction (Van Cuilenburg et al., 1992;

Riffe, Lacy, & Fico, 1998). The following subsections will discuss the main points of the coding instruction. More information about the coding instruction can be found in Appendix D.

The classification system consists of a list of meaning objects[14] per industry. This means that six lists of meaning objects were made: one list for Shell and BP, one list for ABN AMRO and Rabobank, one list for the NS and Schiphol, one list for Albert Heijn and Super de Boer, one list for the Dutch agricultural sector, and one list for the Dutch police. Although the meaning objects may differ between the several lists of meaning objects, they all have the same categories. Each meaning list consist of two parts, one part containing all actors who are of interest to the study, and a second part containing all the relevant issues.

The actors were divided into the next general categories. For each of the categories, examples are given of meaning objects that belong to this category:
-   Organizations: the focal organization/industry, shareholders of the focal company, organizations from the same industry (competitors and partners), and organizations from other industries
-   Politics: national and international governments
-   Powerful actors: the judge, the media, NGOs
-   People: employees, customers

The issues were divided into the following categories:
-   Corporate issues: profits of the organization, news products, staff wages, investments
-   Economic issues: world economy, price of currency, employment
-   Mergers: mergers
-   Societal issues: the environment, human rights, discrimination
-   Image: image, "namedropping"

The meaning object "namedropping" was created to code headlines that consisted only of the name of the focal company (headlines such as "Shell" or "Schiphol"). Three percent of the assertions were coded as namedropping.

The quality of the predicates could take one of the following values: −1, −0.5, 0, 0,5 and 1. The values "+0.5" and "−0.5" were used when a refinement was made to the text (such as "maybe," "slightly," and "perhaps"). In a study on the possibilities and limitations of the NET-method, Van den Berg and Van der Veer (2000) recommended using a three-point scale (positive-neutral-negative) to enhance the reliability of the codings. However, in the experiment by Van den Berg and Van der Veer, the coders were free to use any value between −1 and 1 (e.g., also the value 0.3) to indicate the connection between the two meaning objects. The coders could assign a certain value based on their own insight – clear instructions were not provided. By using clear indications for when the coders should use the values 0.5 and −0.5, it is assumed the recommendations of Van den Berg and Van der Veer were followed. In the case of Shell and BP, the meaning object "gas stations" is not coded, since it is hard to separate this meaning object from the main actors, for example in the headline "Shell stations in Africa." The meaning object "gas stations" occurred rarely, however.

In the media coverage part of the present study, the unit of analysis is the combination of the subject and the object, the "pair of meaning objects." The context unit consisted of the article as a whole. The coders were encouraged to use the context unit only if the headline did not contain enough information to code the headline. For example, in order to code a headline like "Shell profits under pressure, due to Asia and oil price," the coder should use the context of the article to find out if the oil price was high or low. FactLANE did not have the publication rights of a small number (1%) of articles, which meant that the headlines were listed but the full text was not always available. The headlines were coded nevertheless.

The coders used the Excel computer program. We considered using the CETA2 (Computerized Evaluative Text Analysis) computer program, which was developed by De Ridder (1994b). In tests of the reliability of CETA2, though, De Ridder (1994a) found that the reliability of observations of coders who did not use CETA2 was lower than the reliability of observations of coders who did use CETA2. However, De Ridder (1994a) remarked that the coders were not trained extensively, as was also the case in the present study. It was decided to use Excel because it is more user-friendly than the MS DOS program CETA2. A user-friendlier version of CETA2 called, iNET, was not yet available then, although it is being developed at the moment.

Reliability
The previous section addressed three ways of increasing coder reliability: constructing a well-defined list of meaning objects, defining a clear set of predicate types, and using a specialized computer program (Kleinnijenhuis et al., 1997). Another aspect of improving coder reliability is the training of the coders. Most of the seven coders in the present study (all students at the Vrije Universiteit) had taken the course "Communicative Influence," of which the NET-method is a major part. The coders were trained at the Department of Communication Science at the Vrije Universiteit Amsterdam. The researcher instructed the coders (sometimes in presence of a senior researcher) but did not code herself,[15] as recommended by Lombard, Snyder-Duch, and Campanella Bracken (2002). As recommended by Riffe et al. (1998), the training tests were not conducted with the content used for the actual study, but with content about the same focal organizations one month before the research period. The reason for this is that a coder must code independently of both others and of him- or herself. If content is coded several times, prior decisions influence subsequent ones (Riffe et al., 1998). Moreover, Riffe et al. argued that the reliability estimate is inflated by multiple

codings, thus giving false confidence in the study's overall reliability. If the training tests showed that the classification system needed refinement, the coding instruction was adjusted.

In order to prevent the results from being biased by one coder, the media coverage of an organization and the medium were each coded by at least two different coders. The coders coded an additional 10% of the headlines and news items in order to determine *intercoder reliablity*. The headlines and news items were selected at random and randomly returned to the coded set. The coders were not aware of which headlines or news items would be used to measure intercoder reliability.

Overall, intercoder reliability was satisfactory. More detailed information about the calculation of intercoder reliability in the present study is provided in Appendix E.

3.2.4. Measures

The present study focused on the success & failure or support & criticism *received* by the focal organization. It does not provide a theoretical framework for the effects of support & criticism *given* by the focal organization. Therefore, success & failure news and support & criticism news was only used in testing the hypotheses of whether the focal organization was in the object position of the assertion. In the case of issue news, all assertions were selected that contained an issue on the subject and/or object position of the assertion. Although it was coded whether a company was portrayed as having a positive or a negative attitude towards an issue ("the evaluative dimension of the attribute agenda") in order to test the hypotheses, only the *frequency* and not the direction of issue news needed to be taken into account.

The names of the media variables and the variable advertising intensity are presented at the left side of the table below (Table 3.4). They correspond with the left side of the conceptual model (Figure 2.3).

The variable "tenor of success & failure news" was operationalized by two separate items: the direction of success & failure news (which is measured by the *average* direction) and the favorability (which was measured by the *summed* direction) of success & failure news. The same was the case for the variable "tenor of support & criticism news'. The range of the average direction of news is (–1, 1), where 1 indicates all positive coverage, –1 indicates all unfavorable coverage, and 0 indicates a balance between the two over a year. The range of the favorability of news can take larger minus and plus values depending on the amount of media coverage about the organization.

TABLE 3.4.   **Measures of the media variables and advertising intensity**

Variable name	Measures
Amount of television coverage	The cumulated number of times the focal organization or sector was mentioned in *RTL 4* news and *NOS* news.
Amount of print coverage	The cumulated number of times the focal organization or sector was mentioned in the headlines of the five largest Dutch national newspapers.
Tenor of success & failure news:	
Direction of success & failure news	The average direction of success & failure news with the focal organization in the "object position."
Favorability of success & failure news	The summed direction of success & failure news with the focal organization in the "object position."
Tenor of support & criticism news:	
Direction of support & criticism news	The average direction of support & criticism news with the focal organization in the "object position."
Favorability of support & criticism news	The summed direction of support & criticism news with the focal organization in the "object position."
Amount of issue news	The cumulated frequency of news about issues.
Amount of own issue news	The cumulated frequency of news about "owned issues." An "owned issue" means the public perceives an organization to be successfull in handling that particular matter.
Advertising intensity	The cumulated advertising expenditures by the focal organization in the five largest Dutch national newspapers and on television (*RTL 4* and public broadcaster *NOS*).

With regard to the variable "amount of news about owned issues," it should be remarked that the investigator determined whether a certain issue was an owned issue of the focal organization or not (see Appendix F). This is in contrast to the operationalizations used in the studies of Budge and Farlie (1983) and Petrocik (1996), who studied issue ownership in political communication. Budge and Farlie used three coders to determine whether an issue was own or not,[16] while Petrocik asked respondents which political party would be more successful in resolving a problem in several areas like welfare, foreign policy, and economic issues.

An organization's total advertising expenditures was the measure of advertising intensity. This is the same measure as was used by Fombrun and Shanley (1990). Because the respondents were asked if they watched the *NOS* news (which is broadcast on three public channels), advertising data of these three channels were added up.

3.3 CORPORATE REPUTATION

There are several ways to measure corporate images and reputations (see Bouwman, 1998; Van Riel, Stroeker, & Maathuis, 1998). As was indicated in Chapter
1, the concept "overall evaluation" appeared in various definitions of corporate
reputation. Since "evaluation" is the central element in the definition of "attitude" (Ajzen, 1988; Eagly & Chaiken, 1993; Fazio, 1989; Fishbein, 1967; Petty &
Cacioppo, 1986; Pratkanis, 1989), corporate reputation is characterized in the
present study as an attitude towards an organization or sector. In accordance
with this characterization, attitude scales were used to measure reputation. Moreover, attitude scales (or more generally, "closed methods") are preferred when
the objective is to monitor corporate reputation over time, as was the case in the
present study (see Van Riel et al., 1998).

More specifically, in the present study corporate reputation was measured in
the tradition of the expectancy value model (cf. Andreassen & Lanseng, 1997;[7]
Verčič, 2000). The expectancy value model was employed because it is appropriate as a predictive model for measuring corporate reputation (see Van der Pligt &
De Vries, 1995). The expectancy value model and its use as applied in psychology
and in the present study will be discussed in the first three subsections.

3.3.1. Expectancy value model

Expectancy value models are providing a popular framework for understanding
the relationship between attitudes and the evaluative meaning of beliefs. The
foundation of these models was laid by Christiaan Huygens in 1657. Kleine recently translated and explained Huygens' work (Huygens, 1998). He mentioned
that Huygens' publication in 1657 was titled *De ratiociniis in ludo aleae* ("about the
computation in games of chance"), which appeared in a book with mathematical
exercises. Currently, the computations about chance are called "expected value"
estimations. As Van der Pligt and De Vries (1995) indicated in their description
of expectancy value models as used in economics, it is assumed that people are
trying to maximize the average profit of their behavior. For every behavioral alternative, the expected value is calculated by multiplying the possible outcomes (values, V) with the probability (P) that such an outcome will occur. This makes it
possible to calculate for each alternative a_i the expected value.

$$EV_i = \sum_{j=1}^{n} V_i \times P_{ij} \qquad\qquad (i = 1,2....,m) \qquad (1)$$

EV_i stands for the expected value of alternative i. V_i is the value of a specific outcome of a behavioral alternative, and P_{ij} stands for the probability that this outcome will occur. It is assumed that the alternative with the highest expected

value will be chosen (Van der Pligt & De Vries, 1995). For example: If behavioral alternative 1 leads to a certain profit of € 225,-, behavioral alternative 2 leads to a 40% probability of a profit of € 400,- and a 60% probability of a profit of € 80,-, and behavioral alternative 3 offers a 90% probability of a profit of € 150,- and a 10% probability of a profit of € 1000,-. Then, behavioral alternative 3 is the most attractive because of its highest expected value. The expected value of the first alternative is € 225,-, for the second alternative the expected value is 0.4 x € 400,- + 0.6 x € 80,- = € 208,-, and for the third alternative 0.9x € 150,-+ 0.1 × € 1000,- = € 235. Note that this type of expectancy value model is a "one-attribute model;" in this example the attribute is "profit." In the subsections that follow, the focus will be on multi-attribute models.

3.3.2 The expectancy value model as applied by Fishbein

Fishbein (1963; 1967) proposed that attitudes are a function of a) *beliefs* about the attitude object, defined as the subjective probability that the attitude object has each attribute, and b) the evaluative aspect of these beliefs defined as the *evaluation* of each attribute. Since multiple attributes can be combined, this model is a multi-attribute model. The most popular form of this function was given by Fishbein and Ajzen (1975):

$$A_o = \sum_{i=1}^{n} e_i \times b_{io} \tag{2}$$

Where A_o is the attitude towards some object, o; and n is the number of "salient beliefs;" e_i is the evaluation of attribute i (the value component) and b_{io} is the belief i about object o, (i.e., the subjective probability that o is related to attribute i, the expectancy component). Beliefs about an object are formed by associating it with various characteristics, qualities, and attributes (Fishbein & Ajzen, 1980). Evaluation refers to the value or evaluation of the attribute. For example, consumers may associate toothpaste with "fluoride" and evaluate fluoride as positive.

In the case of a consumer's attitude towards a product, the belief strength (or in other words, "the subjective probability that o is related to attribute i") indicates whether a given product has a certain attribute or not. For example, consumers can strongly believe that toothpaste has fluoride or that it does not have fluoride (Fishbein & Ajzen, 1980).

Consider the hypothetical example of a person's attitude towards Royal Dutch Airlines (KLM).[18] Suppose that a person has emitted the following five beliefs (or attributes) with reference to KLM: the Dutch economy, luxury, safety, staff, and environment (see Table 3.5).

To understand the person's attitude towards KLM, it has to be determined how she or he evaluates each of the attributes. This can be done by asking the person to evaluate each of the five attributes on a five-point, good-bad scale:

Do you think it is good or bad if the Dutch economy is stimulated?

good   (+1)   :   (+0.5)   :   (0)   :   (–0.5) : (–1)   bad   *Evaluation*
        extremely   quite   neither/nor   quite   extremely   *(of an attribute)*

Column 2 in Table 3.5 shows the hypothetical results. The person in this example feels it is extremely good if the Dutch economy is stimulated (outcome evaluation of +1), and that he or she does not like luxury (outcome evaluation of -1). Note that the evaluations of attributes are given independent of the relationship of the attributes with the organization.

The next step is to measure the strength of the person's beliefs by asking the person to indicate the likelihood that the company is associated with the attributes. A bipolar scale like the following could measure the strength of each belief:

In your opinion, how likely is it that KLM stimulates the Dutch economy?

likely   (+1) : (+0.5) :   (0)   : (–0.5) :   (–1)   unlikely *Belief strength*
        extremely   quite   neither/nor   quite   extremely

In column 3 of Table 3.5, it can be seen that the respondent in this example thinks it is extremely likely[19] (belief strength of +1) that KLM stimulates the Dutch economy, and that it is extremely unlikely (belief strength of –1) that KLM takes care of the environment. According to Fishbein's development of the expectancy value model, a person's attitude towards an organization (or a behavior) can be predicted by multiplying his or her evaluation of each of the associations with the company by the likelihood that the company is associated with the listed attributes and consequences.

TABLE 3.5.   **A person's attitude towards KLM**

Attributes (beliefs)	Outcome evaluations (e_i)	Belief Strength (b_{io})	Product
1. Dutch economy	1	1	1
2. Luxury	-1	0.5	- 0.5
3. Safety	1	0.5	0.5
4. Staff	1	1	1
5. Environment	0.5	-1	-0.5 +
Attitude towards KLM		Sum	1.5
		Mean	.30

These computations are illustrated in the last column of Table 3.5. It follows from these computations that the reputation of KLM is slightly positive, namely .30, on a scale that ranges from −1 to +1.

The number of salient beliefs

There is one variable of function (2) that has not yet been addressed in the example, namely the variable n, which indicates the number of "salient beliefs." Because people do not have an unlimited information-processing capacity, it is argued that only a relatively small number of beliefs serve as determinants of an attitude at any given moment. Fishbein and Ajzen (1975) stated that under most circumstances, a person's attitude towards an object is primarily determined by no more than five to nine beliefs about the object that are *salient* or *accessible* at a given point of time. Van der Pligt and Eiser (1984) suggested that only three to five beliefs determine the attitude. Fishbein and Ajzen (1975) and Ajzen and Fishbein (1980) mentioned two procedures that can be used to make sure the beliefs are important and meaningful to the respondents.

The first procedure is the use of a free-response format by asking the respondents to list their own beliefs: the attributes they think of when they think of KLM. The advantage of this model is that individuals are able to mention their *own* beliefs. On the other hand, the beliefs of different respondents may differ, which makes it more difficult to compare them. In order to solve this difficulty, Fishbein and Ajzen (1975) proposed composing a list of "modal salient beliefs," beliefs that are salient in a given population. The list of modal salient beliefs can be composed by eliciting beliefs from a representative sample of the population – the beliefs most frequently elicited by this sample make up the modal set for the population in question. The advantage of the assessment of modal salient beliefs is that it makes it easier to compare the beliefs of different individuals than when a free-response method is used.

Fishbein's expectancy value model is integrated into the theory of reasoned action (Fishbein, 1967; Fishbein & Ajzen, 1975). According to this theory, behavior can be predicted by the intention to perform a certain behavior. Behavioral intention can be predicted by the attitude towards the behavior and the social norm (i.e., beliefs that certain referents think the person should or should not perform the behavior in question).

With a view to the purpose of the present study, the focus in the present study is only on the formation of attitude.

3.3.3   Application of the expectancy value model as used in the present study

The method for measuring corporate reputation in the present study should meet three criteria. First of all, as indicated earlier, a closed method should be used, since the objective is to monitor corporate reputation over time (Van Riel et al., 1998). Second, the method should be able to measure various salient beliefs in order to test the second level of agenda-setting-hypothesis. Third, the method should measure the agreement with the policy of an organization and the perceived success of the policy of an organization in order to test hypotheses 2b and 3b. The present study postulates that the reputation of a company is based on the agreement of the people with the policy of the organization, and on their perceived success of the policy of an organization. The following model is used to predict the overall agreement with the policy of an organization:

$$R^{agreement}_{o} = \sum_{i=1}^{n} s_i \times a_{io} \qquad (3)$$

Where $R^{agreement}_{o}$ is the overall agreement with the policy of an organization, o; s_i is a binary variable with value 1 if an attribute i is salient and value 0 if an attribute is not salient; a_{io} is the level of agreement of a respondent with the policy of an organization, o with regard to attribute i. The perceived success of the policy of an organization is measured in the same manner:

$$R^{success}_{o} = \sum_{i=1}^{n} s^i \times s_{io} \qquad (4)$$

Where $R^{success}_{o}$ is the overall success of the policy of the organization, o; s_i is a binary variable with value 1 if an attribute i is salient and value 0 if an attribute is not salient; sio is the success of the policy of an organization o, according to the respondent with regard to attribute i. Overall corporate reputation can be predicted from the agreement with the policy of an organization and the success of the policy of an organization:

$$R_o = b_0 + b_1 \left(\sum_{i=1}^{n} s_i \times a_{io} \right) + b_2 \left(\sum_{i=1}^{n} s_i \times s_{io} \right) + e_i \qquad (5)$$

Where Ro is the reputation of an organization; s_i is a binary variable with value 1 if an attribute i is salient and value 0 if an attribute is not salient; a_{io} is the level of agreement of a respondent with the policy of an organization, o with regard to attribute i; s_{io} is the success of the policy of an organization o, according to the respondent with regard to attribute i; b_0 is the intercept; b_1 and b_2 are the slope coefficients; e_i is the error term.

Whereas in the present study the agreement with the policy of an organization and the success of the policy of an organization are measured, investigators in the marketing area typically measure the extent to which the consumer views the product as satisfactory.[20] In these types of models, the variable a_{io} combines the belief strength component (b_{io} in formula 2) with the evaluative component (e_i in formula 2) in one single measure (which is called *the measure of satisfaction* or *perceived instrumentality*). This is also the case for the variable s_{io}.

The combination of the evaluative component and the strength of belief component in one single measure was criticized by Fishbein et al. They argued that the evaluative component and the strength of belief component should be measured separately (Ajzen & Fishbein, 1980; Cohen, Fishbein, & Ahtola, 1972). Cohen et al. (1972) illustrated this with an example about the carbonation of soft drinks. In Fishbein's model, the strength of beliefs would be high if the brand were carbonated and low if it were not. In contrast, when combining the two components the measure could be high, irrespective of whether or not the brand was carbonated. A consumer could answer he is very satisfied with the carbonation because he likes carbonation and thinks the soft drink is carbonated. On the other hand, he can also be very satisfied because he dislikes carbonation and he believes the soft drink is not carbonated.

In the present study, the attitude towards the organization (R_o) or corporate reputation is more important than its formation, because changes in corporate reputation are observed and not the changes in the actual policy of the organizations/sectors. In addition to this, it can be remarked that the single measure combining the evaluative component and the strength of belief component is often used in the marketing field. Wilkie and Pessemier (1973) investigated 42 empirical studies in the marketing field that used the multi-attribute model. They found the expectancy component was used in only one study. Almost all studies incorporated the expectancy and evaluation component in one measure.

3.3.4 Measures

Table 3.6 lists the operationalization of corporate reputation. The names of the variables are presented at the left side of the table below; they correspond with the right side of the conceptual model (Figure 2.3). The questionnaire was introduced with the following question: For every company or institution, can you state what you think of, i.e., what things you associate it with?

The salience of an association was measured by asking the respondent what he thinks of first and what he thinks of second when he thinks of the focal organization (Shell, for example). He was able to choose two associations from a list of approximately fifteen associations listed on the computer. This meant that only the two most salient beliefs were taken into account. See Appendix G for the list of associations.

TABLE 3.6   **Measures of corporate reputation and policy of the organization. See Appendix G for the complete survey and the answer categories**

Variable name	Measures
Salience	What do you think of first when you think of [name of the focal organization]? What do you think of second when you think of [name of the focal organization]?
Agreement with the policy	Do you agree with the policy of [name of the focal organization] regarding [association mentioned by the respondent]? 1. Strongly disagree 2. Disagree 3. Agree 4. Strongly agree 5. Do not know the policy of the organization 6. Do not want to say
Success of the policy	The policy of [name of the focal organization] is a success regarding [association mentioned by the respondent]. ? 1. Yes, it is a success 2. No, it is not a success 3. Do not know the policy of the organization 4. Do not want to say
Corporate reputation	Can you give a report mark from 1 to 10 for [one of the focal organizations]?

The agreement with the policy of an organization was measured by asking whether the respondent agreed with the policy of the organization or sector regarding their association. For example: if the respondent associated Shell with the environment, the next question was: "Do you agree with Shell's environmental policy?" Answers to this question were scored on a 6-point Likert scale. The success of the policy of an organization was measured by posing the question of whether the policy of the focal organization was a success. For example: if the respondent associated Shell with the environment, the questioned posed was: "Is Shell's environmental policy a success?" Answers to this question were scored on a 4-point Likert scale (see Table 3.6).

The variable reputation was operationalized by asking the respondent to give a report mark for the organization concerned. This means that the report mark was the only item that measured reputation directly. The limitations of this approach will be discussed in Chapter 6.

3.3.5 Sample sizes, media use, and background characteristics of respondents

Table 3.7 lists the number of respondents who took part in the survey. In the columns of the table, the years in which the respondents participated are listed (e.g., there are 100 respondents who took part only in 1998 and filled in the questions about Shell, there are 69 respondents who took part in 1998 and 1999 and filled in the questions of Shell, and so on).

Respondents who took part in the year 2000 filled in all of the questions about the ten focal organizations. Therefore, the number "377" is listed for all ten organizations/sectors in column C. This is in contrast to other years, when the respondents only completed a part of the survey. In 1998 respondents answered the questions for six of the organizations; if they took part in 1999 as well, they answered the questions about the same organizations as the year before.

TABLE 3.7.   Sample sizes of respondents of the survey per organization per period

	A	B	C	D	E	F	Total
	Only in 1998	Only in 1999	Only in 2000	1998-1999 not 2000	1998-2000 not 1999	1998-1999-2000	
Shell	100	0	377	69	84	229	859
BP	86	0	377	68	79	212	822
Albert Heijn	98	0	377	59	89	216	839
Super de Boer	89	0	377	61	87	224	838
NS	91	0	377	70	75	227	840
Schiphol	85	0	377	63	95	222	842
ABN AMRO	85	0	377	62	82	204	810
Rabobank	93	0	377	70	87	190	817
Police	95	0	377	53	93	214	832
Agriculture	89	0	377	67	85	214	832
Total	911	0	3,770	642	856	2,152	8,331

Since respondents answered questions about several organizations, they might appear more than once in the dataset. A total of 1,289 different respondents[21] took part in the survey, once, twice, or three times. The background characteristics of the respondents such as sex, age, number of family members, region, income, and education were measured in 2000. The background characteristics of the respondents who did not participate in 2000 but did participate in 1999 and/or in 1998 could be retrieved from election research that also used the NIPO Telepanel. Data about media use of the respondents were available for 1998 and 2000. In the year 1999, the data of the media use in 2000 were used.

3.4 MEDIA USE AND BACKGROUND CHARACTERISTICS OF THE RESPONDENTS

After removal of the respondents who did not use any of the focal media,[22] on average over the three years it appeared that 45% of the respondents indicated they watch the NOS and RTL 4 news at least once a week; 35% of the respondents watched the NOS news; 14% of the respondents watched the RTL 4 news at least once a week; 6% of the respondents indicated they watched neither NOS nor RTL 4 news.

On average over the three years, the use of print media is as follows: 44% of the respondents did not subscribe to any of the five focal daily newspapers; 22% of the respondents subscribed to *De Telegraaf*; 7% of the respondents subscribed to *Algemeen Dagblad*; 5% of the respondents subscribed to *de Volkskrant*; 2% of the respondents subscribed to *NRC Handelsblad*; 1.5% of the respondents subscribed to *Trouw*, 18.5% of the respondents subscribed to two or more newspapers (6% of the total number of respondents subscribed to *Algemeen Dagblad* and *De Telegraaf*).

The total sample of 1,289 respondents consisted of 49% men and 51% women. The average age of the respondents was 49 years ($SD = 16.26$). The youngest respondent was 18 and the oldest respondent was 90 years old. The educational level of the respondents was as follows: 27% of the respondents had a low level of education,[23] 43% of the respondents had a moderate level of education, and 30% had a high level of education. Of the newspaper readers, 18% had a low level of education, 43% had a moderate level of education, and 39% had a high level of education.[24] Of the viewers of television news, 28% of the respondents had a low level of education, 43% a moderate level of education, and 29% a high level of education.

3.4 DESCRIPTION OF THE MERGED DATASETS AND ANALYSES TECHNIQUES

In order to test the hypotheses, two different datasets were needed. A general description of these datasets will be given in subsection 3.4.1. Subsection 3.4.2 deals with the examination of the data and multiple regression analysis with the Ordinary Least Squares (OLS) method. This method was used to test all the hypotheses of the present study, except for hypothesis 4. In addition, several of the assumptions underlying this technique will be discussed.

In subsection 3.4.3, the analysis technique "Repeated Measures ANOVA" and logistic regression will be elaborated upon. Repeated measures ANOVA was used along with OLS to test hypothesis 6. Logistic regression was needed to test hypothesis 4. In subsection 3.4.4, the last subsection of this chapter, miscellaneous methodological issues will be touched upon.

3.4.1 Description of the datasets with media and corporate reputation variables

In order to test hypotheses 1 to 3, a dataset was needed that related the attitude of a respondent towards a company to the media coverage about that company. The media coverage was assigned on the individual level according to the media use of respondents. This resulted in a "year organization respondent dataset" with 16,662 cases. The effects of advertising intensity (hypothesis 7) were tested together with the models of hypotheses 1 to 3. In this way, the effects of media coverage were controlled for the effects of advertising intensity and vice versa.

In order to test the hypotheses about issue news (hypotheses 4 to 6), a more fine-grained dataset was required to take into account the association of the respondent with a certain organization and the news about that certain association (or issue). This resulted in a "year organization association respondent dataset" with 214,958 cases. This meant that the number of cases was relatively high. Tabachnick and Fidell (2001) explicitly warned for regression analyses with too many cases, since most multiple correlations will be significantly different from zero. However, in the present study, in order to test the hypotheses about the effects of issue news it was necessary to register which issues were salient and which were not at the individual level of the respondents. The construction and description of the two datasets is discussed in more detail in Appendix H.

3.4.2 Multiple regression analysis with the Ordinary Least Squares (OLS) method

Multiple regression analysis was the analysis technique most frequently used in the present study, because the objective of multiple regression is to predict the changes in the dependent variable (reputation) in response to changes in the independent variables (media coverage and advertising intensity). The method was used to test all the hypotheses[25] except for hypothesis 5.

Moreover, multiple regression analysis provides insight into the relative importance of each independent variable in the prediction of the dependent variable (Hair, Anderson, Tatham, & Black, 1998). By using multiple regression, the effect of one independent variable is controlled for the effects of the other variables that are entered in the model. For example, the amount of media coverage on corporate reputation is controlled for the effects of the direction of the news, the favorability of the news, and advertising intensity if the variables are entered simultaneously into the model. There are several methods of constructing the sample regression function. In this study, a sequential method (*Stepwise* estimation) was used to select the variables of the models on an organizational level. Stepwise estimation method was chosen because of the explorative character of this study, with the set of independent variables not closely specified (see Hair et al., 1998). A confirmatory approach was used in the pooled models, and will be elaborated upon in the next subsection.

The classical linear regression model is based on several assumptions (for an overview of the set of assumptions, see for example Gujarati 1995). As mentioned by Gujarati (1995) the set of assumptions may be stringent or called "unrealistic." Nevertheless, the assumptions that may have been violated in this study[26] and also the consequences this might have will be discussed in the subsections below.

Multicollinearity
One of the assumptions of the classical linear regression model is that "there is no perfect multicollinearity." This means there are no perfect relationships among the independent variables. Gujarati (1995) stated that high pair-wise correlations (in excess of 0.8) among the independent variables indicate that multicolline-arity[27] is present. In the case of high multicollinearity, it is likely that the following consequences are encountered: it limits the size of R, because the independent variables are going after much of the same variance on y, it makes the importance of a given independent variable difficult because the effects of the independent variables are confounded due to the correlations among them; it increases the variances of regression coefficients (Stevens, 2002). The greater these variances, the more unstable the prediction equation will be. In the case of the use of sequential search methods, multicollinearity among independent variables can have a substantial impact on the final model specification (see Hair et al., 1998). If two highly correlated independent variables have almost equal correlations with the dependent variable and one of these variables enters the regression model, it is unlikely that the other variable will also enter since the independent variables are multicollinear and there is little unique variance for each variable separately.

The correlations between the independent variables in this study (see Appendix I) were sometimes relatively high. This was the case for the correlations between variables that measured the direction of news, and the variables that measured media favorability. See for example the correlations between the favorability of success & failure TV news and the direction of success & failure TV news of the following organizations: Shell, $r = .96$, $p < .01$ (see Appendix I, Table I1); NS, $r = .91$, $p < .01$ (Table I5); Schiphol, $r = .97$, $p < .01$ (Table I6). It is likely that this was due to the fact that the variables that measured the "direction of news" consisted of the mean direction of success & failure or support & criticism news, while the "favorability of news" consisted of the summed direction of success & failure or support & criticism news. In addition, the intercorrelations between the amount of TV advertising expenditures and the amount of television news were high in the case of ABN AMRO, $r = .94$, $p < .01$ (see Table I7) and in the case of Rabobank, $r = .95$, $p < .01$, (see Table I8). This was probably due to the low number of possible variations in television news (NOS, RTL, or NOS and RTL). High intercorrelations were also present between the variables that measured the amount of issue news and the amount of owned issue news. For example, in the case of Albert Heijn $r_{\text{amount of issue news in print media, amount of owned issues in print media}}$ = .97, p < .01 (see Table I23), and the police $r_{\text{amount of issue news on TV, amount of owned issues on TV}}$ = .98, p < .01 (see Table I29).

The examination of the correlation matrix showed that collinearity was present. The correlations between the variables that measure the amount of issue news and the variables that measure the amount of owned issue news do not have to be a problem, because these variables are not tested together in one model. The variables that measure the direction of the news and the variables that measure the media favorability are tested in one model, however. This is also the case for the variables that measure the amount of TV advertising expenditures and the variables that measure the amount of TV news.

One of the remedial measures to reduce multicollinearity is to omit the least interesting variable from a theoretical point of view (Hair et al., 1998; Pennings, Keman, & Kleinnijenhuis, 1999). Since multicollinearity is a data problem and not a problem of model specification (Gujarati, 1995; Hair et al., 1998), remedial measures are not infallible. It was decided to start with a model that included all the variables, since they are all interesting from a theoretical point of view. In the case of collinearity between a "news variable" and an "advertising variable," it was chosen to give priority to the news variable. This approach was taken because the effects of advertising intensity were found to be robust. In addition, it will be indicated in the text if a significant advertising variable was excluded from the model.

As mentioned by several authors (Gujarati, 1995; Hair et al., 1998) examining a correlation matrix (which shows only simple correlations between two variables) is not enough to identify multicollinearity. Therefore, if multicollinearity was expected, the values of the Variance Inflation Factors (VIF) were examined. If the removal of the advertising variable decreased the VIF values of all the media variables beneath the threshold of 10 (Hair et al., 1998), the advertising variable was removed from the models. In the rare cases that the removal of the advertising intensity variable was not sufficient, other media variables were removed.

Autocorrelation

Another assumption of the classical linear regression model is that there is "no autocorrelation or serial correlation[28] between the disturbances." Gujarati (1995) illustrated this with the following example: the classical model assumes that an effect of an increase of one family's income on its consumption expenditures does not influence the consumption of another family. This may be the case, however, since the two families want to keep up with each other. Similarly, the classical model does not assume that a damaged reputation in a certain year will be carried over to the next year. Due to the stability of reputations this may be the case, however.

Since some of the respondents participated in the study only once, the assumption of independence may be violated for 2,038 respondents of the total of 9,157 cases[29] in the pooled dataset which was used to test hypotheses 1a, 1b, 2a, and 3a. In order to detect whether autocorrelation was present for that group, the Durbin-Watson d statistic was calculated (see Gujarati, 1995). The limits of d are 0 and 4. If d is 2, one may assume that there is no first-order autocorrelation, either positive or negative. If $d = 0$ perfect positive autocorrelation exists; if $d = 4$ perfect negative autocorrelation exists. Therefore, the closer d is to 0, the greater the evidence of positive autocorrelation, and the closer d is to 4, the greater the evidence of negative autocorrelation. The Durbin-Watson d statistic of 0.89 found in this study suggests[30] that positive autocorrelation is present. In other words, units in time t and time $t+1$ are not independent. No remedial measures were taken, since the auto correlation problem concerned only a part of the respondents. Moreover, the problem was solved by the use of autoregressive models in which the autocorrelation between the residuals was modeled. The use of autoregressive models will be gone into in subsection 3.4.4.

3.4.3 Repeated Measures ANOVA and logistic regression

In addition to OLS regression, hypothesis 4 was also tested with Repeated Measures ANOVA. Repeated measures is used if differences between group means are assessed and if the same respondent provides several measures over time. In the case of the testing of hypothesis 4, it was investigated whether respondents who are grouped according to their associations with a certain company differed in the report mark they assigned to the focal organization. The report mark is measured at two points in time. To put it more formally, a Repeated Measures ANOVA with between-subjects factors design was used. The groups of respondents based on the associations of the respondents with the organization in 1999 and in 2000 formed the between-subjects factor, while the report mark was the within-subjects or repeated measures factor.

Stevens (2002) mentioned three assumptions for a single-group univariate repeated measures analysis: independence of the observations, multivariate normality, and sphericity. Stevens mentioned that a violation of the independence assumption is very serious. In order to assess the independence of observations, Glass and Hopkins (1995) stated that, "Whenever the treatments are individually administered, observations are independent. But where treatments involve interactions among persons, such as "discussion" method or group counseling, the observations may influence each other." In this study, the associations and the report mark were individually administered. In addition, in order to test hypothesis 4, the data were examined only at the level of the organization. This means that the observations were independent. Multivariate normality assumes that the joint effect of the variables is normally distributed. Repeated measures ANOVA and MANOVA are fairly robust against violation of multivariate normality (Stevens, 2002). Sphericity is the assumption that the correlations between the scores for all possible pairs of measures are equal in size (Hampton, 2003). If there are only two levels of a within-subjects factor (as is the case of the within-subjects factor report mark in this study) sphericity is not a problem, since there is only one correlation (Hampton, 2003).

Logistic regression

Most of the hypotheses were tested by the method of Ordinary Least Squares regression, due to the ordinal or interval nature of the dependent variable. In the case of hypothesis 5, the dependent variable (salience) was dichotomous (binary). As mentioned by several authors (Hair et al., 1998; Stevens, 2002), this leads to special problems. First, the error terms of a dichotomous variable follow the binomial distribution instead of the normal distribution, so the statistical tests based on the assumptions of normality no longer apply. Second, the variance of a binary variable is not constant, which leads to heteroscedasticity. Logistic regression was developed to specifically deal with these issues.

Logistic regression differs from multiple regression in that it directly estimates the probability of an event occurring, because there are only two possible outcomes for the dependent variable (Hair et al., 1998; Stevens, 2002).

3.4.4 Miscellaneous methodological issues

This subsection deals with miscellaneous methodological issues. Since the impact of missing data can be detrimental because of its potential "hidden" biases of the results and its affect on the generalizability of the results (Hair et al., 1998), the role of missing data in this study will be gone into first. Then it will be discussed why we chose not to center the variable media favorability. Subsequently, it will be addressed how the pooled dataset was controlled for company-specific unobservable factors. Following this, the use of autoregressive and non-autoregressive models and the role of control variables will be elaborated upon. Finally it will be argued why Ordinary Least Squares regression – rather than multilevel analysis – was used to analyze the data of the ten organizations, the pooled data.

Missing values

Missing data played almost no role in the case of the media variables. The media use of the respondents is known just as the content of the media. If a respondent did not use a certain medium title, an "0" was entered for the media variables of this medium. The missing public opinion data were not replaced with the mean, since this would have created artificial relationships in the dataset.

Centering media favorability ... or not?

The variables "favorability of success & failure news" and "favorability of support & criticism news" consisted of the summed scores of the direction of the news concerned. However, statisticians may perceive the term "favorability" as an interaction term, consisting of the amount of success & failure news multiplied by the direction of success & failure news. Therefore, distracting the mean (to center the scores) of the variables "amount of success & failure news" and "direction of success & failure news" before computing the variable "weight," as suggested for example by Aiken and West (1991), was considered. Centering variables prior to the analysis of moderated multiple regression has been advocated to yield a proper interpretation of the resulting regression models and to reduce multicollinearity (Kromrey & Foster-Johnson, 1998).

However, Katrichis (1992) pointed out that this technique produces systematically biased estimates of main effects. Kromrey and Foster-Johnson (1998) demonstrated that the two methods are equivalent, yielding identical hypothesis tests associated with the moderation effect and regressions models that are functionally equivalent. Kromrey and Foster-Johnson (1998) concluded that centering has important implications for both interpretations and hypothesis-testing in hierarchical models, but that in the case of least squares regression analysis, they subscribe to the conclusion reached by Cohen (1978, p. 865): "One might just as well not bother." Furthermore, the interpretation of a regression model, which contains centered variables, is more complex than the interpretation of an uncentered regression model. As explained by Aiken and West (1991), in social science research the predictors are often measured on ordinal scales in which the value zero has no meaning. However, when predictors are centered, then the value of zero is the mean of each predictor. This means that in the following centered regression equation: $\hat{Y} = b_1 X + b_2 Z + b_3 XZ + b_0$, the b_1 coefficient for X represents the regression of Y on X at the mean of the Z variable. Centering creates a value of zero on a scale that is typically meaningful. Due to these drawbacks of centering, it was decided not to center the variable media favorability.

Autoregressive and non-autoregressive models

Two separate regression models will be presented in the chapter about the results, Chapter 5. Models that include the lagged dependent variable, in this case the reputation of the previous year (the autoregressive models), and models without the reputation of the previous year (the non-autoregressive models). As explained by Pennings, Keman, and Kleinnijenhuis (1999), the advantage of entering the lagged dependent variable as an independent variable into the model has the advantage that the independent variables have to explain short-term shifts in order to become significant. Another advantage of the inclusion of lagged dependent variables into the model is that they will tend to reduce the autocorrelation within the residuals.

As mentioned by Pennings, Keman, and Kleinnijenhuis (1999), one should keep in mind that the lagged dependent variable catches all the long-term effects of slowly operating variables. Achen (2000) stated that in many time series applications in the social sciences, the inclusion of lagged dependent variables suppresses the explanatory power of other independent variables due to autocorrelation.

Controlling for unobservable company specific factors in the pooled dataset

In the pooled analyses, reputation was controlled for the influence of the organizations. In other words, it was assumed that respondents would probably assign a different report mark to Albert Heijn than to the NS. This can be caused not only by the effects of media coverage and advertising intensity, but also by factors not observed in this study, e.g., the experiences of the consumers when they shop at Albert Heijn or when they travel by train. Therefore, dummy variables were created for each of the organizations. In the case of the pooled regression models, *confirmatory specification* (Enter) is used to make sure the dummy variables are entered into the pooled models.

It could be a point of discussion as to whether it is useful to use the dummy variables in the autoregressive pooled models; in that case, the lag dependent variable may control for the unobservable company specific factors. Nevertheless, due to reason of comparison of the non-autoregressive and the autoregressive models, the dummy variables were entered in the autoregressive pooled models as well.

Control variables

It is likely that reputation is influenced by media coverage and by advertising intensity. Since both key variables were analyzed simultaneously in one model, the effects of media coverage were controlled for the effects of advertising intensity and vice versa.

Due to reasons of parsimony (see for example Stevens, 2002), not all the background characteristics of the respondents were entered in the models. Some of the models were controlled for the educational level of the respondents, since it might be reasonable to expect that that background variable may explain the effects of media coverage on reputation.

Clustering of the pooled data within respondents

The results in this study were analyzed on an organizational level and on a pooled level, in which the data of the ten organizations were analyzed together in one model (see Chapter 5). The analyses of the pooled dataset may raise questions about the independence of observations. Since respondents answered questions about several organizations, the organizations were clustered "within the respondents." The problem associated with hierarchical data structures is the underestimation of the standard error, and thus, the greater risk of type-1 errors, especially in OLS regressions (Peter, 2003). Therefore, it might be wondered whether multilevel modeling (Snijders & Bosker, 1999) should have been applied.

In our opinion, however, the structure of the pooled dataset used in this study, is not a classical multilevel model. Multilevel data structures result from multistage sample surveys so that pupils (1) are nested within classes (2), and in schools (3). Aitkin, Anderson, and Hinde (1981) demonstrated the importance of taking the grouping of respondents into account. They re-analyzed the study by Bennett

(1976) on the effect of teaching styles on pupils' progress. Bennett (1976) found that formal methods of teaching led to greater progress in the basic skills of the pupils. The data were analyzed using multiple regression techniques, the individual children were the unit of analysis, and the groupings within teachers were ignored. Aitkin et al. (1981) showed that when the analysis accounted for the grouping of children into classes, the significant differences disappeared and the "formally" taught children were no different from the others. The Centre for Multilevel Modelling (2004) mentioned that what happened there was that the children within any one classroom tended to be similar in their performance because they were taught together. As a result, they provided less information than would have been the case if the same number of students had been taught separately by different teachers.

In this study, multilevel analysis was not applied because it was not based on multi-stage sampling. Although the pooled data were clustered within the individual respondents, they were not clustered within *groups* of individuals. Also, the pooled models were in agreement with the models per organization, in which the data were not clustered within groups of individuals.

3.5 SUMMARY

In the present study, three types of data were used: media data, advertising data, and public opinion (reputation) data. In order to gather the public opinion data, TNS NIPO (Dutch Institute for Public Opinion and Market Research), market leader of market research agencies in the Netherlands, kindly conducted our survey among their *panel* of respondents. The panel survey was conducted at three periods in time, namely in the summers of 1998, 1999, and 2000. The survey was filled in by 446 respondents in each of the three years; 606 respondents took part once, and 306 respondents took part twice. Since the media use of each of the respondents was known, it was possible to take the individual media use of the respondents into account when assigning the news and the advertising data to the respondents.

Media coverage about the focal companies in newspapers and on television was analyzed for July 24, 1997 to July 22, 2000. This period was selected in agreement with the public opinion polling data: all the media coverage was analyzed from the year preceding the poll. The newspaper articles were selected from the five largest Dutch daily newspapers. In the Netherlands, subscriptions to daily newspapers are responsible for 85% of the total circulation (Bakker, 2003). This means a respondent's subscription(s) to a certain newspaper is a good predictor of the print news consumption of the respondent, since the impact of single issues is relatively small. Of the television news, the news broadcasts of a public broadcasting company and a commercial channel were analyzed. In total, the media coverage about the focal organizations and sectors consisted of 9,285 newspaper articles and 2,225 news items, which together resulted in 15,883 assertions.

The advertising data were measured in advertising expenditures per focal company per medium. The data were obtained from the Dutch agency BBC. The media and the period from which advertising expenditures data were gathered, were selected to match the media and the period selected for the content analysis.

The Network analysis of Evaluative Texts (NET) method was used to analyze the news, since it provided analytical information about the issues with which the organization was in the news, how the organization was portrayed in the media in relation to its competitors, and whether the organization was depicted as successful or not.

Corporate reputation was measured in the present study as attitude towards an organization. There were two reasons for doing so. The first reason was that reputation was conceptualized in the present study as an attitude towards an organization or sector. The second reason was that closed methods (such as attitude scales) are preferred when the objective is to monitor corporate reputation over time, as was the case in the present study (see Van Riel et al. 1998).

4

MEDIA PROFILES, ADVERTISING INTENSITY, AND REPUTATION OF THE ORGANIZATIONS

This chapter will first describe the media profiles and the advertising expenditures of the ten organizations (the independent variables). The media profiles are described in such a way that the sections match roughly with the order of the hypotheses. This means that the amount of media coverage (hypothesis 1) will be described first, followed by a general overview of the various types of news per organization and a description of the amount of news per medium. In the second section, the focus will be on the success & failure of organizations (hypothesis 2a and 2b). Subsequently, the support & criticism received by the focal organization will be described (hypothesis 3a and 3b). In section 4.4, issue news will be described (hypothesis 4, 5 and 6), together with the associations of respondents (hypothesis 4).

In section 4.5 (hypothesis 7), the advertising expenditures per organization/sector will be described per research period. Finally, in section 4.6 the dependent variables corporate reputation, the agreement with the policy of an organization, and the perceived success of the policy of an organization will be focused upon. All news described in the tables and figures in this chapter is weighted by applying the weight factor (see Box 3.3). Due to the scope of this chapter, only the most important items of the tables will be discussed.

4.1 THE AMOUNT OF NEWS PER PERIOD, NEWS PER MEDIUM, AND THE VARIOUS TYPES OF NEWS

Taken over the entire research period, most news (42%) about the focal companies appeared in the first period (July 24, 1997 to July 25, 1998). This decreased in the subsequent research periods, with 32% of the news in the second research period (July 26, 1998 to August 14, 1999), and 26% of the news in the third research period (August 15, 2000 to July 22, 2000). The total amount of news per organization within a certain year is shown in Figure 4.1. The total amount of news per organization within the whole research period is shown in the row totals of Table 4.1 in the next section. Overall, the Dutch police received most media

FIGURE 4.1. **Amount of news per organization within a period**

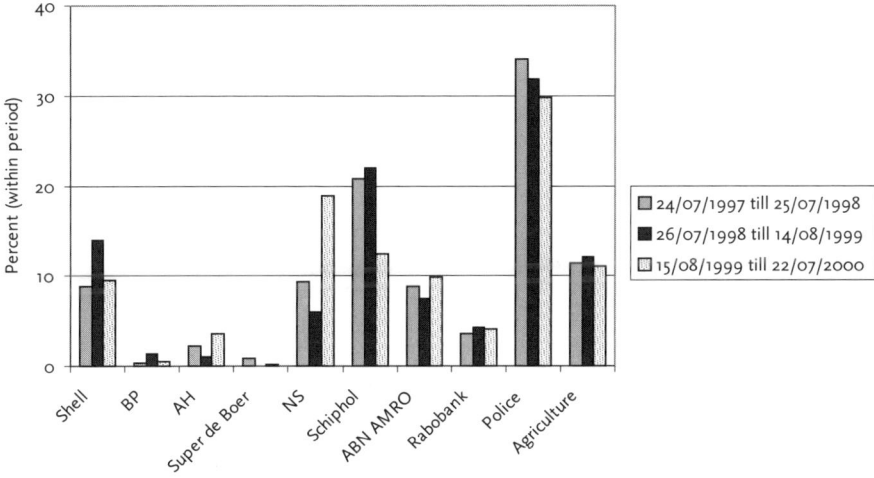

Note. N = 15,884 assertions (weighted)

attention (32%), followed by Schiphol airport (19%), the Dutch agricultural sector (12%), and the NS (11%).

Shell received an average amount of media coverage (11%; see Table 4.1). BP, on the contrary, did not receive much coverage (1%; see Table 4.1). An explanation for this difference in media coverage may be the country of origin. BP is a British company, while Shell is a Dutch-British company. Both companies received most media coverage in the second research period (summer 1998 to summer 1999).

The supermarkets Albert Heijn (2%) and Super de Boer (0.4%) were not often in the media. A reason for this low amount of media attention may be that although their parent companies are quoted on the stock exchange, the supermarket chains themselves are not quoted on the stock exchange. The NS received an average amount of media coverage (11%). Compared to the other organizations, the NS received a great deal of media coverage in the third research period (summer 1999 to summer 2000), when media coverage about the other organizations was low. Compared to the other organizations, Schiphol received a lot of media coverage (19%), especially during the first and second research periods.

Over the total research period, ABN AMRO received 9% of the total amount of news about the focal companies. Within the total amount of news about ABN AMRO, 43% comes from the first research period, the summer of 1997 to the summer of 1998. As compared to ABN AMRO, Rabobank was hardly in the news (4%). This is less than half the amount of the media coverage on ABN AMRO. A reason for this could be that Rabobank is not quoted on the stock exchange. The Dutch police received the most media coverage (32%) of the focal organizations

and sectors. The major part of the news about the police focused on their actions
to combat crime. Compared to the other organizations, the agricultural sector received an average amount of news (12%).

Three different types of news
Figure 4.2 shows the various types of news per organization. Within the ten organizations, "issue news" is the type of news that occurred most often. It made up
38% of the total amount of the news about the ten focal companies. Shell, which
was in the news with the issues profit, economy, and the environment, got relatively more issue news than BP. Due to its introduction of new products, Albert
Heijn received more issue news than Super de Boer. The NS and Schiphol were
both frequently in the news as well, and nearly half of the news about the organizations was issue news. It seems that companies frequently in the news will also
receive a relatively large amount of issue news.

Next to issue news, support & criticism news occurred most (19%) in the
media coverage of the focal companies. BP and Super de Boer received a relatively
large amount of support & criticism news due to the mergers in their sector. In
the case of the other companies, support & criticism news emanated from failed
or successful cooperation with partners from the same industry (Shell, ABN AMRO),
employees (NS), and the government (NS, Schiphol).

The agricultural sector received a lot of support & criticism from the government. In the case of the police, support & criticism came not only from the government but also from the media, as will be elaborated upon later.

At 17%, the total amount of success & failure news came close to support & criticism news. This was mainly due to the fact that within the total amount of media
coverage, the police were covered most, and most (26%) of the news on the police
was about success. Success news about the police reported mainly on their successful crime-fighting actions: "Police in The Hague catches drug baron" (*Trouw*,
December 27, 1997, p. 8); "Police round up drug gang" (*Algemeen Dagblad*, February 25, 1998, p. 3).

As is explained in Chapter 3, these are examples of "results" (RES) predicates.
They illustrate not only the negative attitude of the police towards crime, they also
demonstrate the success of the police. The types of news per organization will be
discussed in more detail in the next subsections.

A large part of the media coverage (26%) about the ten organizations consisted of
"other news." Support & criticism given by the focal organization was categorized
under "other news," since the focus of this study is on the effects of support &
criticism *received* by the focal organization. It does not provide a theoretical framework for the effects of support & criticism *given* by the focal organization. In addition, support & criticism news and success news about an organization *other* than
the ten focal organizations on the object position were categorized in this category.

FIGURE 4.2. **The various types of news per organization**

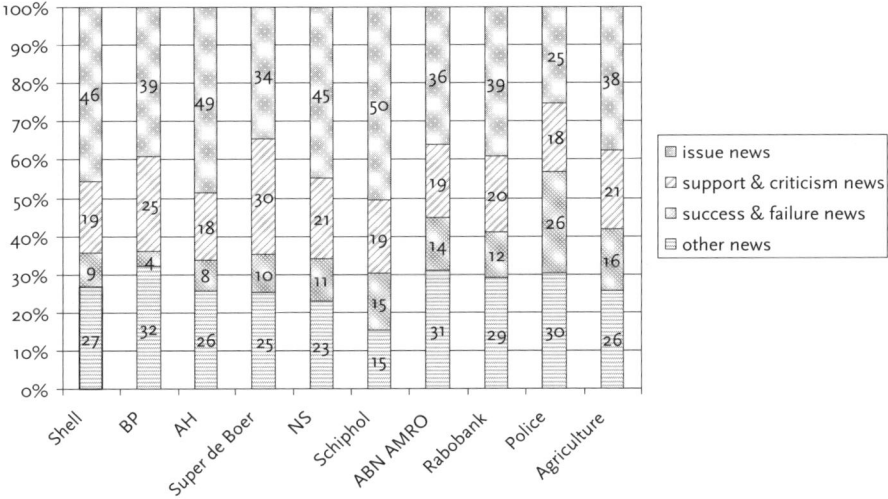

Note. Judgment extrapolation was applied, which means that the evaluation of the focal organizations by the media belongs to the category of support & criticism news. *N* = 15,884 assertions (weighted)

Headlines that covered the name of the focal organization only (such as in the headline "Shell" in *de Volkskrant*, June 10, 1998, p. 2), were included in the category "other news" as well. In the case of issue news, from a theoretical point of view it does not matter whether the focal organization is in the subject or object position as long as the issue is related to the focal organization. This also explains why issue news occurred most.

News about the organization per medium
Table 4.1 shows the news about the organizations per medium. As indicated earlier, the Dutch police accounted for 32% of the total amount of media coverage on the ten focal organizations/sectors. A relatively large amount of the media coverage about the police came from the television broadcasting companies NOS (36%) and RTL (37%). In contrast to the police, Schiphol received less media coverage by the NOS. In general, the media spent 19% of the total amount of media coverage on Schiphol, whereas the NOS spent 17% of the media coverage on the airport. Compared to the other media, *Trouw* paid a great deal of attention to Schiphol (23%). Within the total research period, the agricultural sector received 12% of the total amount of news about the focal organizations. RTL gave a relatively large amount of coverage to the agricultural sector (15%) while the opposite was the case for *Algemeen Dagblad* (9%).

TABLE 4.1.   **News about the organization per medium**

	Alg Dagbl	NRC	Telegraaf	Trouw	Volkskrant	NOS	RTL	Total
	Newspapers					Television		
Police	30%	25%	30%	30%	26%	36%	37%	32%
Schiphol	21%	21%	19%	23%	20%	17%	18%	19%
Agriculture	9%	10%	10%	13%	12%	11%	15%	12%
Shell	11%	17%	8%	11%	12%	9%	10%	11%
NS	14%	11%	8%	11%	12%	11%	9%	11%
ABN AMRO	8%	10%	14%	6%	11%	8%	5%	9%
Rabobank	3%	2%	9%	3%	3%	4%	4%	4%
Albert Heijn	4%	2%	2%	3%	4%	1%	2%	2%
BP	1.1%	2.4%	0.4%	0.9%	0.9%	0.3%	0.5%	0.7%
Super de Boer	0.1%		0.8%	0.1%	0.1%	0.6%	0.5%	0.4%
	100%	100%	100%	100%	100%	100%	100%	100%

Note. N =15,884 assertions (weighted)

Shell was covered relatively frequently by NRC *Handelsblad*. The oil company received 11% of the total amount of news about the ten focal organizations, while it received 17% of the media attention in NRC *Handelsblad*. The NS got a lot of media coverage in *Algemeen Dagblad* (14%) and compared to the other media, was rarely in the news in *De Telegraaf* (8%). ABN AMRO was covered most often by *De Telegraaf* and least often by RTL. Like ABN AMRO, Rabobank also received a lot of media attention in *De Telegraaf* (9%), while it only received 2% of the media attention in NRC *Handelsblad*. *Algemeen Dagblad* and *de Volkskrant* paid most attention (4%) to Albert Heijn and television broadcasting company NOS the least (1%). NRC *Handelsblad* spent 2.4% of its attention on BP. This might not seem like a lot, but it made up 30% of the media coverage about BP. *De Telegraaf* devoted a great deal of media coverage to Super de Boer (.8%) compared to the other media.

4.2 SUCCESS AND FAILURE OF ORGANIZATIONS

This section will describe success and failure per organization; the organizations that received most failure will be described first. The scale of the mean direction of success & failure news ranges from −1, which means that only failure was attributed to the organization, to +1 which means that only success was attributed to the organization. All organizations except the police (.18) and Rabobank (.12) were attributed some failure within the total research period, ranging from −.41 in the case of the agricultural sector to −.08 in the case of ABN AMRO.

TABLE 4.2.   **Mean direction and percentage of success [+1] and failure [-1] news**

	07/24/1997 to 07/25/1998			07/26/1998 to 08/14/1999			08/15/1999 to 07/22/2000			Total research period		
	% of total news	Mean	n	% of total news	Mean	n	% of total news	Mean	n	% of total news	Mean	n
Agriculture	12%	-.28	91	16%	-.37	100	22%	-.58	99	16%	-.41	289
Shell	8%	-.44	46	8%	-.44	56	13%	.12	49	9%	-.26	151
Super de Boer	5%	-1.00	2	25%	-1.00	1	33%	.46	3	11%	-.25	6
NS	9%	-.32	55	13%	-.05	40	12%	-.11	95	11%	-.16	190
BP			0	3%	1.00	2	18%	-.65	4	4%	-.15	5
Schiphol	13%	-.21	188	18%	-.17	203	14%	.10	71	15%	-.15	462
Albert Heijn	8%	-.12	13	4%	-.85	2	9%	.05	13	8%	-.10	28
ABN AMRO	15%	-.15	91	16%	.10	59	10%	-.19	41	14%	-.08	192
Rabobank	7%	.46	17	18%	.20	39	13%	-.34	20	12%	.12	76
Police	28%	.11	647	24%	.15	380	26%	.37	321	26%	.18	1348

Note. The percentages do not add up to 100%, because they describe the percentage of success & failure news about the focal organization compared to the total amount of news about the focal organization. The last column, the percentage of success & failure news within the total research period, matches with the success & failure news section in the stack bars in Figure 4.2. n = 2,745 assertions (weighted).

Fighting to continue farming

Of the ten focal companies/sectors, the tenor of success & failure news over the whole research period was most negative (–.41) in the case of the agricultural sector. Moreover, next to the police, the agricultural sector got the most success & failure news of the ten focal companies (16%). In every research period the agricultural sector received more success & failure news than in the preceding research period (12%, 16%, and 22% respectively), while the tenor of success & failure news was also more negative in each research period (the mean direction is –.28, –.37, –.58 respectively).

In the first period, the success of organic farming and the difficult situation of the farmers got the attention of the media. "Eight farmers quit every day" (*De Telegraaf*, November 1, 1997, p. 7) and "Organic farmers are doing well" (NOS, June 21, 1998) are two headlines that illustrate this. The hog farmers were having a hard time, and then Minister of Agriculture Jozias van Aartsen seized on swine fever to propose a complete restructuring of the hog-farming sector: "Twenty-five percent fewer hogs and 14 million kilos reduction in phosphate discharge" (*NRC Handelsblad*, April 7, 1998, pp. 1, 2). In the second period, heavy rainfall in September 1998 resulted in damage for farmers and market gardeners in the south-western and northern parts of the Netherlands. NOS (October 29, 1998) reported, "The damage for farmers and market gardeners due to the rainfall has run into the hundreds of millions of euros at the moment." In addition, the malaise in the hog sector was still in the news.

In the third research period, the headlines about success & failure were approximately the same in absolute terms (100), but they formed a bigger part of the news about the agricultural sector, since there was less news in the third period. The difficult situation of the farmers was stressed even more (–.58), compared to the previous research period: "Hog farmers quit farming" (NRC Handelsblad, August 16, 1999, p. 3); "Dutch farmers are sick of it" (RTL, October 18, 1999). The mean direction of success & failure news (–.58) indicated that there were also some successes: "Market gardeners happy with suggestion for greenhouses in the Zuidplaspolder" (Trouw, September 21, 1999, p. 4).

Stormy times ahead for Shell, and resistance in Nigeria
Shell got relatively less success & failure news in the first and second research periods (8% in both periods). The tenor was severe, however, and Shell was attributed even more failure (both periods –.44) than the agricultural sector in these periods. The failure news about Shell in the first research period consisted, for example, of the worsening financial expectations for Shell Oil: "Shell Oil sees stormy times ahead" (NRC Handelsblad, January 24, 1998, p. 19). Other failure news about Shell went into the fact that the company missed out on a huge contract in Peru: "Shell missed out on a contract worth billions" (De Telegraaf, September 16, 1997, p. 29). Success news about Shell in the first research period went into Shell's size.

In the second research period, Shell was in the news with failure news about various topics such as the problems with Shell Nigeria: "Kidnapping at Shell Nigeria" (NRC Handelsblad, June 29, 1999, p. 13). The decrease in profits led to failure news about Shell as well: "Shell's profits collapse because of crisis and low oil price" (de Volkskrant, November 6, 1998, p. 15). There was relatively more success & failure news in the third period (13%) and the tenor of success & failure news was positive (.12), contrary to the two preceding periods. NOS news reported about the good financial results of Shell: "Shell had a good year, profits increased by 38%" (February 10, 2000). Failure news about Shell went into the violation of the oil embargo against Iraq: "The Royal Dutch Shell Group has to pay a fine of 2 million dollars for the purchase of illegal oil from Iraq" (RTL, April 25, 2000).

Despite merger, Super de Boer, Vendex, and De Boer not the biggest supermarket chain
In absolute terms, Super de Boer together with BP received the least amount of success & failure news (n = 6) of the ten focal companies/sectors. The success & failure news about Super de Boer in the first research period came out of the merger between Vendex and the Super de Boer's parent company, De Boer Unigro. The supermarkets Super de Boer, Edah, Konmar, and Basismarkt are part of this new company. NOS reported that "the new company will not be the biggest supermarket chain" (May 25, 1998). De Boer Unigro was in the news in

the second research period due to a lower market share (*De Telegraaf*, September 3, 1998, p. 3). In the third research period, Super de Boer was in the news as "victim of an April Fool's Day joke" (*de Volkskrant*, April 10, 2000, p. 2). Unknown persons distributed leaflets containing 12 spectacular bargains available at Super de Boer, and employees in the stores were often asked for these products.

Although the NS fails (delays and quarrels,), it was prosperous and therefore also successful
The tenor of success & failure news about the NS was characterized by ups and downs, although over all the three research periods there was more failure than success attributed to the NS (the tenor is −.32 in the first, −.05 in the second, and −.11 in the third research period). Disruption of train traffic led to failure news about the NS in the first research period: "Heat plays tricks on the NS" (*Algemeen Dagblad*, August 14, 1997, p. 3); "NS train traffic is in disarray today because of falling leaves" (*RTL*, November 7, 1997). In some headlines it was not mentioned *what* was wrong but simply *that* there was something wrong with the NS: "Chaos at NS; "like a dumb blonde" (*NRC Handelsblad*, June 17, 1998); "Chaos again at the NS" (*RTL*, June 17, 1998). There was also some news about the success of the financial results: "NS receives millions of the state thanks to contracts" (*de Volkskrant*, April 4, 1998, p. 3).

In the second research period, headlines reported about the financial success of the NS: "From a financial point of view the NS is doing well" (*NOS*, March 4, 1999). Another NS success was that it was deemed favored to win the operation of the high-speed railway: "NS favored for operation of high-speed railway" (*de Volkskrant*, June 16, 1999, p. 7). As in the previous research period, the NS was in the news with the disruption of train traffic: "Failure of the NS" (*Trouw*, September 9, 1998, p. 3).

In the third research period, NS still had to do its best to acquire the high-speed track, which led to success news: "NS gets a second chance" (*NOS*, November 12, 1999). The acquisition of the high-speed track also led to failure news, however: "NS is sidelined deliberately in the case of high-speed train" (*De Telegraaf*, December 14, 1999, p. 5). Other failure news in this period was the resignation of marketing director Van Eeghen due to his conflict with the chairman of the board, Den Besten: "NS senior executive leaves after conflict about ticket offices" (*Trouw*, September 28, 1999, p. 7). The decision of the marketing director to close down the ticket offices sooner brought him into conflict with the chairman of the board.

BP failed due to forced sale of gas stations, and was successful as the third largest oil giant
Of the ten focal companies/sectors, BP received the least amount of news about success & failure in every respect (4%, *n* = 5). A headline in the second research period went into the takeover of Amoco by BP (*de Volkskrant* August 12, 1998, p. 1). According to the newspaper, next to Shell and Exxon, BP and Amoco together

would be the biggest oil giant in the world. In the third period, BP was in the news with failure because the cabinet and the oil companies agreed that the system of perpetual concessions would come to an end. This meant that 50 gas stations would be sold by auction, and BP was one of the companies whose gas stations would be sold (RTL, February 11, 2000). BP received success news in the third research period as the biggest gas producer in the United States.

Many passengers at Schiphol led to success, and also to claims due to noise nuisance
Of the ten organizations, Schiphol was in the middle group when it came to the amount (15%) and the tenor (–.15) of success & failure news over the whole research period. In contrast to the first two research periods (–.21 and –.17), the tenor of success & failure news about Schiphol was positive in the third period (.10). In the period from mid-1997 to mid-1998, Schiphol was attributed failure due to several topics. One of these was noise nuisance: "350 claims against Schiphol" (NRC Handelsblad, November 28, 1997, p. 3). Another topic was the lack of space for expansion at Schiphol, and the delay in the decision concerning the expansion: "Decision on Schiphol possibly resolved after the elections" (de Volkskrant, March 14, 1998, p. 3). There were also a couple of successes for the airport. One of them was the financing of a second Schiphol airport in the sea by the aviation industry itself: "Schiphol can be financed privately" (de Volkskrant, January 26, 1998, p. 2). Another success for Schiphol was that a judge determined Schiphol was allowed to violate the legal noise conventions: "Schiphol allowed to make too much noise" (Trouw, October 10, 1997, p. 3).

From mid-1998 to mid-1999, the large number of passengers visiting Schiphol contributed to the success news about Schiphol: "Busiest time on record at Schiphol" (RTL, January 1, 1999). The violation of the noise convention continued to lead to confrontations with environmentalists and neighborhood residents: "Action against Schiphol" (de Volkskrant, December 9, 1998, p. 3).

In the period of mid-1999 to mid-2000, Schiphol was in the news with several successes: "Schiphol welcomes 500 millionth passenger" (Trouw, June 28, 2000); "Schiphol best in German test" (De Telegraaf, October 30, 1999, p. 5); "Busiest time ever at Schiphol today" (RTL, July 22, 2000). Another success for Schiphol was the new consultation platform about the future of Schiphol: "New consultations about Schiphol" (Algemeen Dagblad, January 24, 2000, p. 5). Environmental and neighborhood organizations did not join the platform, however. Aside from these successes there were also some setbacks for Schiphol: "Schiphol still has extra rules" (NRC Handelsblad, June 20, 2000, p. 3). Schiphol will be allowed to expand after 2003 as long as it meets the new rules for noise, air pollution, and safety.

Albert Heijn fined, but still number one
On average, Albert Heijn did not do too badly with regard to the tenor of success & failure news compared to the rest of the organizations (–.10). Failure news

about the supermarket in the first research period was related to breaking of environmental laws: "Fines for Albert Heijn for breaking environmental regulations" (*Trouw*, October 10, 1997, p. 4). The supermarket had to pay a fine because it did not follow the "Law for environmentally hazardous fabrics." This regulation obliges supermarkets to check their cooling systems on a quarterly basis. RTL news reported that Albert Heijn was breaking the personal registration laws with its "Bonus Card" (February 19, 1998; February 20, 1998). A merger between Vendex and De Boer-Unigro was success news for Albert Heijn, because Albert Heijn was still number one with regard to turnover: "The new chain of shops comes second to Albert Heijn with regard to turnover" (*RTL*, May 25, 1998). Taken as a whole, there was somewhat more failure news than success news about Albert Heijn (–.12) in the first period.

There was scarcely any success & failure news about Albert Heijn in the second research period, from mid-1998 to mid-1999 (*n* = 2). In the third research period, Albert Heijn was successful because its branch in the South-Eastern Amsterdam district did not close its doors: "Albert Heijn supermarket in Bijlmermeer will actually stay open" (*De Telegraaf*, December 14, 1999, p. 4). The management of the supermarket chain had considered closing the store due to violence in the neighborhood. The leakages due to heavy rainfall in an Albert Heijn store at their Museumplein location in Amsterdam contributed to failure news. Altogether there was somewhat more success news than failure news about Albert Heijn in the third research period (.05). Since Albert Heijn got little media attention compared to the other organizations, there were not many assertions that could be classified as success & failure news about Albert Heijn (*n* = 28).

ABN AMRO financially successful but on its own, and damaged by the IPO of World Online
With regard to the direction of success & failure news over the whole research period, ABN AMRO did not do too badly (–.08) compared to the rest of the ten organizations. However, it did slightly worse than its competitor Rabobank, as will be outlined in the next subsection. In the first period, ABN AMRO was in the news with its failures to take over other companies: "Fortis surpassed ABN AMRO in battle to take over bank" (*NRC Handelsblad*, June 6, 1998, p. 1). ABN AMRO had some coverage about success with the returns of its subsidiary: "More profits for ABN Lasalle" (*NRC Handelsblad*, January 23, 1998, p. 17).

The financial success of ABN AMRO contributed to the success news about ABN AMRO in the second research period: "The bank achieved a net profit of 2,445 million guilders" (*NOS*, August 20, 1998). Although there was a bit more news about the success of ABN AMRO in the second research period, the bank was also in the news with failure, for example with the problems in the ABN branch in Surinam: "Problems for ABN AMRO Surinam not resolved" (*De Telegraaf*, October 14, 1998, p. 2). Due to a conflict between the Central Bank of Surinam and ABN AMRO in that country, the latter closed it doors for some days.

Within the whole research period, the tenor of success & failure news about
ABN AMRO was most negative in the third research period (–.19). This is due in
part to the disaster surrounding the introduction of Internet company World
Online on the stock exchange. In the first half of 2000, ABN AMRO was in the
news with this fiasco regarding the initial public offering (IPO) of World Online,
which was supervised by ABN AMRO. Due to a massive advertising campaign, a
broad public signed up to purchase shares. Moreover, there was a lot of free publicity about Nina Brink, owner of World Online. The media wrote about the sums
of money she was going to make with the IPO of her company. However, from the
very first day (March 17, 2000) World Online was challenged that it had gone to
the stock exchange with major sales of its shares. It appeared that Nina Brink had
an interest in the American company Baystar, which sold the World Online shares.
This accelerated the fall of the prices. ABN AMRO was accused of providing insuffi-
cient information in the prospectus. World Online and ABN AMRO were both
prosecuted by representatives of the thousands of frustrated private investors.
ABN AMRO acknowledged that this affair damaged its reputation (*NRC
Handelsblad*, May 11, 2000, p. 13); Nina Brink resigned.

There was also some news about the successes of ABN AMRO: "ABN AMRO is
getting bigger in the exchange business" (*de Volkskrant*, April 8, 2000, p. 35); "Im-
pulse for business bank ABN AMRO" (*NRC Handelsblad*, May 5, 2000, p. 11).

*Success for Rabobank in the first two periods, failure in the third due to affair in Doe-
tinchem*
In contrast to ABN AMRO, the tenor of success & failure news about Rabobank is
positive over the whole research period (.12). The success in the first research
period stemmed mainly from its 100-year anniversary: "Cycling day at 100-year
old Rabobank" (*Algemeen Dagblad*, June 13, 1998, p. 16). In the second research
period, the media reported about the "D'n Anwas affair." The investment club
"D'n Anwas" was a client of the Rabobank branch in Doetinchem. The members
of the club lost millions of guilders due to mismanagement by the bank. This resulted in the dismissal of the director and the resignation of the board of the
Rabobank branch in Doetinchem, "Board of Doetinchem Rabobank resigns as
well" (*Algemeen Dagblad*, January 4, 1999, p. 11). The arrival of a new chief executive contributed to success news about Rabobank: "Hans Smits, former chairman of the board at Schiphol, to be the new chief executive at Rabobank" (*NOS*,
September 9, 1998). Due to this success, the tenor of success & failure news
about Rabobank was positive (.20) in the second research period.

In the third research period, the tenor of success & failure news was negative
(–.34), however. Rabobank was still in the news with the investment affair in
Doetinchem: "Rabobank under siege again" (*Algemeen Daghlad*, March 2, 2000,
p. 11); "Search of premises at Rabobank and D'n Anwas investors" (*de Volkskrant*,
May 24, 2000, p. 2).

Police more successful when fighting crime
Of the ten companies/sectors, the tenor of success & failure news was most positive in the case of the police, namely .18. In every research period, the police were
more successful (in the first period .11, in the second research period .15, and in
the third research period .37). The police were successful in the first period with
fighting crime: "Amsterdam police eliminate six big drug gangs" (*de Volkskrant*,
November 5, 1997, p. 1); "Police in the 'Delfshaven' area of Rotterdam have confiscated an illegal cannabis nursery" (*RTL*, January 6, 1998). Failure was also attributed to the police in the first research period, however. This was mainly due
to mistakes by the police in their fight against crime. Incidents in Groningen
were an example of this type of news: "Lots of mistakes due to absence of police
in Groningen area" (*NRC Handelsblad*, January 6, 1998, p. 1). The case of the former police officer Lancee, who was wrongfully accused of incest, added to more
news about the failure of the police; "Apologies to Lancee because of blunders by
police and judicial authorities" (*Algemeen Dagblad*, October 8, 1997, p. 7). The
lack of money in the police department also contributed to the failure of the police: "Annual police deficit of 440 million" (*NRC Handelsblad*, May 18, 1998, p. 3).

In the second period, from mid-1998 to mid-1999, a bit more success was attributed to the police (.15). As in the previous research period, there was a lot of
success & failure news that had to do with police work, for instance in the following news item: "Amsterdam police dismantle big international drug network"
(*RTL*, April 28, 1999). There was some failure of the police, too: "Police surprised
by riots" (*Algemeen Dagblad*, February 2, 1999, p. 21); "Police car ends up in the
water in Edam" (*De Telegraaf*, July 30, 1998, p. 1). The bad financial situation of
the police was still in the news: "Deficit of 1.7 billion for police" (*Algemeen Dagblad*, November 5, 1998, p. 1); "Police Gouda goes bankrupt" (*NRC Handelsblad*,
October 14, 1998, p. 3).

In the third period, from mid-1999 to mid-2000, most success was attributed
to the police (.37). The successes of the police when fighting crime contributed substantially to their success: "Success for police in six murder cases" (*De Telegraaf*,
August 28, 1999, p. 3); "The police has confiscated a big bunch of ecstasy pills."
(*RTL*, April 10, 2000); "Police arrests lover boys" (*Trouw*, April 18, 2000, p. 4). Failure was also attributed to the police due to their actions, their workload, and the
lack of money: "Man died following police action" (*de Volkskrant*, November 24,
1999, p. 7); "Amsterdam police in financial problems" (*De Telegraaf*, December 31,
1999, p. 3); "Exodus increases workload of the police of the cities in western Holland" (*de Volkskrant*, January 4, 2000, p. 3). Furthermore, the action of the police
during the European soccer championships in the summer of 2000 contributed to
success & failure news about the police: "According to Minister De Vries of the Department of the Interior, the police are well prepared for the 2000 European soccer
Championships" (*NOS*, June 6, 2000).

4.3 SUPPORT AND CRITICISM RECEIVED BY THE FOCAL ORGANIZATION

This section will describe the support and criticism received by the organization. Support & criticism news was the second most frequently occurring type of news. It made up 19% of the total amount of news about the ten focal companies. Table 4.3 shows that in the case of Super de Boer and BP, the amount of support & criticism news was far above the average amount of support & criticism news (30% and 25% respectively). In BP's case this was due to its merger with Amoco. The merger of the parent company of Super de Boer, De Boer Unigro, with Vendex Food resulted in support & criticism news about Super de Boer. This will be elaborated upon later in this subsection.

TABLE 4.3.   **Mean direction and percentage of support [+1] & criticism [-1] received by organization** [1]

	07/24/1997 to 07/25/1998			07/26/1998 to 08/14/1999			08/15/1999 to 07/22/2000			Total research period		
	% total amount news	Mean	n	% total amount news	Mean	n	% total amount news	Mean	n	% total amount news	Mean	n
Albert Heijn	17%	-.15	29	13%	-.24	7	20%	-.46	26	18%	-.29	62
NS	21%	-.20	129	17%	-.05	51	22%	-.29	175	21%	-.22	355
Police	20%	-.25	449	17%	-.20	266	18%	-.17	221	18%	-.22	936
Schiphol	19%	-.19	261	20%	-.21	211	19%	.08	90	19%	-.15	561
Agriculture	16%	-.23	122	25%	-.02	150	25%	-.13	118	21%	-.12	391
ABN AMRO	20%	-.13	121	13%	-.30	50	22%	.01	89	19%	-.12	260
Shell	19%	.08	111	14%	-.23	102	25%	-.17	93	19%	-.10	306
Rabobank	23%	.54	56	20%	.05	45	15%	.27	25	20%	.31	126
BP	42%	-.13	9	19%	.81	14	18%	.17	6	25%	.38	29
Super de B.	34%	1.00	16		.	0.	25%	-.63	3	30%	.75	19

Note. Only support & criticism news with the focal organization in the object position is selected. The percentages do not add up to 100%, because they describe the percentage of support & criticism news about the focal organization compared to the total amount of news about the focal organization. The last column, the percentage of support & criticism news within the total research period, matches with the top of the stack bars in Figure 4.2. $n = 3,046$ assertions (weighted)

The rest of the organizations received approximately the same amount of support & criticism news as the other focal organizations. The scale of the mean direction of support & criticism news ranges from –1, which means that the organization received criticism only, to +1, which means that the focal organization received only support and no criticism. Super de Boer (.75), BP (.38), and Rabobank (.31) received more support than criticism news. For the rest of the organizations, the opposite was the case. Albert Heijn received most criticism (–.29), followed by the police (–.22), and the NS (–.22).

Albert Heijn under siege by Montignac and the media

Albert Heijn received the most severe criticism of the ten focal companies (–.29). The tenor of support & criticism news about Albert Heijn was more negative in the third research period (–.46) than in the first (–.15) and second period (–.24). Within the whole research period, the industry criticized or supported Albert Heijn most frequently (56%, Table 4.4). Part of the support and criticism of Albert Heijn came from "diet guru" Montignac. Frenchman Michel Montignac's diet – "the Montignac method" – was a hype in the Netherlands at the time. The basic idea of the Montignac method is that people do not gain weight because they eat too much, but because they eat the "wrong" products. Montignac developed a list of products that were approved according to the Montignac method. Because the Montignac method was a hype, it was clear why supermarkets wanted their products to be approved for the Montignac method. Albert Heijn labeled its products accordingly, saying the products "were suitable for the Montignac method." As a reaction to this, the publisher of the Montignac books threatened to initiate proceedings against Albert Heijn. Montignac considered it to be an illegal use of his name. On February 12, 1998, Albert Heijn gave up and removed the "suitable for Montignac" labels. The national newspapers paid attention to this.

Both Lidl and Laurus supermarkets launched the attack on Albert Heijn by taking its customers away (see for example *Algemeen Dagblad*, March 3, 1999, p. 17). There was 15% of support & criticism from the media, which was mainly negative (–.48). Examples of media criticism are: "Albert Heijn lets oldest customers down" (*de Volkskrant*, March 13, 1999, p. 53); "Albert Heijn demands too much" (*Trouw*, September 25, 1998, p. 7); "'Free' chopping board appears to be a dirty affair" (*de Volkskrant*, June 3, 2000, p. 31).

ABN AMRO supported and attacked by other banks, and criticized by the media

Competitor Rabobank did better than ABN AMRO when it came to the tenor (–.15) of support & criticism news over the whole research period, as was also the case with the tenor of success & failure news.

Most support & criticism came from the industry (37%), which mildly criticized ABN AMRO (–.09). For example, ABN AMRO was supported by Rabobank because they were going to cooperate in telephony. The battle between Fortis and ABN AMRO to take over the Belgian Generale Bank led to much criticism news for ABN AMRO, because a battle is a "mutual relationship": "The battle between Fortis and ABN AMRO for the Belgian Generale Bank will probably be decided to the advantage of Fortis" (*NOS*, June 5, 1998). ABN AMRO received a lot of support & criticism from the media (28%), with more criticism than support (–.30). The following headlines are examples of criticism from the media: "ABN AMRO wrong" (*NRC Handelsblad*, December 18, 1997, p. 29); "ABN AMRO is bluffing" (*Trouw*,

TABLE 4.4.   Actors who supported or criticized the focal organization in the media

	Internal		Media		Industry		Government		Countries		Experts, NGO		Empl. union		Other actors	
	Row %	Mean	Row %	Mean	Row %	Mean	Row %	Mean	Row %	Mean	Row %	Mean	Row %	Mean	Row %	Mean
Albert Heijn	4%	-0.17	15%	-0.48	56%	-0.22	2%	-0.15	2%	1.00	4%	-1.00	5%	0.10	11%	-0.69
ABN AMRO	12%	0.44	28%	-0.30	37%	-0.09	6%	-0.11	2%	-0.40	8%	-0.42	3%	0.56	3%	-0.47
Shell	25%	-0.40	10%	-0.29	27%	0.44	5%	-0.61	19%	-0.05	12%	-0.33	2%	-0.63	0%	1.00
Rabobank	17%	0.45	19%	-0.14	43%	0.62	3%	0.69	0%	1.00	4%	-0.19	2%	0.23	11%	-0.23
BP	25%	0.04	3%	-1.00	51%	0.92	5%	1.00	2%	-1.00	14%	-0.71		.		.
Super de Boer		.		.	100%	0.75		.		.		.		.		.
NS	8%	-0.68	8%	-0.54	20%	0.27	29%	-0.29	1%	0.22	4%	-0.80	25%	-0.21	4%	-0.34
Police	10%	0.00	20%	-0.73	15%	0.22	22%	0.08	2%	-0.39	4%	-0.39	3%	-0.34	24%	-0.37
Schiphol	4%	0.32	10%	-0.39	26%	0.03	34%	-0.20	4%	0.09	16%	-0.35	3%	-0.55	4%	-0.01
Agriculture	7%	-0.16	23%	-0.52	11%	-0.12	46%	-0.03	4%	0.75	3%	-0.19	2%	0.97	3%	-0.20

Note. n = 3,021 assertions (weighted)

June 8, 1998, p. 4). An example of support from the media is: "ABN AMRO's offer for Generale Bank very smart" (*Algemeen Dagblad*, May 27, 1998, p. 1).

The internal support (12%) came mainly from senior executives who were going to work at ABN AMRO: "Banker Joost Kuiper to ABN AMRO" (*Algemeen Dagblad*, August 21, 1998, p. 11); "Groenink new man at ABN AMRO" (*NRC Handelsblad*, November 12, 1999, p. 1). Due to the arrival of the new senior executives, there was more support than criticism for ABN AMRO (.42). The bank received 8% of support & criticism in the category "experts and NGOs." A great deal of this news was about the conflict between the central bank of Surinam and ABN AMRO.

Shell supported by the industry, and criticized internally (due to the sale of Shell units)
Shell belonged to the middle group of the ten companies/sectors when it came to the tenor of support & criticism news; the oil company was criticized slightly over the entire research period (–.10). The tenor of support & criticism in the case of Shell was positive only in the first research period. Competitor BP was scarcely in the news, but was doing better as will be focused upon later.

Like most companies, Shell received most support & criticism from other companies (27%); there was more support than criticism from other companies (.44). This was caused by the cooperation with other companies. Since cooperation is a mutual relationship, it is assumed that the companies who cooperated with the oil company have a positive affective relationship with Shell. An example of Shell's cooperation is: "Once before, Albert Heijn and Shell decided to cooperate together by setting up supermarkets in gas stations" (*RTL*, September 17, 1997). The unsuccessful attempts to get interests in other companies led to criticism news of other companies of Shell: "Woodside puts Shell offer aside" (*De Telegraaf*, July 1, 2000, p. 7). Compared to the other organizations, Shell received a lot of internal support & criticism (25%), which consisted more of criticism than of support (–.40). This was mainly due to the sale of business units: "Shell close to selling PVC branch" (*Algemeen Dagblad*, March 3, 1998, p. 13).

Compared to the other organizations, Shell received a lot of support & criticism from the category "country" (19%). This type of support & criticism news, which is nearly balanced (–.05), had to do with the business activities of Shell in foreign countries: "The three countries gave Shell the order to see if it is possible to extend the gas line 1,500 kilometers to Turkey" (*NOS*, December 28, 1997); "Shell wants to go to Iran despite threat from US" (*De Telegraaf*, April 28, 1998, p. 33). Shell also received a lot of support & criticism from NGOs: "Anti-Shell demonstration" (*NRC Handelsblad*, November 11, 1997, p. 15); "Greenpeace argues with Shell about leaflet on the environment" (*NRC Handelsblad*, March 21, 2000, p. 2). Because Greenpeace bought Shell shares, which according to the NET-method indicates a positive affective relationship of Greenpeace with Shell, the mean direction was –.30 and not entirely negative (–1). The media gave 10% of the support & criticism to Shell, and as was also the case with the other companies, they give more criticism than support (–.29): "Shell scared" (*Trouw*, December 28,

1998, p. 3); "Shell now the ugly duckling of the Seven Sisters" (*de Volkskrant*, August 8, 1998, p. 2). There was also support from the media for Shell: "Shell is on the right track" (*Trouw*, March 25, 1998, p. 7); "Shell still has the most beautiful assets" (*NRC Handelsblad*, June 5, 1999, p. 18).

Rabobank had more friends than enemies due to support by the industry and its senior executives

In contrast to competitor ABN AMRO, the tenor of support & criticism news about Rabobank was positive on average over the entire research period (.31). The relative amount of support & criticism news about the two banks was nearly equal.

Rabobank received most support & criticism from the industry (43% Table 4.4), which consisted more of support than of criticism (.62). The cooperation of the bank with other organizations led to a great deal of support for Rabobank: "ABN AMRO and Rabobank together in telephony" (*de Volkskrant*, September 18, 1997, p. 33). The announced merger between Rabobank and Achmea was called off in December 1998, however: "The merger announced in July between Rabobank and bank insurance company Achmea is off" (*NOS*, December 17, 1998).

Rabobank was again accused of wrongful conduct by investment club "D'n Anwas." The media gave 19% of the support & criticism. Compared to the other organizations, the media criticized Rabobank mildly (–.14): "Stayer Hans Smits crown prince at Rabobank" (*De Telegraaf*, September 11, 1997, p. 3); "New chief executive Smits scores a bull's eye" (*Trouw*, September 10, 1998, p. 7). Rabobank received 17% internal support & criticism, which was mainly the result of the coming and going of men at the top: "Smits becomes new chief executive at Rabobank" (*NRC Handelsblad*, September 9, 1998, p. 20); "Manager at Doetinchem Rabobank resigns" (*De Telegraaf*, December 5, 1998, p. 4).

BP received support because of merger with Amoco and cooperation with Elf

Due to its merger with Amoco, BP got a relatively large amount of support & criticism news (25%). Moreover, the tenor of support & criticism news about BP was for the most part positive over the whole research period (.38). This was in contrast with the tenor of support & criticism news about Shell.

On balance, in the first research period BP was mildly criticized (–.13), as shown in Table 4.3. The number of assertions was low, however (*n* = 9). In the second research period, BP mainly received support (.81) from the other organizations. This was due to the fact that BP was going to merge with Amoco: "Merger BP and Amoco" (*Algemeen Dagblad*, August 12, 1998, p. 1). Furthermore, BP was supported by the cooperation with Elf: "Elf and BP Amoco together in polypropylene" (Trouw, June 16, 1999, p. 6). Again, the number of assertions was low (*n* = 14).

In the third research period, BP received more support than criticism news (.17). Table 4.4 shows that BP received a lot (51%) of support & criticism of the industry. The direction of .92 shows that the companies mainly supported BP. This is due to the news about the cooperation of BP with other companies in the industry (such as Amoco and Elf). The oil company also received internal (25%) support & criticism that is nearly balanced (.04). Internal support & criticism was due to the sale of parts of BP to Internatio-Müller and other companies. If BP was commented on by NGOs (14%) it was by Greenpeace, who mainly criticized BP (–.71).

Super de Boer had the most friends
Like BP, Super de Boer was scarcely in the news, and got a relatively large amount of support & criticism news due to mergers (30%). Moreover, the tenor of support & criticism news was predominantly positive (.75), as was also the case with BP. Super de Boer received support in the first research period, because it would then belong to one company together with other supermarkets. NOS (May 25, 1998) reported: "This means that the supermarkets Edah, Basismarkt, Konmar, and Super de Boer will belong to one concern." There was no support & criticism news about Super de Boer in the second research period.

In the third research period, Super de Boer received "criticism" (–.63) from Albert Heijn: "Super de Boer reduced the price to 99 cents per liter, after which Albert Heijn reduced it to just under that amount" (NOS, January 28, 2000). Super de Boer was supported and criticized by supermarkets ("Industry") only, which is expressed in Table 4.4.

NS criticized by the government, quarreled with unions, and received support of industry
The tenor of support & criticism news about the NS was relatively negative (–.22) compared to the support & criticism news about other organizations/sectors. Only Albert Heijn was criticized more severely than the NS. The criticism was more severe in the first (–.20) and the third (–.29) periods than in the second period (–.05).

In the first period, the NS received most support & criticism (31% Table 4.5) from the employees and the labor union. In the beginning of the research period, NOS (September 10, 1997) reported that the labor union turned against the commercial policy of the railway company. In April 1998 there was a conflict between the NS and the labor unions about a new labor agreement. Six months later, *de Volkskrant* reported (June 10, 1998, p. 1) there was an agreement between the NS and the labor unions.

The industry gave the NS 28% of the support & criticism, which was nearly in balance (.05). A lot of criticism came from competitor Lovers Rail: "Manager Lovers angry at NS because of passenger count" (*de Volkskrant*, August 11, 1997, p. 3). The cooperation with other companies led to support for the NS: "NS, Schiphol, and KLM want to cooperate" (*NRC Handelsblad*, July 23, 1998, p. 13). The amount of support & criticism from the government was 24% in the first period, and rather negative (–.22). Most of the support and criticism came from then Minister of Transport and Public Works Annemarie Jorritsma. She supported the NS with an allowance of 32 million guilders, according to *De Telegraaf* (September 11, 1997, p. 7). Jorritsma criticized the NS because they did not want to talk with her about reducing the fare increases: "Jorritsma criticizes NS attitude" (*Algemeen Dagblad*, November 4, 1997, p. 4).

The NS received considerably less support & criticism news in the second period. Due to the installation of a new government, Jorritsma was succeeded by Tineke Netelenbos as Minister of Transport and Public Works. Most support & criticism came from the industry (23%) and the government (21%). Both actors gave more support than criticism (.28 and .26 respectively). An example of support from the industry is: "Albron catering firm on NS trains" (*NRC Handelsblad*, December 22, 1998, p. 20). The NS was supported by the new Minister of Transport and Public Works in its decision to close down ticket booths: "Netelenbos: closing down ticket booths is the business of NS" (*Trouw*, June 2, 1999, p. 7). The political parties CDA and PvdA supported the NS, because they wanted the NS to operate part of the high-speed railway to Paris (*Trouw*, June 16, 1999, p. 4).

The employees and labor unions gave 19% of the support and criticism. They gave much more criticism than support (–.66). NOS (January 19, 1999) reported that engine drivers and ticket inspectors were dissatisfied with the NS's day-off regulation. The labor union federation FNV threatened the NS with strikes (*Trouw*, April 17, 1999, p. 7).

In the third period, the NS got the most support & criticism from the government (36%), as in the previous two periods. The criticism of the government was more severe (–.41) than in the preceding two periods. NOS news (November 17, 1999) reported that Minister Netelenbos refused to negotiate further with the NS about the operation of the Dutch part of the high-speed railway; the offer by the NS did not meet her expectations. In December, several newspapers noticed that the minister wanted the government to get more of a hold on the NS. Netelenbos argued that the government did not have much of a hold on the privatized NS, while at the same time it was not exposed to the "discipline of the market." Another example of criticism news by the minister of the NS was her demand for the return of 80 million guilders by the NS. Netelenbos demanded the return of this amount from contractors working on the Schiphol railway tunnel, who were suspected of committing fraud by using fake invoices. The minister reprimanded NS senior executive Den Besten, because his managers knew or could have known about the manipulation (*NOS*, May 10, 2000). Dutch parliament wanted measures to be taken against the management of the NS if it appeared they knew about the misuse of subsidies on the construction of the Schiphol railway tunnel. Den Besten resigned in July 2000; he denied being pressured by Minister Netelenbos (*Trouw*, July 17, 2000, p. 1).

In the third period, employees and labor unions gave 21% of the support & criticism, which was nearly in balance (–.08). In September 1999, the labor unions accused NS senior executive Den Besten of bad management (*de Volkskrant*, September 29, 1999, p. 2). The NS received the support of the labor unions because they reached an agreement on a social plan for the NS in December 1999. More support from the labor unions for the NS came in January, when the labor unions and the NS talked and reached an agreement about tackling the violence in trains. A new plan by the NS to tackle the delays was not well-received by the employees, however. NOS news (June 2, 2000) reported that engine drivers and conductors organized an unofficial strike because they were against the plans of the NS to put them on fixed routes. Labor union federation FNV was against the reorganization plans of the NS despite the agreement reached earlier: "FNV against NS plans" (de Volkskrant, June 13, 2000, p. 17). The NS received 10% internal support & criticism news due to a conflict at the top and the departure of NS marketing manager Van Eeghen and NS senior executive Den Besten: "Disagreement among NS leadership" (NRC Handelsblad, September 9, 1999, p. 2). There is far more criticism than support (–.81).

The police had a cool relationship with the government, and is crazy in the eyes of the media
The tenor of support & criticism news about the police was relatively negative (–.22) compared to the support & criticism news about other organizations/sectors. The police was criticized as severely as the NS, and only Albert Heijn was criticized more severely than the police. There was more criticism in the first period (–.25) than in the second (–.20) and the third periods (–.17).

As shown in Table 4.4, the media gave a total of 20% of the support & criticism of the police. Table 4.5. shows that in the first research period, the media gave 16% of the support & criticism of the police. The police was criticized by the media in the first research period in headlines such as: "Police are guilty as well" (NRC Handelsblad, November 6, 1998, p. 3); "Police have gone nuts" (De Telegraaf, December 1, 1997, p. 3); "Police doesn't know enough about Moroccan youth" (de Volkskrant, April 28, 1998, p. 3).

The police was criticized a great deal (27% see Table 4.5) in the first period by members of the government, mainly by then Minister of the Interior Hans Dijkstal and Minister of Justice Winnie Sorgdrager: "Sorgdrager angry about failure of police" (NRC Handelsblad, September 16, 1997, p. 3); "Dijkstal concerned about the deterioration of standards and values of the police" (De Telegraaf, December 5, 1997, p. 7). In part due to the criticism of the two ministers, the police got some support from other actors in the government: "Field Officer Daverschot is annoyed about the image of the police and judicial authorities in Groningen" (de Volkskrant, January 20, 1998, p. 3); "Wallage: more power for the police" (De Telegraaf, February 25, 1998, p. 6); "Top justice official: leave police alone" (Trouw, May 8, 1998, p. 9). There was nearly a balance between support and criticism on the police (–.07, see Table 4.5).

TABLE 4.5. **Actors who supported or criticized the police, NS, the agricultural sector, and Schiphol per research period**

Organization	Period	Internal		Media		Industry		Government		Countries		Experts, NGOs		Empl. union		Other actors	
		Row %	Mean	Row %	Mean	Row %	Mean	Row %	Mean	Row %	Mean	Row %	Mean	Row %	Mean	Row %	Mean
Police	I	14%	-0.08	16%	-0.74	13%	0.27	27%	-0.07	1%	-1.00	3%	-0.20	3%	-0.11	22%	-0.54
Police	II	7%	0.12	24%	-0.75	11%	0.30	20%	0.14	1%	1.00	5%	-0.22	2%	-0.73	29%	-0.23
Police	III	2%	0.53	23%	-0.67	24%	0.11	17%	0.47	5%	-0.50	4%	-1.00	3%	-0.51	21%	-0.24
NS	I	4%	-0.83	8%	-0.54	28%	0.05	24%	-0.22	1%	1.00	2%	-0.75	31%	-0.22	3%	-0.63
NS	II	10%	-0.09	15%	-0.36	23%	0.28	21%	0.26	7%	-0.13	1%	-1.00	19%	-0.66	4%	1.00
NS	III	10%	-0.81	6%	-0.65	15%	0.59	36%	-0.41		.	7%	-0.80	21%	-0.08	4%	-0.57
Farmers	I	5%	-0.20	24%	-0.57	5%	0.30	54%	-0.24	3%	0.80	1%	-0.22	2%	0.89	6%	-0.23
Farmers	II	9%	0.03	23%	-0.57	7%	-0.37	48%	0.24	4%	0.45	4%	-0.37	2%	1.00	3%	-0.23
Farmers	III	7%	-0.45	23%	-0.41	22%	-0.12	36%	-0.15	5%	1.00	3%	0.16	2%	1.00	2%	-0.02
Schiphol	I	7%	0.26	6%	-0.52	25%	-0.05	36%	-0.19	3%	-0.30	16%	-0.48	4%	-0.45	3%	0.54
Schiphol	II	1%	0.61	15%	-0.46	23%	-0.09	37%	-0.31	3%	0.53	14%	-0.06	2%	-0.81	5%	-0.12
Schiphol	III	3%	0.45	7%	0.33	38%	0.34	26%	0.13	6%	0.14	17%	-0.56		.	4%	-0.82

n = 1,829 assertions (weighted)

In the second research period, the media criticized the police even more often (24% see Table 4.5). The criticism of the media was severe, as in the first research period (–.75): "Police lax in Zandvoort porn case" (*Algemeen Dagblad*, December 31, 1998, p. 1); "Nervous police take tough action" (*Trouw*, February 17, 1999, p. 1). In the second period, the government and political parties gave 20% of the support & criticism of the police. Contrary to the first period, the government mildly supported the police (.14). In the beginning of the second research period (early August 1998), a new government was installed. The Minister of the Interior was then Bram Peper and the Minister of Justice was Benk Korthals. Examples of support & criticism from the government and political parties are: "CDA: police should conduct searches more often" (*Trouw*, November 3, 1998, p. 3); "Interior Minister Peper wants extra measures to combat increasing violence against the police" (*RTL*, June 21, 1999). In the second period, the police received a lot of support & criticism from "other actors" (29%, Table 4.5). In the case of the police, the category "other actors" consisted mainly of civilians and petty criminals (such as rioters).

In the third research period, these actors gave the most support & criticism (24%, see Table 4.5) of the police, but they gave less support (.11) than in the previous re search period The police's quarrel with crime reporter Peter R. de Vries was one of the examples of criticism of the police: "Police and Peter R. de Vries keep quarreling away" (*de Volkskrant*, March 25, 2000, p. 3). Other examples are: "Companies keep police away in cases of fraud" (*Trouw*, January 22, 2000, p. 4); "Taxi federation threatens police with legal proceedings" (*De Telegraaf*, February 15, 2000, p. 3); "Veilig Verkeer Nederland critical of alcohol checks by police" (*RTL*, March 11, 2000). The media gave nearly the same amount of support & criticism (23%) as the "industry." They still gave severe criticism, although it was a bit less than in the previous research period, namely –.67 (see Table 4.5). Foreign countries gave 5% of the support & criticism in the third research period.

Actions by the Rotterdam police led to an enormous outrage in Italy. Rotterdam police arrested seven journalists from the public broadcasting company *RAI* during a soccer match. The journalists had tried to film how a group of Italian disabled persons were brought into the stadium only with difficulty. The police summoned the journalists to stop filming, which caused them to come to blows. This incident was taken seriously in Italy: "The Italian ambassador lodged an official protest in The Hague against the actions of the police before the championship" (*NOS*, July 3, 2000); "The riot has to do with the way the Italian ambassador is treated by the Rotterdam police" (*RTL*, July 4, 2000). "Italy is furious about the actions of the Rotterdam police towards the television crew filming in the Kuip stadium" (*RTL*, July 9, 2000).

Schiphol played along by the government, and had a neutral relationship with the industry

The amount of support & criticism news about Schiphol (19%) did not differ much from the amount of support & criticism news about the other companies/sectors. The tenor of support & criticism news about Schiphol was negative on average over the whole research period (–.15), but not so negative as in the case of Albert Heijn, NS and the police.

In the first two periods, the tenor of support & criticism news was negative (–.19 and -.21 respectively), while this was reversed in the third period (.08). In the first period, Schiphol got most support & criticism from the government (36%). One of the major subjects of criticism by the government was the noise nuisance caused by Schiphol; Schiphol's plan to increase the number of flights divided the politicians. The new plan by Schiphol would have made 40,000 extra flights possible in 1998. The legal regulations for noise nuisance would have been violated on six of the 235 points of measurement (NRC *Handelsblad*, January 14, 1998, p. 2). Minister of Transport and Public Works Jorritsma said in the NOS news (January 20, 1998) that she supported Schiphol and thought it was desirable to change the noise law, despite her earlier claim in August 1997 when she stated that Schiphol should stick to this. This led to a conflict with the Minister of Environment De Boer. *Algemeen Dagblad* reported (February 13, 1998, p. 16) that the government postponed the decision concerning the expansion of the number of lights at Schiphol. At the end of the first research period, NRC *Handelsblad* reported (June 24, 1998, p. 3) that the negotiators of the new government did not reach an agreement concerning the expansion of Schiphol and a potential new location for a new airport. In the first period, a quarter of the support & criticism came from the industry. There was slightly more criticism than support (–.05). The category "experts and NGOs," which included judges and neighbors, gave 16% of the support & criticism of Schiphol. They gave far more criticism than support (–.48) due to the exceeding of the noise nuisance levels.

In the second research period, Schiphol received 37% of the support & criticism from the government. There was more criticism from the government than in the first research period (–.31). The new Minister of Transport and Public Works Netelenbos rejected Schiphol's "1998-1999 users' plan," because more than 12,000 homes would suffer from noise nuisance. Then Prime Minister Wim Kok got into a foul mood about Schiphol (*Algemeen Dagblad*, September 18, 1998, p. 5). According to the newspaper, Kok was irritated because time after time the airport came with new figures to put politicians under pressure, and because it did not keep its promises with regard to its agreements to restrict noise nuisance. Within two weeks, NOS (September 29, 1998) reported that parliament was going to hold a hearing to find out which figures about the growth and noise nuisance of Schiphol were correct. Former minister Jorritsma was under siege because she was accused of holding back information about the growth of Schiphol. Minister of the Environment Jan Pronk and Minister Netelenbos disagreed about Schiphol, because according to Pronk it had not yet been decided that Schiphol would be allowed to grow. This had been announced by Minister Netelenbos the night before. A day later, *Trouw* (November 19, 1998, p. 8) reported that the government had postponed the decision about the location of Schiphol.

The industry gave 23% of the support & criticism of Schiphol, which is slightly more negative (–.09) compared to the previous research period. In the second period, compared to the first and the third research periods, the media gave a lot of support & criticism (15%), which consisted more of criticism than of support (–.46), "Schiphol is a big shame" (*Trouw*, 23 September 1998, p. 7). In the second period, the "experts and NGOs" gave 14% of the support & criticism (–.06). Nearly all support & criticism came from the environmental movements. The environmental group Milieudefensie (Friends of the Earth Netherlands) bought a piece of land (the Bulderbos) to block the plans of Schiphol concerning a new runway.

In the third research period, most support & criticism came not from the government as in the previous two research periods but from the industry (38%), which gave more support than criticism (.34). In the *NOS* news (June 14, 2000), the airport received support from KLM because the airline wants to have its own terminal at Schiphol. KLM is Schiphol's biggest customer. Schiphol also received support from NS and KLM because of their cooperation with Schiphol on the high-speed railway (*NRC Handelsblad*, April 12, 2000, p. 2; *de Volkskrant*, April 13, 2000, p. 2).

In the third research period, the government gave 26% of the support & criticism, which in this period consisted more of support than of criticism (.13). Several print media reported in November that Minister Netelenbos would be strict with Schiphol, and would not tolerate any more breaking of the noise regulations. In December 1999, Schiphol received the support of the Amsterdam city council, because it wanted to keep its influence after the privatization of Schiphol. The city council wanted to keep at least 10% of the shares. Experts and NGOs gave 17% of the support and criticism in the third research period. They gave more criticism than support (–.56). *NOS* and *RTL* (September 5, 1999) reported on demonstrators from Milieudefensie who attached a banner onto a control tower at Schiphol. At the end of November, the environmental movement left the consultation platform because environmental regulations had been broken.

Farmers supported and criticized by the government in turn, and criticized by the media
The tenor of support & criticism news about the agricultural sector in the whole research period was average (–.12) compared to the rest of the organizations. Of the three research periods, the tenor of support & criticism news about the agricultural sector as the least negative in the second research period (–.02).

The government and political parties gave by far the most support & criticism (54% Table 4.5) on the agricultural sector in the first period. This was due to government regulations put in place to eradicate swine fever. Minister of Agriculture Van Aartsen reprimanded the hog farmers and haulers for not following the government guidelines that severely limited transport between hog farms. The minister seized on swine fever to restructure the hog-farming sector and reduce the number of hogs, which led to discussions with the hog farmers. In March 1998 Van Aartsen supported the hog farmers by introducing a regulation to help them with difficulties resulting from reductions in the Dutch hog-farming sector.

The hog farmers also received some support from political parties: "Due to pressure by the VVD and CDA, the government adjusted its plan to reduce the number of hogs" (RTL, 12 December, 1997). Nevertheless, the farmers were criticized more than supported by the government (–.24). The media were more negative than the government (–.57). In the first research period, they were responsible for 24% of the support & criticism news about the farmers. Examples of support & criticism of the media are: "Not all hog farmers are immoral" (NRC Handelsblad, January 8, 1998, p. 6); "Incompetent farmer Pauw is mostly to blame for cattle mortality" (de Volkskrant, March 10, 1998, p. 6).

In the second period, from mid-1998 to mid-1999, the agricultural sector again received most support & criticism from the government (48%). Contrary to the previous period, the government gave more support than criticism (the mean direction was .24). The new Minister of Agriculture Apotheker supported the farmers and market gardeners with a visit. The farmers also received support from members of parliament: "Encouraging pat on the back by members of parliament for sad farmer" (Algemeen Dagblad, September 17, 1998, p. 6). Minister Apotheker was forthcoming to those farmers and market gardeners who suffered from the heavy rainfall: "Farmers and market gardeners who are suffering because of the bad weather received Minister Apotheker's promise that they will be reimbursed for damages" (NOS, October 28, 1998).

In the second period, the media gave approximately the same amount of support & criticism (23%) as in the first research period. The tone was the same, and there was more criticism than support (–.57): "Cow chip gives hope for intractable farmers" (Algemeen Dagblad, January 5, 1999); "Farmers protected against capricious market for too long" (Trouw, October 7, 1998, p. 11).

In the third period, the farmers received most support & criticism news from the government (36%), which was less than in the two previous research periods. There was more criticism than support (–.15) from the government. The plan of then Minister of Agriculture Laurens-Jan Brinkhorst to reduce the manure surplus led to criticism news by the government of the farmers: "Minister Brinkhorst expects that the plans to drive back the manure problem will cost 6,000 farmers their jobs" (RTL, September 10, 1999). Contact was restored between Brinkhorst and the farmers so they could talk about the restructuring of the hog sector: "Farmers and Brinkhorst overcome the deadlock" (Trouw, January 28, 2000, p. 7). In February, the minister came to terms with the farmers about a buy-out regulation for their hogs, chickens, and calves: "Brinkhorst reaches an agreement with the farmers" (*de Volkskrant*, February 21, 2000, p. 6).

The media criticized the farmers somewhat less compared to the previous research period (–.41). In the third research period, problems between the pharmaceutical company Bayer and the farmers led to a great deal of support & criticism news from the industry (22%). Despite the earlier headlines in March, it appeared in September that Bayer was not willing to pay damage reimbursement to the farmers: "Bayer rejects farmers' claims" (*Algemeen Dagblad*, September 17, 1999, p. 13). According to *de Volkskrant* (January 5, 2000, p. 7), Bayer did not want to pay damage reimbursement because independent research showed that the causal relationship between the polluted vaccine and the illness of the cows was not demonstrable in all cases. The conflict went on until May 2000: "Bayer challenges farmers to prove their criticism" (*Trouw*, May 23, 2000, p. 7).

4.4   NEWS ABOUT ISSUES IN THE MEDIA AND ASSOCIATIONS OF RESPONDENTS

This section describes the associations mentioned by respondents, together with all the issues the organization was related to in the headlines in print media and news items. These variables are used in the next chapter in order to test hypothesis 4 about agenda-setting.

People associated Shell with gas stations and pollution, and the media associated Shell with profit and pollution
Table 4.6 shows that most people (46% in 1998 and 2000; 45% in 1999) associated Shell with the general association "gas stations." The meaning object "gas stations" was not coded, however, since it was hard to separate from the main actors, as was mentioned in the previous chapter. Environmental pollution (13% in 1998; 11% in 1999; 10% in 2000) and mini- supermarkets (10%) were the second and third associations of the respondents. The topic of the environment was also covered by the media when they wrote about Shell (17% of the issue news in the first, 10% in the second, and 13% in the third research periods). Note that the evaluation of the issue (the environment or environmental pollution) was not taken into account in the present study, as indicated in subsection 3.2.4. The topic mini-supermarkets did not receive a lot of media attention.

Shell's profits were most frequently covered by the media: "Shell had a good year, profits increased by 38%" (*NOS*, February 10, 2000). This issue was mentioned by the respondents as the fifth association. The media also wrote a lot about the economy, which was the fourth association of respondents. Examples of economic issues in the media are: "Asia and oil price spoil Herkströters last year" (*de Volkskrant*, February 13, 1998, p. 2); "Liberalization of energy market took Shell by surprise" (*Trouw*, August 7, 1998, p. 7).

TABLE 4.6.   **Public associations and media issues connected with Shell**

	Opinion 1998	Opinion 1999	Opinion 2000	Media 1998	Media 1999	Media 2000
Gas stations	46%	45%	46%	-	-	-
Environmental pollution	13%	11%	10%	17%	10%	13%
Mini-supermarkets	10%	10%	10%	2%	0%	5%
(Dutch) economy	8%	9%	8%	17%	17%	16%
Profitability	4%	5%	6%	29%	31%	26%
Advertising	3%	2%	2%	1%	0%	1%
Employment	3%	3%	2%	1%	3%	2%
(Lack of) materials	2%	4%	4%	12%	11%	17%
Values	2%	1%	1%	8%	1%	1%
Efficiency	1%	0%	1%	3%	17%	5%
Mergers, takeovers	0%	0%	0%	4%	8%	8%
Other	2%	2%	2%	6%	2%	8%
Do not know/Shell	6%	7%	8%	-	-	-
Total	100%	100%	100%	100%	100%	100%
	$n = 500$	$n = 308$	$n = 727$	$n = 254$	$n = 394$	$n = 171$

Note. assertions are weighted. With regard to the media coverage, durable goods such as solar energy belong to the category "environment." The category "do not know/Shell" consists of the respondents who did not have associations with or were not familiar with Shell.

BP was a meaningless company when it came to associations and issues

As was the case with Shell, the most frequently occurring association with BP was "gas stations" (46%). It was remarkable, however, that subsequently 18% of the respondents did not know what to think of when they were asked what they thought of when they thought of BP. In general, BP did not receive a lot of media coverage, and this was also the case for issues mentioned in the media and related to BP. In the second research period, the oil company received most issue news due to its merger takeovers: "British oil concern is going to merge with the American Amoco" (NOS, August 11, 1998); "Oil giant BP/Amoco takes over the American Atlantic Richfield and pushes Shell out of its second place among the world's ranking list of oil companies" (NOS, April 1, 1999).

Although BP received most of its issue news about mergers and takeovers in the second research period, it did not show up in the associations of respondents. In the second research period, still only 1% of the respondents thought about mergers and takeovers when they thought about BP. This was the same amount as in the first and the third research periods. In the first period, a relatively large number of media (which is less in absolute numbers) reported on the profits of BP. However, only 1% of the respondents thought about profit when they thought of BP.

TABLE 4.7.   Public associations and issues in the media connected with BP

	Opinion 1998	Opinion 1999	Opinion 2000	Media 1998	Media 1999	Media 2000
Gas stations	46%	46%	46%	-	-	-
Do not know	18%	19%	18%	-	-	-
Mini-supermarkets	12%	10%	10%	10%	0%	0%
Environmental pollution	8%	6%	7%	14%	6%	23%
Advertising	5%	3%	3%	0%	0%	6%
(Lack of) materials	3%	3%	3%	0%	6%	0%
Dutch economy	2%	4%	3%	0%	9%	0%
Employment	2%	1%	2%	0%	4%	0%
Profitability	1%	1%	2%	61%	7%	12%
Mergers, takeovers	1%	1%	1%	7%	57%	45%
Other, do not know BP	3%	4%	4%	8%	10%	14%
Total	100%	100%	100%	100%	100%	100%
	n = 464	n = 292	n = 702	n = 7	n = 53	n = 9

Note. assertions are weighted. With regard to the media coverage, durable goods such as solar energy belong to the category "environment."

ABN AMRO: mergers, economy, and profit mentioned by the respondents and the media

The issues that were covered most frequently in the media (mergers, the economy, and profits of ABN AMRO) were also mentioned most often by the respondents after they mentioned the general associations (see Table 4.8). The associations not often mentioned by the respondents (such as values and employment) were only infrequently in the news.

The fluctuations in media attention for an issue per research period were not precisely reflected in the fluctuations of the associations mentioned by the respondents, however. ABN AMRO was most in the news with profit in the second research period (31%), but the association "profit" was mentioned slightly less by the respondents in this period (3%). The same was the case for the issue and association "Dutch economy".

Note that more than half of the respondents (around 70%) associated ABN AMRO with a general association (such as bank accounts, bank cards, or financial services) in each of the three years. As will be explained in the next chapter, these permanent issues will be excluded in the analysis of the agenda-setting hypothesis.

TABLE 4.8.   **Public associations and issues in the media connected with ABN AMRO**

	Opinion 1998	Opinion 1999	Opinion 2000	Media 1998	Media 1999	Media 2000
Bank accounts, bank cards	38%	38%	39%	-	-	-
Financial services	30%	32%	29%	-	-	-
Mergers and takeovers	7%	5%	7%	40%	9%	15%
Dutch economy	4%	3%	4%	27%	30%	18%
Profitability	4%	3%	4%	16%	31%	23%
Values	1%	0%	1%	7%	9%	9%
Advertising	4%	4%	3%	2%	0%	3%
Employment	1%	1%	1%	0%	1%	3%
Efficiency (and inefficiency)	1%	1%	1%	1%	2%	10%
Other	2%	2%	2%	8%	18%	8%
Do not know/ABN AMRO	9%	9%	9%	-	-	-
Total	100%	100%	100%	100%	100%	100%
	n = 459	n =277	n = 698	n = 246	n = 184	n = 178

Note. assertions are weighted. The category "do not know/ABN AMRO" consists of the respondents who did not have associations with or were not familiar with ABN AMRO.

Examples in the media in which ABN AMRO was related to economic issues are: "ABN AMRO pushes stock exchange to a new peak" (*de Volkskrant*, July 9, 1998, p. 21) and "Crisis in Asia good for ABN AMRO" (*Algemeen Dagblad*, September 24, 1997, p. 10). An example of news in which ABN AMRO is related to an intended takeover is: "ABN AMRO one of the last three bidders for French CIC" (*de Volkskrant*, March 24, 1998, p. 17).

Rabobank associated with advertising, and the media wrote about the"Rabo racing team"

In contrast to ABN AMRO, in the case of Rabobank the media reported most about the category "advertising." This was the result of the cycle racing team sponsored by Rabobank, and which was in the news frequently: "Rabobank cyclists fight for selection" (*De Telegraaf*, June 16, 1998, p. 2). Rabobank's cycling team got a lot of media attention, while at the same time advertising is the association mentioned most often by the respondents after general associations such as bank account and financial services. The second issue the media wrote about is the Dutch economy, for example in headlines like: "Rabobank grows due to stock exchange euphoria?" (*NRC Handelsblad*, March 4, 1998, p. 15); "Rabobank: small chance of recession" (*Algemeen Dagblad*, September 25, 1998, p. 11). The Dutch economy was also the second association of respondents after the general associations.

The media spent less attention on the profits of Rabobank than on the profits of ABN AMRO. Note that the respondents also associated Rabobank less with profit

(Table 4.9) compared to ABN AMRO (Table 4.8). Although by far the most respondents associated Rabobank with general associations such as bank accounts and financial services, the rest of the issues that the media paid attention to seemed to be reflected by the associations of the respondents.

The fluctuations in the issue news about Rabobank were not reflected directly in the associations of respondents, however. The decline in media coverage about the Dutch economy in the third research period (from 30% in the second to 7% in the third period) did not immediately show up in the number of respondents who associated Rabobank with the Dutch economy, which was still 4%, as in the previous research period. An increase in the news about the efficiency of Rabobank (from 6% in the second research period to 19% in the third research period), did not show up in the percentage of respondents that associated Rabobank with efficiency, which was 2% in both periods.

TABLE 4.9.   **Public associations and issues in the media connected with Rabobank**

	Opinion 1998	Opinion 1999	Opinion 2000	Media 1998	Media 1999	Media 2000
Bank accounts, bank cards	37%	38%	40%	-	-	-
Financial services	32%	29%	32%	-	-	-
Advertising	4%	6%	4%	28%	32%	23%
Dutch economy	3%	4%	4%	21%	30%	7%
Profitability	4%	4%	2%	11%	8%	12%
Mergers and takeovers	2%	1%	1%	20%	8%	8%
Values	1%	1%	1%	6%	5%	2%
Employment	1%	1%	1%	4%	3%	1%
Efficiency (and inefficiency)	2%	1%	2%	3%	6%	19%
Other	5%	4%	3%	7%	8%	29%
Do not know/Rabobank	9%	10%	10%	-	-	-
	100%	100%	100%	100%	100%	100%
	n = 463	n = 272	n = 691	n = 108	n = 78	n = 82

Note. assertions are weighted. The category "do not know/Rabobank" consists of the respondents who did not have associations with or were not familiar with the Rabobank.

In the third period, the category "other" news was large (29%) since there was a lot of news about corporate issues and new products of Rabobank. The collective labor agreement, for example, belonged to the category corporate issues: "Rabobank enters into its own collective labor agreement for the first time" (NRC *Handelsblad*, May 25, 2000, p. 18). An example of a "new product" of Rabobank was: "Rabobank gets counter in supermarkets" (*Algemeen Dagblad*, April 12, 1999, p. 11).

Albert Heijn associated with "a wide selection," and the media wrote about its new products

Most respondents (around 30%) associated Albert Heijn with its wide selection of products. This matched with the issue "Albert Heijn's new products," which was covered more than any other issue in the media in the first and the third research periods (16% in 1998 and 31% in 2000). Examples of Albert Heijn's new products are: "New Spice Girl single only at Albert Heijn" (*Algemeen Dagblad*, August 8, 1997, p. 2); "Albert Heijn is going to sell books in supermarkets" (*NRC Handelsblad*, May 13, 1998, p. 21). However, in the second research period, there was no media coverage on the new products of Albert Heijn.

TABLE 4.10.   **Public associations and issues in the media connected with Albert Heijn**

	Opinion 1998	Opinion 1999	Opinion 2000	Media 1998	Media 1999	Media 2000
Wide selection of products	31%	31%	30%	16%	0%	31%
Price level	15%	17%	18%	30%	3%	8%
Opening hours	16%	13%	13%	-	-	-
Service	7%	7%	8%	-	-	-
Dutch economy	2%	2%	3%	9%	11%	11%
Mergers, takeovers	3%	1%	3%	2%	7%	2%
Environmental pollution	0%	0%	0%	4%	9%	0%
Standards and values	0%	0%	0%	21%	0%	6%
Advertising	9%	9%	8%	5%	0%	0%
Profitability	2%	2%	2%	6%	24%	17%
Efficiency	1%	1%	1%	3%	7%	2%
Other	6%	8%	5%	3%	40%	22%
Do not know /Albert Heijn	8%	8%	8%	-	-	-
	100%	100%	100%	100%	100%	100%
	n = 488	n = 289	n = 719	n = 75	n = 25	n = 69

Note. assertions are weighted. The category "do not know/Albert Heijn" consists of the respondents who did not have associations with or were not familiar with Albert Heijn.

The second and third associations mentioned by the respondents was the price level (17% of the respondents) and the opening hours (14% of the respondents). The association "price level" was matched with media coverage about Albert Heijn's "Bonus Card," which gives a discount on its products: "Albert Heijn introduces Bonus Card for regular customers" (*de Volkskrant*, February 9, 1998, p. 2). There was no news about the opening hours of Albert Heijn. The category "other" news was large in the second period, due to the headlines about product recalls at Albert Heijn: "AH recalls meat" (*NRC Handelsblad*, May 28, 1999, p. 13).

Super de Boer was a meaningless company except for the issue and association "mergers"
Most respondents (16% in 1998, 18% in 1999 and 18% in 2000) associated
Super de Boer with "price level." The second association with Super de Boer was
"the wide selection of products," which occurred equally as often as the association "price level," except for in the year 2000.

TABLE 4.11. **Public associations and issues in the media connected with Super de Boer**

	Opinion 1998	Opinion 1999	Opinion 2000	Media 1998	Media 1999	Media 2000
Price level	16%	18%	18%	0%	0%	78%
Wide selection of products	16%	18%	17%	0%	0%	12%
Do not know	15%	14%	18%	-	-	-
Do not know Super de Boer	14%	13%	11%	-	-	-
Mergers, takeovers	10%	4%	2%	56%	63%	0%
Advertising	10%	8%	9%	0%	0%	0%
Dutch economy	1%	1%	1%	18%	0%	0%
Openings hours	7%	10%	8%	-	-	-
Service	5%	6%	8%	-	-	-
Returns	0%	0%	1%	23%	0%	0%
Other	6%	8%	6%	2%	37%	10%
	100%	100%	100%	100%	100%	100%
	$n = 480$	$n = 293$	$n = 722$	$n = 30$	$n = 2$	$n = 6$

Note. assertions are weighted.

A relatively high percentage of the respondents mentioned that they did not have
an association with Super de Boer (15% in 1998, 14% in 1999 and 18% in 2000)
or that they did not know the supermarket (14% in 1998, 13% in 1999 and 11% in
2000). The third association mentioned by the respondents was "mergers and
takeovers." This association was mentioned in 1998 by 10%, in 1999 by 4%, and
in 2000 by 2% of the respondents. There was a relatively Odette amount of
media coverage (56%) in 1998 about the mergers of the parent company of
Super de Boer, De Boer Unigro, with the Vendex concern. In 1999 and 2000
there was scarcely any media coverage of Super de Boer and issues ($n = 2$ and $n = 7$ respectively).

Schiphol associated with KLM and in the news about expansion, but there were some matches
The media paid by far the most attention to the expansion of Schiphol (43% in
1998, 37% in 1999 and 28% in 2000). The headlines reported about the growth
of Schiphol or the fifth runway: "Fifth Schiphol runway close at hand" (*De*

Telegraaf, February 13, 1998, p. 2); "Schiphol allowed to handle 20,000 extra flights this year" (*RTL,* February 5, 1998). This association was not listed by the respondents, however.

TABLE 4.12.   **Public associations and issues in the media connected with Schiphol**

	Opinion 1998	Opinion 1999	Opinion 2000	Media 1998	Media 1999	Media 2000
KLM	34%	33%	32%	3%	2%	6%
Holiday trips	15%	17%	20%	-	-	-
Dutch economy	12%	10%	8%	4%	4%	7%
Environment. pollution	10%	11%	10%	31%	27%	22%
Mergers and takeovers	0%	0%	1%	1%	0%	0%
Profitability	0%	1%	0%	1%	1%	2%
Employment	3%	4%	3%	1%	1%	0%
Efficiency	1%	0%	0%	1%	1%	0%
Other	21%	19%	23%	6%	10%	13%
Do not know/Schiphol	3%	4%	3%	-	-	-
Expansion of Schiphol	-	-	-	43%	37%	28%
Societal issues	-	-	-	10%	16%	21%
	100%	100%	100%	100%	100%	100%
	n = 487	*n* = 294	*n* = 728	*n* = 768	*n* = 520	*n* = 282

Note. assertions are weighted. The category "do not know/Schiphol" consists of the respondents who did not have associations with or were not familiar with Schiphol.

Next to the expansion of Schiphol, the media paid the most attention to environmental pollution, in most cases the noise nuisance of the airport. In 1998, 31% of the news about issues examined environmental pollution; this was 27% in 1999 and 22% in 2000. Examples of headlines in the media in which Schiphol was related to environmental pollution are: "Schiphol produced less noise in 1998" (*De Telegraaf,* February 10, 1999, p. 2); "Environmental limit does not matter to growth of Schiphol" (*Trouw,* July 29, 1998, p. 7). This issue was mentioned by approximately 10% of the respondents. Furthermore, there was some news about the economy, which was mentioned by the respondents as the third association.

The most frequently mentioned association by the respondents, the airline KLM, did not receive a lot of media attention in the news about Schiphol.

NS meant "delays" to the respondents; there was some media coverage about the delays
Most respondents associated the NS with delays (28% in 1998, 33% in 1999, and 29% in 2000). The issue of delays was the sixth most frequently occurring issue in the media (9% in 1998, 10% in 1999, and 6% in 2000). As shown in Table 4.13,

there were more associations ("service in trains and stations," "inefficiency," and "strikes") that broadly matched with the media coverage about the NS. Examples of headlines about strikes are: "Possible NS strike on May 25" (NRC Handelsblad, May 14, 1998, p. 5); "NS strike is off" (De Telegraaf, March 26, 1998, p.1).

TABLE 4.13.   **Public associations and issues in the media connected with NS**

	Opinion 1998	Opinion 1999	Opinion 2000	Media 1998	Media 1999	Media 2000
Delays	28%	33%	29%	9%	10%	6%
Transport	23%	22%	25%	-	-	-
Service in trains and stations	17%	15%	16%	5%	16%	9%
Efficiency (and inefficiency)	5%	6%	5%	8%	6%	5%
Environment (noise, air)	4%	3%	3%	2%	1%	1%
Strikes	4%	2%	6%	14%	14%	12%
Dutch economy	3%	3%	3%	15%	8%	8%
Employment	2%	5%	2%	1%	2%	3%
Profitability	1%	2%	1%	5%	9%	4%
Other	1%	4%	5%	9%	11%	16%
Do not know/NS	6%	5%	5%	-	-	-
Corporate issues	-	-	-	26%	12%	7%
Societal issues	-	-	-	7%	10%	30%
	100%	100%	100%	100%	100%	100%
	n = 484	n = 308	n = 716	n = 297	n = 155	n = 341

Note. assertions are weighted. The category "do not know/NS" consists of the respondents who did not have associations with or were not familiar with the NS.

The NS received the most media coverage about societal issues, with 30% in the third research period and 17% within the entire research period. Most news in this category was on safety and crime in trains and around stations: "Train on fire at NS station in Leiden" (NRC Handelsblad, January 22, 1998, p. 3). The corporate issues were the second most frequently occurring issue in the media related to the NS (15% within the entire research period). Much this news was on the price of train tickets. However, the issues of safety and the price of train tickets were not offered to the respondents.

The police meant "crime" to the media, and was the second association of respondents
Table 4.14 indicates there could be a global match between the associations mentioned by the respondents and the issues reported in the media, and this will be tested in the next chapter. "Crime" was the most frequently occurring issue in the media and the second association of the respondents. Examples of the police and

crime in the media are: "Police picked up ten people for blackmail and receiving stolen goods" (NRC Handelsblad, December 27, 1997, p. 3); "Police arrests 25 squatters" (Algemeen Dagblad, May 1, 1999, p. 5).

"Safety" was the second most frequently occurring issue in the media and the third association mentioned by the respondents. Media attention for this issue dropped in the second research period, which was not reflected in the "safety association" of the respondents. An example of news about the police and safety is: "Experienced police officers must bring peace and quiet back to Western Amsterdam district" (RTL, May 12, 1998).

TABLE 4.14. **Public associations and issues in the media connected with the Dutch police**

	Opinion 1998	Opinion 1999	Opinion 2000	Media 1998	Media 1999	Media 2000
Maintaining authority	28%	27%	28%	-	-	-
Crime	22%	24%	21%	45%	43%	54%
Safety	15%	16%	17%	17%	9%	19%
More police on the streets	14%	14%	16%	-	-	-
Weak actions by police	10%	10%	8%	-	-	-
Standards and values	3%	3%	3%	5%	3%	4%
Employment	1%	0%	0%	6%	5%	9%
Inefficiency	2%	2%	2%	12%	4%	3%
Other	3%	2%	2%	4%	10%	2%
Do not know	2%	2%	2%	-	-	-
Corporate issues	-	-	-	11%	25%	10%
	100%	100%	100%	100%	100%	100%
	n = 479	n = 278	n = 719	n = 495	n = 477	n = 307

Note. assertions are weighted.

Most respondents associated the police with the general association "maintaining authority." The second most frequently described issue in the media, "corporate issues," could not be matched with the associations of respondents, either. The following headlines are examples of corporate issues: "Police allowed to try out pepper spray" (de Volkskrant, June 26, 1999, p. 1); "Police or dealer can use a scanner to see if a scooter is stolen" (RTL, November 6, 1997).

The environment came second when thinking or reporting about the agricultural sector
If the media coverage about issues and the associations mentioned by the respondents are compared, it appears that respondents mentioned "the environment" as their second association, while it was also the second most frequently covered issue by the media. Examples of the issue of the environment in the

media are: "Too much nitrate worries farmers, too" (*NRC Handelsblad*, June 8, 1998, p. 7); "Farmers: get manure surplus under control on our own" (*Algemeen Dagblad*, July 7, 1999, p. 12).

TABLE 4.15.   **Public associations and issues in the media connected with the agricultural sector**

	Opinion 1998	Opinion 1999	Opinion 2000	Media 1998	Media 1999	Media 2000
Fresh products	33%	33%	33%	-	-	-
Environment (pollution)	32%	28%	31%	6%	22%	32%
Dutch economy	13%	14%	13%	3%	2%	1%
EU subsidy	7%	10%	8%	-	-	-
Employment	4%	3%	4%	7%	8%	10%
Strikes	2%	2%	3%	5%	7%	11%
Efficiency	0%	1%	1%	4%	6%	1%
Other	3%	3%	2%	5%	10%	18%
Do not know/agric. sector	5%	6%	7%	-	-	-
Swine fever	-	-	-	51%	6%	5%
BSE	-	-	-	8%	28%	5%
Cattle/harvest	-	-	-	12%	12%	17%
	100%	100%	100%	100%	100%	100%
	n = 472	*n* = 288	*n* = 710	*n* = 371	*n* = 199	*n* = 119

Note. assertions are weighted. The category "do not know/agricultural sector" consists of the respondents who did not have associations with or were not familiar with the agricultural sector.

There was an increase in the news about the environment, which went from 6% in the first research period to 22% in the second and 32% in the third research period. The increase in news about the environment from the first to the second period was caused by the heavy rainfall ("environment"), which hurt the farmers and market gardeners. In the third period, there was a lot of news about the manure surplus. These changes in the media coverage were not directly reflected in the associations of the respondents. There was no match between the first and more general association of the respondents ("fresh products") and the issues reported in media. The same was the case for the most frequently covered issue in the media, "swine fever." This issue was not presented to the respondents.

The third issue that was frequently reported in the media was the mad cow disease "BSE," especially in the second research period, when BSE made up 28% of the issue news. This was due to the new cases of BSE, which were reported in the media: "New case of BSE discovered" (*Algmeen Dagblad*, August 27, 1998, p. 1). This issue was not presented to the respondents. The third association of the respondents, "the Dutch economy," did not receive a lot of media attention.

4.5 ADVERTISING EXPENDITURES PER ORGANIZATION/SECTOR PER RESEARCH PERIOD

In Table 4.16, the advertising expenditures are presented for every organization per medium and per research period. Compared to the other ten organizations and sectors, the total amount of advertising expenditures of Shell and BP were relatively low. Except for the police, BP had the lowest amount of advertising expenditures for the total research period.

The advertising expenditure patterns of Shell and BP were rather different. Over the three research periods, Shell spent by far the highest amount on advertising expenditures in the third research period, namely 1.2 million euros (these figures have been rounded off, see Table 4.16 for the exact figures). BP, on the other hand, spent the highest amount on advertising expenditures in the first research period, namely 1.1 million euros. In addition, in the second and the third research periods, BP did not spend a single euro on television advertising with either a public broadcaster or RTL 4. In all three research periods, Shell spent money on television advertising. In the last research period, when the advertising expenditures of Shell were the highest of the three research periods, Shell spent half its advertising budget on advertising on public television channels and – compared to the previous two research periods – a relatively low amount on advertising on the commercial channel RTL 4.

Next to supermarket Albert Heijn, the two banks together had the highest total amount of advertising expenditures over the three research periods. Rabobank had the highest amount of advertising expenditures per research period, namely 11.9 million euros in the second research period. Both banks spent more on advertising on public television channels than on RTL 4. Compared to Schiphol, the advertising expenditures of the NS were relatively high. The advertising expenditures of the airport came to 1.9 million euros in the total research period, while the advertising expenditures of the NS came to 13.1 million euros within the total research period. Schiphol did not spend any money on advertising on television within the total research period, whereas the NS spent most of its money on television advertising.

At 27.6 million euros, Albert Heijn had the highest advertising expenditures of the ten focal organizations and sectors within the total research period. The amounts do not vary that much between the three research periods. Albert Heijn spent more money on advertising in newspapers (mainly in *De Telegraaf*) than on advertising on television. With 10.3 million euros within the total research period, Super de Boer did not even have half of the advertising budget of Albert Heijn. Like Albert Heijn, Super de Boer spent more money on advertising in newspapers than on advertising on television. The advertising expenditures of the agricultural sector (1.6 million euros) and the police (1 million euros) were relatively low compared to the other organizations. The police did not spend money on television advertising in the first two research periods.

TABLE 4.16   **Advertising expenditures per organization per period, amounts in € x 1000**

Organization and period	Public television	RTL 4	Het Parool	Alg Dagblad	De Telegraaf	De Volkskr	NRC	Trouw	Total
Shell period 1	78.0	142.8	0.0	11.0	68.0	6.0	27.0	2.0	334.8
Shell period 2	40.0	95.3	0.0	25.0	102.0	29.0	56.5	7.0	354.8
Shell period 3	638.0	76.0	0.0	77.0	229.0	78.0	79.5	0.0	1177.5
BP period 1	264.0	130.0	20.0	135.8	230.5	132.0	132.0	28.0	1072.3
BP period 2	0.0	0.0	12.0	49.3	85.5	46.0	54.0	10.0	256.8
BP period 3	0.0	0.0	29.0	7.0	37.0	8.0	0.0	0.0	81.0
ABN period 1	2352.0	258.0	37.0	929.3	2217.8	724.0	1049.0	33.0	7600.0
ABN period 2	2558.0	190.5	39.5	544.5	1464.5	182.3	351.3	20.5	5351.0
ABN period 3	4112.5	570.3	32.5	566.8	1953.3	176.5	569.0	5.5	7986.3
Rabobank per. 1	1288.3	616.3	185.5	794.0	1567.3	739.3	876.8	265.8	6333.0
Rabobank per. 2	6047.5	1394.5	94.0	705.3	1780.3	755.0	909.8	174.5	11860.8
Rabobank per. 3	1466.5	359.0	24.3	421.0	1050.3	430.0	577.8	137.0	4465.8
NS period 1	1690.5	1626.5	55.0	310.5	526.8	344.0	300.5	96.3	4950.0
NS period 2	1453.3	1439.5	54.0	234.5	379.0	224.0	194.5	55.8	4034.5
NS period 3	1730.8	1462.0	37.5	162.8	270.3	176.3	209.0	72.8	4121.3
Schiphol per.1	0.0	0.0	1.5	154.3	196.0	141.0	3.8	1.5	498.0
Schiphol per. 2	0.0	0.0	7.5	246.8	380.8	236.8	123.3	6.5	1001.5
Schiphol per. 3	0.0	0.0	1.0	79.0	142.0	85.5	68.0	1.0	376.5
Albert Heijn per. 1	213.0	283.0	592.3	1491.0	2434.8	1463.0	1676.0	528.5	8681.5
Albert Heijn per. 2	748.5	1015.0	427.8	1383.0	2326.3	1159.3	1442.5	475.5	8977.8
Albert Heijn per. 3	501.5	842.0	157.0	1545.8	2801.5	1540.3	1878.3	643.8	9910.0
Super de B. per. 1	207.0	512.0	0.0	323.3	511.0	266.5	0.0	0.0	1819.8
Super de B. per. 2	93.0	210.0	0.0	918.8	1599.0	859.0	0.0	0.0	3679.8
Super de B. per. 3	349.0	679.0	1.0	952.0	1772.5	1002.3	1.0	0.0	4756.8
Agric. sector per. 1	53.0	80.0	0.0	10.0	78.0	11.0	12.0	5.0	249.0
Agric. sector per. 2	90.0	69.0	16.0	79.0	275.0	27.0	33.0	10.0	599.0
Agric. sector per. 3	171.5	450.3	8.0	39.0	16.0	6.0	10.0	2.0	702.8
Police period per. 1	0.0	0.0	8.0	12.0	88.8	29.0	4.0	0.0	141.8
Police period per. 2	0.0	0.0	16.0	149.0	122.3	70.0	31.0	9.0	397.3
Police period per. 3	55.0	93.0	8.0	128.3	56.0	90.0	0.0	0.0	430.3

Note. Source: *BBC* De Media en Reclame Bank

Research period 1: July 24, 1997 to July 25, 1998;
Research period 2: July 26, 1998 to Aug 14, 1999;
Research period 3: Aug 15, 2000 to July 22, 2000.

It should be remarked that the column "public television" is the total of advertising expenditures of the focal companies and sectors on three public television channels ("Nederland 1," "Nederland 2," and "Nederland 3"). Because the NOS news is broadcast on these three channels, the advertising expenditures of all three channels were taken into account, as was mentioned in the previous chapter.

4.6 CORPORATE REPUTATION AND THE POLICY OF THE FOCAL ORGANIZATIONS

This section will describe the public opinion data and the dependent variables. The report mark used as overall measure of reputation in the hypotheses 1, 2a, 3a, 5, 6, and 7 will be elaborated upon first. Subsequently, the perceived success the policy of a certain organization (hypothesis 2b) will be focused upon. Finally, the agreement with the policy of an organization will be described.

Albert Heijn had the best and the NS the worst reputation
Table 4.17 shows that Albert Heijn had the best reputation in 1999 (M = 6.9, SD = 1.52) and in 2000 (M = 6.8, SD = 1.39). In 1999 and 2000, Schiphol had the second-best reputation (M = 6.6, SD = 1.48 in 1999; M = 6.7, SD = 1.46 in 2000). Of the ten focal companies, the NS had the worst reputation in 1999 (M = 5.5, SD = 1.39) and 2000 (M = 5.7, SD = 1.44). In 1999 the police had the second-worst reputation (M = 6.0, SD = 1.54). In the year 2000 the police was doing better (M = 6.2, SD = 1.32).

TABLE 4.17.   **Mean report marks of the focal organizations in 1999 and 2000**

	1999			2000			*Paired t-test*	
	Mean mark	SD	n	Mean mark	SD	n	t	n
Albert Heijn	6.9	1.52	268	6.8	1.39	668	1.01	209
Schiphol	6.6	1.48	279	6.7	1.46	687	2.82**	216
ABN AMRO	6.5	1.34	251	6.3	1.44	642	-.9	188
Rabobank	6.5	1.59	244	6.4	1.44	633	.59	174
Agriculture	6.4	1.25	274	6.5	1.27	655	1.84	204
Super de Boer	6.3	1.31	209	6.3	1.37	541	1.28	149
BP	6.2	1.28	270	6.2	1.24	643	-1.58	199
Shell	6.2	1.45	298	6.2	1.46	678	-.42	226
Police	6.0	1.54	266	6.2	1.32	683	1.65	213
NS	5.5	1.39	294	5.7	1.44	674	1.96	223

Note. Respondents who do not use media were removed from the dataset when testing the hypotheses and are therefore not included in this table.
*p<.05; **p<.01

If the focus was on the respondents who participated at least in 1999 and in 2000, it seems there was only one significant difference in the change of report marks of the focal organizations. In 2000, Schiphol's report mark was significantly higher than in 1999, $t(215) = 2.82$, $p = .005$ (two tailed). The report mark of the NS was nearly significantly higher in 2000 than in 1999, $t(222) = 1.96$, $p = .051$ (two tailed). Note that the number of respondents was decreased in the paired t-tests, because only respondents who participated twice were selected. In the analyses of the hypotheses, all respondents using media will be included in the non-auto-regressive models.

Albert Heijn had the most and the police the least successful policy
Albert Heijn had the most successful policy, according to the respondents. The scale of success of the policy ranged from –1, which means that all the respondents thought the organization was unsuccessful to +1, which means that all the respondents thought the policy of the focal organization was a success. According to the respondents, Albert Heijn had the second most successful policy in 1998 ($M = .67$, $SD = .55$), and the best policy in 1999 ($M = .61$, $SD = .63$) and in 2000 ($M = .59$, $SD = .61$). Rabobank had the most successful policy in 1999 ($M = .68$, $SD = .58$), the fourth most successful policy in 1999 ($M=.59$, $SD = .62$), and the second/third most successful policy in 2000 ($M = .57$, $SD = .67$). In all the three years, the police had the least successful policy ($M = −.48$, $SD = .72$ in 1998, $M = −.55$, $SD = .69$ in 1999; and $M = −.32$, $SD = .78$ in 2000).

TABLE 4.18.   **Mean success of the policy of the focal organizations in 1998, 1999, and 2000 scale ranging from -1 (no success) to +1 (success)**

	1998			1999			2000			RM Anova	
	Mean	SD	n	Mean	SD	n	Mean	SD	n	F	n
Albert Heijn	.67	.55	380	.61	.63	234	.59	.61	549	2.44	142
Rabobank	.68	.58	260	.59	.62	150	.57	.67	347	.50	76
ABN AMRO	.62	.59	242	.60	.62	130	.56	.67	362	.39	61
Super de B.	.59	.65	215	.62	.62	132	.57	.63	332	.171	66
BP	.49	.63	159	.54	.58	98	.50	.66	248	1.59	26
Schiphol	.49	.70	334	.40	.74	178	.45	.73	449	1.75	96
Shell	.45	.71	316	.31	.73	176	.36	.75	417	3.24*	82
Agriculture	.20	.76	330	.12	.76	203	.16	.78	452	2.29	115
NS	-.23	.77	359	-.36	.76	218	-.29	.79	492	.20	118
Police	-.48	.72	416	-.55	.69	227	-.32	.78	591	1.08	159

Note. Respondents who do not use media were removed from the dataset when testing the hypotheses and are therefore not included in this table.
*p<.05

In order to determine whether the success of the policy of an organization varied per year, repeated-measures analyses were conducted per organization with "success of the policy of the organization in a certain year" as a within-subjects variable. The results were significant only in the case of Shell, Wilks' Lambda =. 93, $F(2, 79) = 3.23$, $p = .044$. As shown in Table 4.18, there was a sharp decrease in the perceived success of Shell's policy from 1998 ($M = .45$, $SD = .71$) to 1999 ($M = .31$, $SD = .73$). In 2000 the success of Shell's policy increased again to $M = .36$ ($SD = .75$). For all the other organizations, no significant difference was found between the success of the policy in 1998, in 1999, and in the year 2000.

Albert Heijn was agreed with the most, and the NS and the police agreed with the least
Table 4.19 shows that Albert Heijn is the organization the respondents agreed with most. The scale of agreement ranges from –1, which means that all the respondents only strongly disagreed with the policy of the focal organization to +1, which means that all the respondents only strongly agreed with the policy of the focal organization. In 2000, the respondents agreed just as much with Albert Heijn ($M = .25$, $SD = .42$) as with Super de Boer ($M = .25$, $SD = .36$). Super de Boer was second in line as the organization respondents most agreed with ($M = .22$, $SD = .38$ in 1998; $M = .26$, $SD = .35$ in 1999; $M = .25$, $SD = .36$ in 2000). The third and fourth organizations whose policies were most agreed with were Rabobank and ABN AMRO. The NS and the police (except for the year 2000) were the only two focal organizations with which the respondents slightly disagreed. In 1998 the respondents slightly disagreed with the policy of the police ($M = –.07$, $SD = .49$). The police were the least agreed with in 1998. In 1999 and 2000 the NS was the organization least agreed with.

TABLE 4.19. **Mean agreement with the policy of the focal organizations in 1998, 1999, and 2000 scale ranging from -1 (*strongly disagree*) to +1 (*strongly agree*)**

	1998			1999			2000			RM Anova	
	Mean	SD	n	Mean	SD	n	Mean	SD	n	F	n
Albert Heijn	.31	.40	416	.30	.40	255	.25	.42	627	2.17	181
Super de B.	.22	.38	276	.26	.35	176	.25	.36	438	.76	89
Rabobank	.21	.37	403	.20	.35	238	.18	.34	601	2.8	154
ABN AMRO	.20	.30	398	.19	.30	247	.15	.33	617	3.26*	175
Schiphol	.19	.39	453	.16	.35	273	.18	.36	678	.35	204
Agriculture	.16	.37	431	.12	.37	266	.12	.37	627	1.1	192
BP	.10	.24	419	.09	.23	263	.10	.26	624	.35	185
Shell	.11	.35	474	.06	.31	291	.07	.36	664	.97	215
Police	-.07	.49	450	-.06	.47	262	.00	.44	674	.18	203
NS	-.05	.45	440	-.08	.41	283	-.04	.39	655	.22	204

Note. Respondents who do not use media were removed from the dataset when testing the hypotheses and are therefore not included in this table.
*p<.05

In order to determine if the agreement with the policy of an organization varied per year, repeated-measures analyses were conducted per organization with "agreement with the policy of the organization in a certain year" as a within-subjects variable. The results were significant only in the case of ABN AMRO, Wilks' Lambda = .96, $F(2, 172) = 3.26$, $p = .04$. For all the other organizations, no significant difference was found between the agreement with the policy in 1998, in 1999, and in the year 2000.

5

RESULTS

This chapter will focus on the testing of the hypotheses. In order to examine the data, the correlations of the main study variables incorporated in the analyses will be discussed first. Then it will be examined whether corporate reputation can be predicted from the perceived success of the policy of an organization and the agreement with the policy of an organization, as was postulated in section 3.3.3. Subsequently, in section 5.3 the influence of the total amount of news (hypothesis 1), the tenor of success & failure news (hypotheses 2a), and advertising intensity (hypothesis 7) on corporate reputation will be estimated. In section 5.4, the empirical results of the impact of the tenor of support & criticism news (hypotheses 3a), the total amount of news and advertising intensity on reputation will be presented. In section 5.5, a combined model with the effects of the total amount of news, the tenor of success & failure, support & criticism news, and advertising intensity on corporate reputation will be estimated. The effects of success & failure news on the perceived success of organizational policy, hypothesis 2b, will be tested in subsection 5.6. Hypothesis 3b, the effects of support & criticism news on the agreement with organizational policy, will be focused upon in subsection 5.7.

In section 5.8, the results of the tests of the agenda-setting hypothesis (hypothesis 4) will be described. Then the results of the testing of the priming hypothesis will be presented (hypothesis 5). In section 5.10, the impact of news about owned issues on corporate reputation (hypothesis 6) will be assessed. Finally, an overview will be given of the results per hypothesis (section 5.11). If applicable, a comparison will be made between the different models.

5.1 CORRELATIONS BETWEEN THE MEDIA VARIABLES, ADVERTISING AND REPUTATION

In Appendix I, the tables with the correlations among the main study variables are displayed in order to get a first impression of the results. There are three different sets of correlations tables. The tables in Part I of Appendix I present the correlations between and within the variables that are used to test hypotheses 1a, 1b, 2a, 3a, and 7. The tables in Part II of Appendix I present the correlations between and within the variables that are used to test hypotheses 2b and 3b. The tables in Part III of Appendix I present the correlations between the variables, which are used to test hypotheses 4, 5, and 6. Part I of Appendix I will be discussed first.

The amount of television news correlated for three organizations positively with report mark (Shell, $r = .07$, $p < .05$, see Table I1; NS, $r = .07$, $p < .05$, see Table I5; police, $r = .18$, $p < .001$, see Table I9). These correlations indicated that the reputation of these companies and professional groups improve if they appear in the television news. For none of the organizations did the amount of print news correlate significantly with report mark.

If the tenor of success & failure television news correlated significantly with report mark (in the case of Schiphol and the police), it was a positive correlation, except for the agricultural sector ($r_{\text{favorability success \& failure TV news, reputation}} = -.07$, $p < .05$, see Table I10). Being successful in the television news may go together with a good reputation.

The opposite was the case for the tenor of support & criticism television and print news. The tenor of support & criticism news correlated negatively with report mark for four organizations (Shell, Albert Heijn, NS, and Schiphol). If criticism in the media increases, this seems to go together with a better reputation.

In accordance with the mere exposure theory, positive correlations were found between the advertising intensity of an organization and its reputation. This was the case for the following six organizations: BP (see Table I2), Albert Heijn (see Table I3), Super de Boer (see Table I4), Rabobank (see Table I8), the police (see Table I9), and the agricultural sector (see Table I10). In the subsection below, the correlations between the favorability of success & failure news and the perceived success of the policy will be discussed (Part II of Appendix I).

The correlations between the favorability of success & failure news and the perceived success of the policy were somewhat mixed. In agreement with the expectations, positive correlations were found between the favorability of success & failure news and the perceived success of the policy. This was the case for three organizations: NS (see Table I15), ABN AMRO (see Table I17), and the police (see Table I19). In the case of two organizations, Albert Heijn and the NS, however, a negative correlation was found between the two variables. The results from the regression analyses may provide more insights into these findings.

A negative correlation was expected between "the favorability of support & criticism news" and "the agreement with the policy of the organization," since a lot of the news comes from competitors. Indeed, a negative relationship between these two variables was found in the case of Shell (Table I11), Albert Heijn (Table I13), and ABN AMRO (Table I17). A positive correlation between the two variables was found in the case of the police (Table I19) and the agricultural sector (Table I20). This may suggest that criticism of an impartial actor such as the government may influence the agreement with the policy of the organization accordingly.

The subsection below will discuss the correlations between the independent and dependent variables that were used to test the hypotheses about agenda-setting, priming issue ownership (hypotheses 4, 5, and 6 respectively).

For eight organizations there was a significant correlation between the amount of issue news and the salience of an issue (hypothesis 4). Note that the correlations were negative for five organizations. This was due to the fact that the general associations of the respondents (such as "gas stations" in the case of Shell and BP, and "transport" in the case of the NS) were included in the correlation tables. Since these associations could not be matched with the media coverage about the focal companies (see Appendix F), the general associations were excluded in the testing of hypothesis 4 (section 5.8). As was expected on the basis of hypothesis 5, the correlations between the interaction variable "the frequency of issue news the policy of news" and the reputation variable were positive. This was the case for eight organizations. In the regression models, these effects were controlled for the main effects. The amount of news about owned issues correlated with report mark in the case of BP (see Table I22) and the police (see Table I29). There were no significant correlations between the amount of owned issue news and report mark for the rest of the organizations. The low number of significant correlations indicates that it may be questionable whether the amount of owned issue news predicts report mark.

Overall, these results suggest it is likely that effects of media coverage on corporate reputation will be found, although some hypotheses will receive stronger support than other hypotheses.

5.2 SUCCESS AND AGREEMENT WITH ORGANIZATIONAL POLICY AS PREDICTORS OF REPUTATION

In Chapter 3 it was argued that corporate reputation can be predicted from the agreement with the policy of an organization and the success of the policy of an organization. This was tested by using Ordinary Least Squares regression. The results of this model are presented in Table 5.1.

As shown in Table 5.1, both variables are significant predictors of corporate reputation. The more the respondents agreed with the policy of the organization, the more positive they were about the reputation of the organization ($\beta = .42$, $p < .001$). Accordingly, respondents also assigned a higher reputation to the organization the more they thought the policy of the organization was a success ($\beta = .24$, $p < .001$). The two independent variables explained 37% of the variance.

TABLE 5.1.   Stepwise regression analysis of the success and agreement with organizational
              policy on reputation

Predictors	
Agreement with the policy of an organization	.42***
Success of the policy of an organization	.24***
Adjusted R^2	.37
df	(2, 5982)
F	1751.18***
N	5,985

Note. Beta coefficients are shown.
*** $p \leq .001$

In Table 5.1, one regression model is shown for the data of the ten organizations. If regression models were made for each organization separately, it appeared that both predictors were significant in all ten models (see Appendix J). In other words, the two variables "the agreement with the policy of the organization" and "the perceived success of the policy of an organization" predict the reputation of an organization.

5.3   THE EFFECTS OF THE TENOR OF SUCCESS & FAILURE NEWS, THE AMOUNT OF NEWS, AND ADVERTISING INTENSITY ON REPUTATION (H. 1A, 1B, 2A, AND 7)

In this section, the effects of the total amount of news, the favorability and the direction of success & failure news, and advertising intensity on reputation will be examined. As mentioned in Chapter 3, hypotheses 1a, 1b, 2a, and 7 will be tested together. By examining several hypotheses together, the effects of the amount of news on reputation are controlled for the effects of the tenor (which consists of the favorability and direction) of success & failure news and for the effects of advertising intensity and vice versa. This results in the following equation:

$\text{Reputation}_{i,t,o} = b_o + b1$ (Amount of TV news$_o$) + b2 (Amount of print news$_o$) + b3 (Direction success & failure TV news$_o$) + b4 (Direction success & failure print news$_o$) + b5 (Favorability of success & failure TV news$_o$) + b6 (Favorability of success & failure print news$_o$) + b7 (Advertising intensity TV$_o$) + b8 (Advertising intensity newspapers$_o$) + e_i

The models will be presented for each organization separately and at the level of all the organizations (the pooled dataset). The results will also be reported when the regression model is controlled for the educational level of the respondents.

Testing of the model for each of the organizations separately rendered the following results (see Table 5.2). In the last column, the lagged dependent variable (the reputation of the previous year) was entered in the models. In the second to last column, the lagged dependent variable was not included in the models; these are what are known as "non-autoregressive models." These models will be discussed first, starting with the hypothesis about the effects of the amount of news, hypothesis 1.

The amount of television news had a positive effect on the reputation of Shell ($\text{ß} = .10$, $p < .01$), the NS ($\text{ß} = .07$, $p < .05$), Schiphol ($\text{ß} = .14$, $p < .001$), and the police ($\text{ß} = .15$, $p < .001$). This means that the reputations of these organizations improved if they received more coverage on the television news. The amount of print news was significant in the case of Super de Boer and Schiphol. The reputation of Super de Boer was improved by the amount of print news ($\text{ß} = .14$, $p < .01$), while it worsened the reputation of Schiphol ($\text{ß} = -.19$, $p < .001$).

This may indicate that an organization that is in the news relatively rarely benefits from coverage in the print media, while it may harm organizations that are frequently in the print news. For five organizations, the amount of either print or television news was not found to significantly influence reputation.

TABLE 5.2.   **Stepwise regression analyses of the effects of the tenor of success & failure news and the amount of news and advertising intensity on reputation**

Models per organization[a]	Predictors (news and advertising, previous year)	Cross-sectional	Auto-regressive
Shell	Amount of TV news	.10**	ns
	Favorability of success & failure TV news	.07*	ns
	Reputation,1999	-	.52***
	Adjusted R^2	.01	.27
	df	(2, 973)	(1, 224)
	F	4.06*	82.41***
	n	976	226
BP	Advertising intensity newspapers	.09**	ns
	Direction of success & failure newspapers	ns	.15*
	Reputation,1999	-	.45***
	Adjusted R^2	.01	.23
	df	(1, 911)	(1,197)
	F	7.43**	54.61***
	n	913	199
Albert Heijn	Advertising intensity newspapers	.09**	ns
	Reputation,1999	-	.72***
	Adjusted R^2	.01	.52
	df	(1, 934)	(1, 207)
	F	8.29**	227.63***
	n	936	209

Note. Cell entries are Betas from Ordinary Least Squares regression.
For ABN AMRO, no significant effects were found.
* $p \leq .05$; ** $p \leq .01$; *** $p \leq .001$

TABLE 5.2. **(Continued)**

Models per organization[a]	Predictors (news and advertising, previous year)	Cross-sectional	Auto-regressive
Super de Boer	Amount of print news	.14**	ns
	Favorability of success & failure print news	.15**	ns
	Advertising intensity TV	.10**	ns
	Reputation$_{1999}$	-	.40***
	Adjusted R^2	.02	.15
	df	(3, 746)	(1, 147)
	F	6.06***	27.65***
	n	750	149
NS	Amount of TV news	.07*	ns
	Reputation$_{1999}$	-	.57***
	Adjusted R^2	.00	.32
	df	(1, 966)	(1, 221)
	F	4.66*	106.91
	n	968	223
Schiphol	Amount of TV news	.14***	ns
	Amount of print news	-.19***	ns
	Favorability of success & failure TV news	.23***	.14**
	Favorability of success & failure print news	-.22***	ns
	Reputation$_{1999}$	-	.61***
	Adjusted R^2	.03	.41
	df	(4, 961)	(2, 213)
	F	9.48***	76.91
	n	966	216
Rabobank	Advertising intensity newspapers	.08*	ns
	Reputation$_{1999}$	-	.65***
	Adjusted R^2	.00	.41
	df	(1, 875)	(1, 172)
	F	4.91*	122.30
	n	877	174
Police	Amount of TV news	.15***	ns
	Direction of success & failure TV news	.11***	ns
	Advertising intensity newspapers	ns	.17**
	Reputation$_{1999}$	-	.53***
	Adjusted R^2	.04	.32
	df	(2, 946)	(2, 210)
	F	20.91***	50.06
	n	949	213
Agricultural sector	Favorability of success & failure TV news	-.07*	ns
	Advertising intensity TV	ns	.15*
	Reputation$_{1999}$	-	.52***
	Adjusted R_2	.00	.31
	df	(1, 927)	(2, 201)
	F	4.13*	45.78***
	n	929	204

Hypothesis 2a predicts that the more positive the favorability or direction of success & failure television news, the better the reputation of the organization will be. This was the case for the model of Shell[1] (ß = .07, $p < .01$), Schiphol (ß = .23, $p < .001$), and the police (ß = .11, $p < .001$). In other words, the more television news there is about the increase in the profits of Shell, the better the reputation of Shell will be. The reputation of Schiphol will improve as well if there is television news about the success of Schiphol, such as the increase in the number of passengers. The reputation of the police will be improved if they are in the television news more often about their success in fighting crime. The favorability of success & failure TV news[2] is negatively related to the reputation of the agricultural sector (ß = –.07, $p < .05$). This negative relationship indicates that more television news about the struggle of the farmers to keep farming improves their reputation.

For print news, significant effects of the favorability of success & failure news were found for Super de Boer (ß = .15, $p < .01$) and for Schiphol (ß = –.22, $p < .001$). Although there was very little success & failure news in the print media about Super de Boer, the reputation of the supermarket seemed to get worse nevertheless if there was more news about its failure, such as the loss of market share. The negative relationship between the favorability of success & failure news in newspapers and the reputation of Schiphol indicates that the newspaper readers will, in contrast to the television viewers, become more negative towards Schiphol if the airport is in the news with its success (such as how busy it is). For five organizations, no effects of the favorability (or direction) of success & failure news on reputation were found.

Hypothesis 7 poses that the higher the advertising intensity of an organization, the better its reputation. The significant effects were all in the predicted direction. Higher advertising expenditures in newspapers improved corporate reputation in the models of four organizations (BP, Albert Heijn, and Rabobank). In addition, the advertising expenditures on television improved the reputation of Super de Boer.

In some of the non-autoregressive models (for example NS, Rabobank, and the agricultural sector) the Adjusted coefficient of determination, the Adjusted R^2, was relatively low, namely 0.5%. This will be dealt with in the discussion of the results in Chapter 6.

Autoregressive models per organization
In the last column of Table 5.2, Beta coefficients are shown only if in addition to the lagged dependent variable, a media or advertising variable was significant. Since this model includes a lag dependent variable, the number of respondents was limited to the respondents who participated in both 1999 and 2000. It was examined whether this decrease in the number of respondents influenced the

results described in the previous section. This was done by estimating the non-autoregressive model with respondents who participated both in 1999 and in 2000. It appears that only three media variables remain significant. These three variables were the total amount of television news in the model of Schiphol, the favorability of success & failure television news in the model of Schiphol, and the direction of success & failure television news in the model of the police. This means that in the rest of the models per organization, the media variables – nine in total – disappeared solely due to the decrease in the number of respondents. Seeing that the decrease of the number of respondents leads to a major decrease in the number of significant media variables, it can be expected that only a few of the media variables will be significant in the autoregressive models, which was also the case.

The reputation of the previous year is by the far the best predictor of the reputation of the current year. Except for the favorability of success & failure TV news that still positively influenced the reputation of Schiphol, all the other media variables significant in the non-autoregressive models have disappeared. In the case of BP, the direction of success & failure news about BP in the print media ($\text{ß} = .15, p < .05$) became significant. This variable was not significant in the non-autoregressive model. Advertising expenditures in newspapers turned out to improve the reputation of the police ($\text{ß} = .17, p < .01$), whereas the reputation of the agricultural sector was improved by advertising expenditures on television[3] ($\text{ß} = .15, p < .05$).

Controlling for the educational level of respondents in the models per organization
In order to investigate whether the effects of television news and print news can be explained by the educational level of the respondents, the models were controlled for this background characteristic (see Table K1, Appendix K). Overall, the non-autoregressive models of the organizations that received a great deal of media coverage (NS, Schiphol, police, and the agricultural sector) changed very little after controlling for educational level of the respondents.

For three of these organizations, the education variable was added to the model (NS, Schiphol, and the agricultural sector). The model of the police remained the same as the model that was not controlled for educational level. In the model of Schiphol, two media variables remained significant, while the amount of print news was no longer significant and the variable favorability of success & failure print news was replaced by the variable direction of success & failure print news ($\text{ß} = -.09, p < .05$). The models of Shell and Super de Boer changed substantially. In the model of Shell, neither the media variables nor the advertising variables were significant any longer. When the model of Super de Boer was controlled for education, the amount of print news and the favorability of success & failure print news were no longer significant due to a small decrease in the number of cases.[4] This means that the non-autoregressive model of Super de Boer is relatively unstable.

In sum, the models of organizations that received a great deal of media coverage (Schiphol, NS, police, and the agricultural sector) changed very little after controlling for education. For three of these organizations, the education variable was added in the model. If the educational level of the respondent was significant, the s were negative. This means that more highly educated people have a more negative attitude towards the focal organizations than those people with a lower level education.

When the autoregressive models of the effects of the total amount of news and the direction and favorability of success & failure news on corporate reputation were controlled for education, the two media variables (direction of success & failure print news in the case of BP, and favorability of success & failure TV news for Schiphol) and the two advertising variables remained significant.

The pooled models of the effects of the amount of news and success & failure news on reputation
In Table 5.3, the pooled dataset is used to examine the effects of the total amount of news and the favorability (or direction) of success & failure news on corporate reputation. The last column presents the autoregressive pooled model, while the second to last column presents the non-autoregressive model that will be discussed first.

In the non-autoregressive pooled model, the amount of television news ($ß = .11$, $p < .001$) about the ten focal organizations improved the reputation of the organizations. This is in accordance with the results of the models per organization that were presented earlier. The amount of print news about the focal organizations had a negative impact on the reputation of the organizations ($ß = -.04, p < .001$). This means that the reputation of the organization will worsen the more they are covered in the print media.

The pooled model also shows that the more successful an organization is portrayed in the television news, the better its reputation will be ($ß = .07, p < .01$). The favorability of success & failure news in the print media was not significant. Advertising expenditures in newspapers improved corporate reputation ($ß = .06$, $p < .001$), whereas in the pooled model, no effect was found for advertising expenditures on television.

In the case of the autoregressive pooled model,[5] the same procedure was applied as on the autoregressive models on organizational level. Before the autoregressive pooled model was estimated, it was examined whether the decrease in the number of respondents influenced the results found earlier. Therefore, the autoregressive pooled model was estimated with the reduced number of respondents. It appears that the effects of the amount of print news and the favorability of success & failure TV news were no longer significant, due to the decrease in the number of cases.

TABLE 5.3. **Pooled confirmatory regression analysis of the impact of the tenor of success & failure news and the amount of news and advertising on corporate reputation**

Predictors (news and advertising, previous year)	Cross-sectional	Auto-regressive
Key variables		
Amount of TV news	.11***	.10
Amount of print news	-.04***	-.03
Direction of success & failure TV news	-.01	.02
Direction of success & failure print news	.01	-.01
Favorability of success & failure TV news	.07**	-04
Favorability of success & failure print news	.01	.03
Advertising intensity newspapers	.06***	.04
Advertising intensity TV	.03	-.04
Control variables		
Shell	-.10***	-.05
BP	-.06***	-.05
Albert Heijn	.03	.02
Super de Boer	-.05**	-.01
NS	-.23***	-.08
Schiphol	.02	.04
ABN AMRO	-.05*	.01
Rabobank	-.03	.01
Police	-.19***	-.07
Reputation, $_{1999}$	-	.57***
Adjusted R^2	.06	.37
df	(17, 9139)	(18, 1982)
F	34.20***	65.76***
N	9,157	2001

Note. Cell entries are Betas from Ordinary Least Squares regression.
The agricultural sector was the base category. Variables were selected by confirmatory specification (using "Enter").
$p \leq .05$; ** $p \leq .01$; *** $p \leq .001$

None of the key variables were significant in the autoregressive pooled model. The amount of television news and advertising expenditures in newspapers were not significant, due to the addition of the lagged dependent variable. The amount of television news was nearly significant, however ($\beta = .10$, $p = .079$). The reputation of 1999 was the only significant predictor of the reputation of the organizations in 2000.

If the non-autoregressive pooled model was controlled for the educational level of the respondents (see Table K2, Appendix K), the model did not change except for the fact that the variable education became significant ($\beta = -.05$, $p < .001$).

5.4 THE EFFECTS OF SUPPORT & CRITICISM NEWS, THE AMOUNT OF NEWS, AND ADVERTISING ON REPUTATION (H. 1A, 1B, 3A, AND 7)

The following subsections will discuss the models that examine the effects of the total amount of news and the effects of support & criticism news on corporate reputation. Hypotheses 1a, 1b, 3a, and 7 will be tested together so that the influence of one variable is controlled for the influence of the other two variables. For example: the *amount* of news is controlled for the effects of the *favorability* and *direction* of support & criticism news and *advertising* intensity. This results in the following equation:

Reputation$_{i,t,o}$ = b$_0$ + b1 (Amount of TV news$_o$) + b2 (Amount of print news$_o$) + b3 (Direction success & failure TV news$_o$) + b4 (Direction success & failure print news$_o$) + b5 (Favorability of success & failure TV news$_o$) + b6 (Favorability of success & failure print news$_o$) + b7 (Advertising intensity TV$_o$) + b8 (Advertising intensity newspapers$_o$) + e$_i$

Testing of the model for each of the organizations separately rendered the following results (see Table 5.4). In the last column, the lagged dependent variable (the reputation of the previous year) was entered in the models. The non-auto-regressive models are presented in the second to last column. These models will be discussed first, starting with the hypothesis about the effects of the amount of news on reputation, hypothesis 1.

The amount of television news had a positive impact on the reputation of Shell (ß = .10, $p <$ .01), the NS (ß = .07, $p <$ .05), and the police (ß = .18, $p <$ .001), after controlling for support & criticism news. This means that the reputation of these organizations improves if they receive more coverage in the television news. These results are in line with the results of the model in which the effects of the amount of television news were controlled for success & failure news. The amount of print news had a negative impact on the reputation of the police (ß = - .07, $p <$ .05).

Hypothesis 3a predicts that the tone of support & criticism news is negatively related to reputation if the organization is criticized by its competitors. The direction of support & criticism TV news had a negative influence on the reputation of Shell (ß = - .09, $p <$ .05). This means that the more NGOs criticize Shell, the better Shell's reputation will be. The direction of support & criticism print news also had a negative influence on Albert Heijn (ß = - .08, $p <$ .05) and Schiphol (ß = - .08, $p <$ .05). This indicates that the reputation of Albert Heijn will improve the more it is criticized by its competitor Montignac. If Schiphol is supported in the newspapers by the airlines and the government in the third research period, the reputation of Schiphol will get worse. Only for the agricultural sector did a positive relationship exist between the favorability of support & criticism news and reputation (ß = .11, $p <$ .01) . This means that the more support for the agricultural sector, the better its reputation and vice versa, and the more criticism for the agricultural sector, the worse its reputation.

Thus, the direction of the Beta coefficients with regard to the tenor of support & criticism news are different for Shell, Albert Heijn, and Schiphol on one hand, and the agricultural sector on the other hand. The negative influence of support & criticism news on the reputations of Shell, Albert Heijn, and Schiphol may be explained by the fact that the subjects of criticism were "traditional opponents" or "traditional friends." Their criticism or support had the opposite reaction. This means that if traditional opponents (such as the NGOs in the case of Shell) criticize the organization, the reputation of the criticized organization will improve. Support from traditional friends results in a more negative reputation. Schiphol received the support of the industry, but the industry may be considered to be a suspicious partner when it comes to protecting noise regulations.

TABLE 5.4.   **Stepwise regression analyses of the effects of the tenor of support & criticism news and the amount of news and advertising on reputation**

Models [a]	Predictors (news and advertising, previous year)	Cross-sectional	Auto-regressive
Shell	Amount of TV news	.10**	ns
	Direction of support & criticism TV news	-.09*	ns
	Reputation, $_{1999}$	-	.52***
	Adjusted R^2	.01	.27
	df	(2, 973)	(1, 224)
	F	5.40**	82.41***
	n	976	226
BP	Advertising intensity newspapers	.09**	ns
	Amount of print news	ns	.15*
	Reputation, $_{1999}$	-	.453***
	Adjusted R^2	.01	.23
	df	(1, 911)	(2, 196)
	F	7.43**	30.59***
	n	913	199
Albert Heijn	Direction of support & criticism print news	-.08*	ns
	Advertising intensity newspapers	.09**	ns
	Reputation, $_{1999}$	-	.72***
	Adjusted R^2	.01	.52
	df	(2, 933)	(1, 207)
	F	6.98***	227.64***
	n	936	209
Super de Boer	Advertising intensity TV	.08*	ns
	Reputation, $_{1999}$	-	.40***
	Adjusted R^2	.01	.15
	df	(1, 748)	(1, 147)
	F	4.42*	27.65***
	n	750	149

TABLE 5.4.   (Continued)

Models[a]	Predictors (news and advertising, previous year)	Cross-sectional	Auto-regressive
NS	Amount of TV news	.07*	ns
	Reputation, 1999	-	.57***
	Adjusted R^2	.00	.32
	df	(1, 966)	(1, 221)
	F	4.66*	106.91***
	n	968	223
Schiphol	Direction of support & criticism print news	-.08*	-.10*
	Favorability of support & criticism print news	ns	-.17**
	Reputation, 1999	-	.58***
	Adjusted R^2	.01	.43
	df	(1, 964)	(3, 212)
	F	6.31*	54.70***
	n	966	216
Rabobank	Advertising intensity newspapers	.08*	ns
	Direction of support & criticism TV news	ns	-.18**
	Reputation, 1999	-	.65***
	Adjusted R^2	.00	.44
	df	(1, 875)	(2, 171)
	F	4.91*	69.07***
	n	877	174
ABN AMRO	Direction of support & criticism print news	ns	.16**
	Reputation, 1999	-	.61***
	Adjusted R^2		.39
	df		(2, 185)
	F		61.88***
	n	893	188
Police	Amount of TV news	.18***	ns
	Amount of print news	-.07*	ns
	Advertising intensity newspapers	ns	.17**
	Reputation, 1999	-	.53***
	Adjusted R^2	.03	.32
	df	(2, 946)	(2, 210)
	F	17.56***	50.06***
	n	949	213
Agricultural sector	Favorability of support & criticism print news	.11**	ns
	Advertising intensity TV	.07*	.15
	Advertising intensity newspapers	.10**	ns
	Reputation, 1999	-	.52***
	Adjusted R^2	.01	.31
	df	(3, 925)	(2, 201)
	F	5.41***	45.78***
	n	929	204

Note. Cell entries are Betas from Ordinary Least Squares regression.

* p ≤ .05; ** p ≤ .01; *** p ≤ .001

The positive influence of support & criticism news on the reputation of the agricultural sector may be explained by the fact that most criticism came from the government. The government may be perceived as neutral or even as an ally of the farmers in a globalizing food market. Therefore, the reputation of the agricultural sector will be influenced in the same direction as the support & criticism given by the government. This means that the reputation of the agricultural sector will get worse the more the sector is criticized by the government, and better the more the sector is supported by the government.

With regard to hypothesis 7, if the advertising expenditures on television and in newspapers have an effect on reputation, it is a positive effect. Higher advertising expenditures led to a better reputation in the models of five organizations. The reputation of organizations that were rarely in the news (BP, Super de Boer, Rabobank), was influenced by the amount of advertising expenditures, but not by the amount of media coverage or by the tenor of support & criticism news. In addition, higher advertising expenditures led to a better reputation for Albert Heijn and the agricultural sector.

In some of the non-autoregressive models (e.g., NS, Rabobank) the Adjusted coefficient of determination, the Adjusted R^2, was relatively low, namely 0.5%. This will be dealt with in the discussion of the results in Chapter 6.

Autoregressive models per organization
In the last column of Table 5.4, Beta coefficients are shown only if a media or advertising variable was significant in addition to the lagged dependent variable. Since this model includes a lag dependent variable, the number of respondents was limited to the respondents who participated in 1999 and in 2000. Hypothesis 1 (about the effects of the amount of media coverage on corporate reputation) will be discussed first.

Table 5.4 shows that in three models (Shell, NS, and the police), the amount of television news no longer has a significant impact on the reputation if the lagged dependent variable is included in the model. The amount of print news significant in the non-autoregressive model of the police is no longer significant in the autoregressive model. As in the previous subsection, it is examined whether the disappearance of the variables that measure the amount of media coverage was due to the *lower number of cases* in the autoregressive models.

A total of four of the media variables that measured the amount of media coverage were significant in the non-autoregressive models, but not in the autoregressive models. Three of them disappeared to the lower number of cases in the autoregressive models (amount of television news about Shell and the NS, amount of print news about the police), while one of them (amount of television news about the police) disappeared due to the inclusion of the lagged dependent variable in the model. Due to the lower number of cases, the amount of print news became significant in the autoregressive model of BP (ß = .15, $p < .05$).

Hypothesis 3a predicts that the tone of support & criticism news is negatively related to reputation if the organization is criticized by its competitors. As in the non-autoregressive model, the support Schiphol received in the print media worsened its reputation (ß = - .10, $p < .05$). There also appears to be a negative relationship between the favorability of support & criticism television news about Schiphol and the reputation of Schiphol (ß = - .17, $p < .01$). This variable was not significant in the non-autoregressive model. This means that the reputation of the airport will improve according to the amount of criticism it gets in the television news from the NGOs and vice versa: the more support the airport receives in the television news, the worse its reputation will become. The reputation of Rabobank got worse according to the amount of support it got on television (ß = −.18, $p < .01$) after including the lagged dependent variable in the model. For ABN AMRO the opposite was the case, however, and the more print news about organizations willing to cooperate with ABN AMRO, the better the reputation of the bank (ß = .16, $p < .01$). It is not exactly clear why the effects of support & criticism news on reputation differed for the two banks. Rabobank received a lot of support in the third research period due to cooperation with other organizations. Rabobank also received new accusations by the investment club, however. The accusations by the investment club may evoke an opposite reaction among the respondents. This resulted in a reverse effect, and the more criticism Rabobank received by the investment club, the better its reputation became.

In addition to the support & criticism variables that become significant in the autoregressive model, there were also variables that were significant in the non-autoregressive model, but were not significant in the autoregressive model (for example, in the case of Shell, Albert Heijn, and the agricultural sector). Due to the lower number of cases, the direction of support & criticism television news was no longer significant in the case of Shell, and the favorability of support & criticism print news no longer influenced the reputation of the agricultural sector. For Albert Heijn, that the direction of support & criticism print news is not significant in the autoregressive model is due not to the lower number of cases but to the addition of the lagged dependent variable.

In the autoregressive model, only advertising expenditures on TV about the agricultural sector remained significant, while advertising expenditures in newspapers in the case of the police become significant after including the lagged dependent variable. The effects of the advertising expenditures of BP, Albert Heijn, Rabobank, and the agricultural sector in newspapers disappeared due to the lower number of cases. Only in the case of Super de Boer did the effects of advertising expenditures on TV disappear because of the inclusion of the lagged dependent variable and not because of a lower number of cases.

Controlling for the educational level of respondents
In order to investigate whether the different effects of the amount of television news and the amount of print news can be explained by the educational level of the respondents, the models will be controlled for this background characteristic. This results in the following models (see Appendix L). In the case of four models (Albert Heijn, BP, Schiphol, and Rabobank), there was only a minor change in the regression coefficients after controlling the non-autoregressive models for the educational level of the respondents. Two models (Shell and the police) changed due to a decrease in the number of cases, because the educational level of all the respondents was not known. These regression models were still in agreement with the previous findings. Support & criticism news worsened the reputation of Shell, and the amount of television news improved the reputation of the police.

Three models (Super de Boer, NS, and the agricultural sector) changed due to the addition of the educational level of the respondents. In all three models, the educational level of the respondents was added. In the case of Super de Boer, the educational level of respondents was the only significant variable in the model. The findings of the models of the NS and the agricultural sector were still in agreement with the previous findings. The more criticism and the more television news, the better the reputation of an organization. See Appendix L for a more detailed description of the models.

When the autoregressive models of the effects of the total amount of news and the tenor of support & criticism news on corporate reputation were controlled for education, all six models presented in the last column of Table 5.4 remained the same. See Appendix L for a more detailed overview of the changes after controlling for education.

Pooled models of the effects of the amount of news and support & criticism news on reputation
Now that the models have been discussed per organization, the focus will be on the model in which the data of all ten organizations are included. Table 5.5 presents the effects of the total amount of news, the favorability and direction of support & criticism news on corporate reputation, for the pooled dataset. In the non-autoregressive pooled model, the amount of television news had a positive impact ($\beta = .12$, $p < .001$) on corporate reputation, while the amount of news in the print media negatively ($\beta = -.05$, $p < .01$) influenced corporate reputation after controlling for support & criticism news. Except from the significance level of the amount of print news, these findings do not differ from the model in which the amount of news is controlled for news about success & failure that is presented in the second to last column of Table 5.3. The same goes for advertising expenditures in newspapers that had a positive influence on corporate reputation ($\beta = .05$, $p < .001$). Contrary to the models per organization, the tenor of support & criticism news (both direction and the favorability) did not have a significant effect on corporate reputation in the pooled model, however.

The estimation of the autoregressive pooled model[7] of the effects of the total amount of news and support & criticism news on corporate reputation are presented in the last column of Table 5.5.

TABLE 5.5.   **Pooled confirmatory regression analysis of the effects of the tenor of support & criticism news and the amount of news and advertising on reputation**

Predictors (news and advertising, previous year)	Cross-sectional	Auto-regressive
Key variables		
Amount of TV news	.12***	.06
Amount of print news	-.05**	-.02
Advertising intensity newspapers	.05***	.04
Favorability of support & criticism TV news	-.01	-.03
Favorability of support & criticism print news	-.01	.02
Direction of support & criticism TV news	-.02	-.02
Direction of support & criticism print news	.00	-.03
Advertising intensity TV	.02	-.03
Control variables		
Shell	-.07***	-.07**
BP	-.02	-.04
Albert Heijn	.05**	.01
Super de Boer	-.03	.00
NS	-.21***	-.08**
Schiphol	.05***	.05
ABN AMRO	-.03	.02
Rabobank	.01	.03
Police	-.12***	-.06
Reputation, $_{1999}$	-	.57***
Adjusted R^2	.06	.37
df	(17, 9139)	(18, 1982)
F	33.55***	65.98***
N	9,157	2,001

Note. Cell entries are Betas from Ordinary Least Squares regression.
* $p \le .05$; ** $p \le .01$; *** $p \le .001$

None of the media and advertising variables were significant. The influence of the amount of print news disappeared in the autoregressive model due to the lower number of cases. This was examined by estimating the non-autoregressive model with the same number of cases as the autoregressive model. It appears that in that case, the amount of print news was no longer significant. The disappearance of the effects of the amount of television news and advertising expenditures in newspapers was not caused by the addition of the lagged dependent variable.

If the non-autoregressive pooled model was controlled for the educational level of the respondents, the amount of television news and the advertising expenditures in newspapers remained significant. The amount of print news was nearly significant ($\beta = -.03$, $p = .09$).

If the autoregressive pooled model was controlled for education, none of the media and advertising variables were significant, as in the autoregressive pooled model in which education was omitted. The educational level of respondents was not significant, either.

5.5 THE EFFECTS OF THE AMOUNT OF NEWS, SUCCESS & FAILURE AND SUPPORT & CRITICISM NEWS, AND ADVERTISING ON CORPORATE REPUTATION (H. 1A, 1B, 2A, 3A, AND 7)

In this final section about the effects of success & failure and support & criticism news on reputation, the effects of both types of news are estimated together for each organization and for the pooled dataset. In this way the effects of success & failure news are controlled for support & criticism news and vice versa. This means that regression models will be presented that include the impact of the total amount of news (hypothesis 1), the favorability (or direction) of success & failure news (hypothesis 2a), the favorability (or direction) of support & criticism news (hypothesis 3a), and advertising (hypothesis 7) on corporate reputation. This results into the following equation:

$Reputation_{i,t,o} = b_o + b1$ (Amount of TV news$_o$) + b2 (Amount of print news$_o$) + b3 (Favorability of success & failure TV news$_o$) + b4 (Direction success & failure TV news$_o$) + b5 (Direction success & failure print news$_o$) + b6 (Favorability of success & failure print news$_o$) + b7 (Favorability of support & criticism TV news$_o$) + b8 (Direction support & criticism TV news$_o$) + b9 (Direction support & criticism print news$_o$) + b10 (Favorability of support & criticism print news$_o$) + b11 (Advertising intensity TV$_o$) + b12 (Advertising intensity newspapers$_o$) + e_i

Testing this model for each of the organizations separately rendered the following results (see Table 5.6). The table will be discussed per hypothesis, starting with the non-autoregressive models. Hypothesis 1 (about the effects of the amount of media coverage on corporate reputation) will be discussed first.

The amount of television news had a positive influence on the reputation of Shell ($\beta = .10$, $p < .01$), the NS ($\beta = .07$, $p < .05$), and the police ($\beta = .15$, $p < .001$). This was in agreement with the previous results. The amount of television news also had a positive influence on the reputation of the agricultural sector ($\beta = .07$, $p < .05$), if the model was controlled for success & failure news and for support & criticism news. This means that the reputation of this sector improves the more it receives attention in the television news. The amount of print news was significant in the case of Super de Boer and the agricultural sector. The reputation of Super de

Boer was improved by the amount of print news ($\beta = .14$, $p < .01$), while it worsened the reputation of the agricultural sector ($\beta = -.31$, $p < .001$). As suggested earlier, this may indicate that an organization that is in the news relatively infrequently benefits from coverage in the print media, while it may harm organizations that are frequently in the print news.

TABLE 5.6.   Stepwise regression analyses of the effects of the amount of news, the tenor of success & failure news, the tenor of support & criticism news, and advertising on reputation

Models per organization[a]	Predictors (news and advertising, previous year)	Cross-sectional	Auto-regressive
Shell	Amount of TV news	.10**	ns
	Direction of support & criticism television news	-.09*	ns
	Reputation, 1999	-	.52***
	Adjusted R^2	.01	.27
	df	(2, 973)	(1, 224)
	F	5.40**	82.41***
	n	976	226
BP	Advertising intensity newspapers	.09**	ns
	Direction of success & failure print news	ns	.15*
	Reputation, 1999	-	.45***
	Adjusted R^2	.01	.23
	df	(1, 911)	(2, 196)
	F	7.43**	30.83***
	n	913	199
Albert Heijn	Advertising intensity newspapers	.09**	ns
	Direction of support & criticism print news	-.08*	ns
	Reputation, 1999	-	.72***
	Adjusted R^2	.01	.52
	df	(2, 933)	(1, 207)
	F	6.98***	227.64***
	n	936	209
Super de Boer	Amount of print news	.14**	ns
	Favorability of success & failure print news	.15**	ns
	Advertising intensity TV	.10**	ns
	Reputation, 1999	-	.40***
	Adjusted R^2	.02	.15
	df	(3, 746)	(1, 147)
	F	6.06***	27.65***
	n	750	149
NS	Amount of TV news	.07*	ns
	Reputation, 1999	-	.57***
	Adjusted R^2	.00	.32
	df	(1, 966)	(1, 221)
	F	4.66*	106.91***
	n	968	223

TABLE 5.6. **(Continued)**

Models per organization[a]	Predictors (news and advertising, previous year)	Cross-sectional	Auto-regressive
Schiphol	Favorability of success & failure TV news	.33***	ns
	Favorability of support & criticism TV news	-.22***	-.17**
	Direction of success & failure print news	-.07*	ns
	Direction of support & criticism print news	-.12***	-.10*
	Reputation, 1999	-	.58***
	Adjusted R^2	.04	.43
	df	(4, 961)	(3, 212)
	F	11.96***	54.70***
	n	966	216
Rabobank	Advertising intensity newspapers	.08*	ns
	Direction of support & criticism television news	ns	-.18**
	Reputation, 1999	-	.65***
	Adjusted R^2	.00	.44
	df	(1, 875)	(2, 171)
	F	4.91*	69.07***
	n	877	174
Police	Amount of TV news	.15***	ns
	Direction of success & failure TV news	.13***	ns
	Direction of support & criticism print news	.09**	ns
	Advertising in newspapers	ns	.17**
	Reputation, 1999	-	.53***
	Adjusted R^2	.05	.32
	df	(3, 945)	(2, 210)
	F	16.59***	50.06***
	n	949	213
Agricultural sector	Amount of television news	.07*	ns
	Amount of print news	-.31***	ns
	Favorability of success & failure print news	-.27***	ns
	Advertising in newspapers	.10**	ns
	Advertising television news	ns	.15*
	Reputation, 1999	-	.52***
	Adjusted R^2	.03	.31
	df	(4, 924)	(2, 201)
	F	7.69***	45.78***
	n	929	204
ABN AMRO	Direction of support & criticism print news	ns	.16**
	Reputation, 1999	-	.61***
	Adjusted R^2		.39
	df		(2, 185)
	F		61.88***
	n	893	188

Note. Cell entries are Betas from Ordinary Least Squares regression.

* p ≤ .05; ** p ≤ .01; *** p ≤ .001

Hypothesis 2a predicts that the more positive the favorability or direction of success & failure news, the better the reputation of the organization will be. The more favorable the success & failure television news, the better the reputation of Schiphol ($ß = .33$, $p < .001$) and the police ($ß = .13$, $p < .001$). This is in accordance with the results found earlier.

For print news, significant effects of the favorability or direction of success & failure news were found for Super de Boer ($ß = .15$, $p < .01$), Schiphol ($ß = -.07$, $p < .05$), and the agricultural sector ($ß = -.27$, $p < .001$). Although there was not much news about Super de Boer, the reputation of the supermarket worsened if there was more news about its failure, such as the loss of market share. Contrary to television viewers, newspaper readers reacted critically to the success of Schiphol.

These results were more or less the same[8] as in the model, which consisted only of the amount of news and success & failure news (Table 5.2). The favorability of success & failure print news about the agricultural sector had a negative influence on the reputation of the sector. This means that newspaper readers, like television viewers, assigned higher report marks to the agricultural sector the more the sector was in the news with their failure. The more the farmers were in the news with their struggle to keep farming, the larger the sympathy of the respondents for the sector. This suggests that an underdog effect occurred.

Hypothesis 3a proposes that the reputation of an organization will improve if it is criticized by its competitors. The favorability (or direction) of support & criticism TV news had a negative influence indeed on the reputation of Shell ($ß = -.09$, $p < .01$) and the reputation of Schiphol ($ß = -.22$, $p < .001$). In addition, the favorability (or direction) of support & criticism print news also had a negative impact on the reputation of Albert Heijn ($ß = -.08$, $p < .05$) and Schiphol ($ß = -.12$, $p < .001$). This means that the reputation of these organizations will improve the more they are criticized by their opponents, and worsen the more they are supported by their friends.

For the police, the effect of the direction of support & criticism print news on their reputation appeared to be positive ($ß = .09$, $p < .01$). Remember that in a previous model (Table 5.4) it was found that the favorability of support & criticism print news had a positive significant effect on the reputation of the agricultural sector9 as well. This suggests that the government is perceived as a more reliable actor than the actors who support and criticize the companies. Therefore, the reputations of the sectors changed in the same direction as the support & criticism they received.

Hypothesis 7 postulates that the higher the advertising intensity, the better the reputation of the organization or sector. For the models of five organizations (BP, Albert Heijn, Super de Boer, Rabobank, and the agricultural sector), the regression

coefficient of advertising intensity was significant and positive, thus in the predicted direction.

Again, the Adjusted coefficient of determination, the Adjusted R^2, was relatively
low, especially for the non-autoregressive models of the NS and Rabobank, namely
0.5%. As remarked earlier, this will be dealt with in the discussion of the results in
the next chapter.

Autoregressive models per organization
The last column of Table 5.6, the autoregressive models, will be discussed per hypothesis. The amount of print news and the amount of television news did not
have a significant impact on the reputation of the focal organizations. In all
cases, this was caused by the reduced number of respondents.

Hypothesis 2a states that the more favorable the success & failure news, the
better the reputation of the organization will be. Only the direction of success &
failure print news in the case of BP ($\beta = .15$, $p < .05$) turned out to be significant
in the autoregressive model. This variable was not significant in the non-autoregressive model. Three variables were not significant in the autoregressive model
due to the lower number of cases in this model, compared to the number of cases
in the non-autoregressive model. These three variables were the favorability of
success & failure print news in the models of Super de Boer and the agricultural
sector, and the favorability of success & failure television news in the model of
Schiphol. Two variables disappeared due to the inclusion of the lagged dependent variable in the model. These variables were the favorability of success & failure print news in the case of Schiphol, and the direction of success & failure
television news in the model of the police.[10]

Hypothesis 3a predicts that the reputation of an organization will improve the
more it is criticized by its competitors. As shown in the autoregressive models in
Table 5.6, this was the case for two organizations. Support & criticism about
Schiphol in the newspapers ($\beta = -.17$, $p < .01$) and support & criticism about
Schiphol on television ($\beta = -.10$, $p < .05$) had a negative impact on the reputation
of Schiphol. This means that the more the airport is criticized, the worse its reputation will be and vice versa. Support & criticism television news had a negative
influence on the reputation of Rabobank ($\beta = -.18$, $p < .01$) in the autoregressive
model. The autoregressive model of ABN AMRO showed that, surprisingly, support & criticism print news had a positive impact on the reputation of ABN AMRO
($\beta = .16$, $p < .01$). These results of the banks do not differ from the results found
earlier (see Table 5.4).

Three variables that were significant in the non-autoregressive models were
not significant in the autoregressive models. Two variables disappeared due to
the lower number of cases in the autoregressive models: the direction of support
& criticism television news in the model of Shell and the direction of support &

criticism print news in the model of the police. In the model of Albert Heijn, the direction of support & criticism print news was no longer significant due to the addition of the lagged dependent variable.

Hypothesis 7 assumes that the higher the advertising intensity, the better the reputation of the organization or sector. In the autoregressive model of the police, the advertising expenditures in newspapers appeared to be significant, while it was not significant in the non-autoregressive model. Instead of the advertising expenditures in newspapers, advertising expenditures on television became significant in the model of the agricultural sector.

In the models of three organizations (BP, Albert Heijn, and Rabobank) the effects of advertising expenditures were no longer significant in the autoregressive models due to the lower number of cases. In the case of Super de Boer, the effects of advertising expenditures on television were no longer significant due to the addition of the lagged dependent variable.

In sum, when the models were controlled for both success & failure and for support & criticism news, the results did not change substantially. The amount of television news had a positive impact on reputation. The amount of print news had a negative influence on reputation, except for organizations that are rarely in the news, such as Super de Boer. Moreover, the models show that the more successful the organization is depicted in the television news, the higher the report marks given by the respondents. In other words, there was a positive impact of success and failure television news on reputation. Newspaper readers reacted negatively to success of Schiphol, however, and felt sorry if the agricultural sector failed.

Support and criticism about the focal company in television news and in the print media had a negative impact on the reputation of the companies. This means that the more the organizations are criticized by their competitors, the better their reputations will be and vice versa. In the case of the police, there was a positive impact of support & criticism news on the reputation of the police. This can be explained by the subject of criticism. The police received a great deal of support & criticism from the government, which may be perceived as an impartial actor by the respondents. Therefore, the reputation of the police changed in the same direction as the support & criticism the sector they received.

Pooled models of amount of news, success & failure and support & criticism news, and advertising intensity on reputation
The next subsection will examine the model that includes the data of all ten organizations. Table 5.7 presents the effects of the total amount of news, the favorability and the direction of success & failure news, the favorability and the direction of support & criticism news, and advertising intensity on corporate reputation, based on a pooled model.

First, the non-autoregressive pooled model will be discussed per hypothesis. The amount of television news had a positive impact on the reputation of the organizations ($ß = .10$, $p < .001$) in the pooled model, while the amount of print news had a negative impact on the reputation of the organizations ($ß = -.04$, $p < .05$).

Hypothesis 2a predicts that the more favorable success & failure news, the better the reputation of the organizations will be. This was the case for the favorability of success & failure TV news ($ß = .08$, $p < .001$). The favorability of success & failure print news, and the direction of success & failure television news and print news were not significant.

TABLE 5.7. **Pooled confirmatory regression model of the effects of the amount of news, success & failure news and support & criticism news, and advertising intensity on reputation**

Predictors (news and advertising, previous year)	Cross-sectional	Auto-regressive
Key variables		
Amount of TV news	.10***	.04
Amount of print news	-.04*	-.02
Favorability of success & failure TV news	.08***	.03
Favorability of success & failure print news	.01	.03
Direction of success & failure TV news	-.03	-.07
Direction of success & failure print news	.01	-.00
Favorability of support & criticism TV news	-.01	-.04
Favorability of support & criticism print news	.00	.03
Direction of support & criticism TV news	-.03	-.07
Direction of support & criticism print news	-.00	-.04
Advertising intensity newspapers	.05***	.03
Advertising intensity TV	.03	-.02
Control variables		
Shell	-.09***	-.05
BP	-.05**	-.03
Albert Heijn	.03	.01
Super de Boer	-.04*	.02
NS	-.23***	-.06
Schiphol	.03	.07
ABN AMRO	-.06**	.02
Rabobank	-.02	.03
Police	-.18***	-.05
Reputation, $_{1999}$	-	.57***
Adjusted R^2	.06	.37
df	(21, 9135)	(22, 1978)
F	27.85***	54.05***
N	9,157	2,001

Note. Cell entries are Betas from Ordinary Least Squares regression.

* $p \le .05$; ** $p \le .01$; *** $p \le .001$

Hypothesis 3a proposes that the tone of support & criticism news is negatively related to reputation if the organization is criticized by its competitors. As in the
previous pooled model (Table 5.5), no impact of support & criticism news (neither the favorability nor the direction) was found.

Hypothesis 7 postulates that the higher the advertising intensity, the better
the reputation of the organization or sector. The pooled model shows that higher
advertising expenditures of the focal organizations in print media led to a better
the reputation of these organizations. The effects of advertising expenditures on
television were not significant.

In the autoregressive pooled model, none of the media and advertising variables
were significant. The influence of the amount of print news disappeared in the
autoregressive model due to the lower number of cases. This was examined by
estimating the non-autoregressive model with the same number of cases as the
autoregressive model. It appears that in that case, the amount of print news was
no longer significant. The disappearance of the effects of the amount of television
news and advertising expenditures in newspapers was not caused by the lower
number of cases, but by the addition of the lagged dependent variable. Furthermore, it should be noted that the VIF values exceeded the threshold of 10. When
multicollinearity was brought down to an acceptable level (and all the independent variables have VIF values below 10), the model did not change substantially.

In sum, when the non-autoregressive pooled and the autoregressive pooled model
were controlled both for success & failure and support & criticism news, they did
not differ substantially with the pooled models presented earlier that were controlled for one type of news only (Table 5.3 and Table 5.5). The amount of television news had a positive influence on reputation, while the amount of print news
had a negative influence on reputation in the non-autoregressive model. Moreover, it was found that, in agreement with the hypothesis, the favorability of TV
news had a positive impact on the reputation of organizations. The advertising
intensity in newspapers had a positive impact on corporate reputation as well,
which is in agreement with the mere exposure hypothesis.

5.6 THE EFFECTS OF SUCCESS & FAILURE NEWS AND ADVERTISING ON SUCCESS OF POLICY (H. 1, 2B, AND 7)

In this subsection, the effects of the amount of news (hypothesis 1), the favorability
of success & failure news (hypothesis 2b), and advertising intensity (hypothesis
7) on the perceived success of the policy of the organization are estimated. Separate variables were made for television and print media. This results in the following equation:

Success of policy$_{i, t, o}$ = b$_o$ + b1 (Amount of TV news$_o$) + b2 (Amount of print news$_o$) + b3 (Favorability of success & failure TV news$_o$) + b4 (Favorability of success & failure TV news$_o$) + b5 (Advertising intensity TV$_o$) + b6 (Advertising intensity newspapers$_o$) + e$_i$

Testing this model for each of the organizations separately rendered the following results (see Table 5.8). For each organization two different models were estimated, a cross-sectional and an autoregressive model.

The table will be discussed per hypothesis, starting with the non-autoregressive models. Hypothesis 1 (about the effects of the amount of media coverage on corporate reputation) will be discussed first.

In none of the models did the amount of TV news lead to a more successfully perceived policy of the organizations. However, in the case of Super de Boer, the amount of TV news led, remarkably, to a less successfully perceived policy of the organizations ($\beta = -.09$, $p < .05$). In agreement with the expectations, the amount of print news led to a more successfully perceived policy of organizations that are not frequently in the news, such as Super de Boer ($\beta = .12$, $p < .01$). The amount of print news had a negative impact on the perceived success of the policy of the organizations that are in the news frequently, as in the case of the NS ($\beta = -.07$, $p < .01$), Schiphol ($\beta = -.28$, $p < .001$), and the agricultural sector ($\beta = -.18$, $p < .001$).

In the non-autoregressive models of ABN AMRO and the police, the favorability of success & failure television news had a positive impact on the perceived success of the policy of these organizations. This means that the more successful ABN AMRO is depicted in the television news, the more successful its policy according to the respondents ($\beta = .07$, $p < .05$). This is the same for the police: the more successful the police are portrayed on television, the more successful the policy of the police is judged by the respondents ($\beta = .14$, $p < .001$). These models are in line with the models of the favorability of success & failure television news on reputation presented earlier. The model of Albert Heijn is an exception, however: the more success was attributed to the supermarket, the less successful its policy was judged by the respondents ($\beta = -.07$, $p < .05$).

The favorability of success & failure print news had a negative impact on the perceived success of the policy of Schiphol and of the agricultural sector. Newspaper readers judged the policy of Schiphol as less successful the more the airport was in the news with its successes, such as business. In addition, print and other news about the hard times for the agricultural sector not only improved its reputation, but also the perceived success of the policy of the agricultural sector. The media effects disappeared in the autoregressive models. This was due to the lower number of cases. When the non-autoregressive models were estimated with the same number of cases as in the autoregressive models – or in other words, with respondents who have participated at least twice – none of the media variables were significant.

TABLE 5.8.   Stepwise regression analyses of the effects of amount of news, success & failure news, and advertising on success of the policy

	Predictors (news and advertising, previous year)	Cross-sectional	Auto-regressive
Albert Heijn	Favorability of success & failure TV news	-.07*	ns
	Advertising intensity newspapers	.08**	.10*
	Success of policy, $_{t-1}$	-	.48***
	Adjusted R^2	.01	.25
	df	(2, 1160)	(2, 353)
	F	5.87**	58.46***
	n	1,163	356
Super	Amount of print news	.12**	ns
de Boer	Amount of TV news	-.09*	ns
	Success of policy, $_{t-1}$	-	.41***
	Adjusted R^2	.01	.17
	df	(2, 676)	(1, 179)
	F	4.23*	36.51***
	n	679	181
NS	Amount of print news	-.07*	ns
	Advertising intensity TV	.07*	ns
	Success of policy, $_{t-1}$	-	.41***
	Adjusted R^2	.01	.17
	df	(2, 1066)	(1, 314)
	F	4.64**	63.45***
	n	1,069	316
Schiphol	Amount of print news	-.28***	ns
	Favorability of success & failure print news	-.22***	ns
	Success of policy, $_{t-1}$	-	.40***
	Adjusted R^2	.02	.16
	df	(2, 958)	(1, 263)
	F	10.79***	49.97***
	n	961	265
ABN AMRO	Favorability of success & failure TV news	.07*	ns
	Success of policy, $_{t-1}$	-	.25***
	Adjusted R^2	.00	.06
	df	(1, 732)	(1, 171)
	F	4.08*	11.09
	n	734	173
Police	Favorability of success & failure TV news	.14***	ns
	Advertising intensity TV	ns	.1*
	Success of policy, $_{t-1}$	-	.33***
	Adjusted R^2	.02	.11
	df	(1, 1232)	(2, 380)
	F	25.73***	25.28
	n	1,234	383

TABLE 5.8.   (Continued)

	Predictors (news and advertising, previous year)	Cross-sectional	Auto-regressive
Agricultural	Amount of print news	-.18***	ns
sector	Favorability of success & failure print news	-.11*	ns
	Success of policy, $_{t-1}$	-	.29***
	Adjusted R^2	.01	.08
	df	(2, 982)	(1, 302)
	F	6.54**	28.5***
	n	985	304

Note. Cell entries are Betas from Ordinary Least Squares regression.
For Shell, BP, and Rabobank, no significant effects were found.
$* p \leq .05; ** p \leq .01; *** p \leq .001$

In the models of Albert Heijn and the NS, higher advertising expenditures led to a better reputation of these organizations. In the autoregressive models, the effects of advertising expenditures became significant in the model of the police. They remained significant in the model of Albert Heijn, but they disappeared in the model of the NS.

The low Adjusted coefficients of determination, the Adjusted R^2, in some of the non-autoregressive models will be dealt with in the discussion of the results in the next chapter.

In Table 5.9, the pooled dataset is used to examine the impact of the favorability of success & failure news on the perceived success of the policy of an organization. The model was controlled for the impact of advertising.

Table 5.9 shows that in the non-autoregressive pooled model, the amount of print news had a negative impact on the perceived success of the policy of the organizations ($\beta = -.06, p < .001$). Moreover, the favorability of success & failure television news had a positive influence on the perceived success of the organizations ($\beta = .05, p < .001$). This means that in general an organization will be perceived as having a more successful policy if it is attributed success in the media and vice versa. In addition, advertising in newspapers also had a positive influence on the perceived success of the policy of the organization ($\beta = .04, p < .05$). This implies that the higher the advertising expenditures of an organization in the print media, the more successful its policy will be perceived.

In the autoregressive pooled models, the negative influence of the amount of print news and the positive effect of the advertising expenditures in newspapers remained significant. The favorability of success & failure television news was no longer significant in the autoregressive pooled model due to the decrease in the number of cases. This was examined by estimating the non-autoregressive pooled model with the same number of cases as in the autoregressive pooled model, or in other words with respondents who participated at least twice. In that case, the favorability of success & failure television news was no longer significant.

TABLE 5.9.   Confirmatory regression analysis of the effects of the amount of news, success & failure news, and advertising on success of the policy

Predictors	Cross-sectional	Auto-regressive
Key variables		
Amount of TV news	.01	.02
Amount of print news	-.06***	-.04*
Favorability of success & failure TV news	.05***	.02
Favorability of success & failure print news	.02	.01
Advertising intensity TV	.002	.03
Advertising intensity newspapers	.04*	.05*
Control variables		
Shell	.07***	.04
BP	.08***	.04*
Albert Heijn	.14***	.07*
Super de Boer	.11***	.08***
NS	-.20***	-.14***
Schiphol	.10***	.06*
ABN AMRO	.13***	.08**
Rabobank	.13***	.08**
Police	-.30***	-.19***
Success of policy, $_{t-1}$	-	.38***
Adjusted R^2	.24	.38
df	(15, 8980)	(16, 2499)
F	190.13***	98.46***
N	8,996	2,516

Note. Cell entries are Betas from Ordinary Least Squares regression.
* $p \leq .05$; ** $p \leq .01$; *** $p \leq .001$

5.7 THE EFFECTS OF SUPPORT & CRITICISM NEWS AND ADVERTISING ON AGREEMENT WITH POLICY (H. 1, 3B, AND 7)

This subsection will estimate the effects of the amount of news (hypothesis 1), the favorability of support & criticism news (hypothesis 3b), and advertising (hypothesis 7) on the agreement with the policy of an organization. This results in the following equation:

Agreement with policy$_{i,t,o}$ = b$_o$ + b1 (Amount of TV news$_o$) + b2 (Amount of print news$_o$) + b3 (Favorability of support & criticism TV news$_o$) + b4 (Favorability of support & criticism TV news$_n$) + b5 (Advertising intensity TV$_n$) + b6 (Advertising intensity newspapers$_o$) + e$_i$

Testing this model for each of the organizations separately rendered the following results (see Table 5.10). For each organization two different models were estimated, a non-autoregressive and an autoregressive model.

TABLE 5.10. **Stepwise regression analyses of the effects of the amount of news, the tenor of support & criticism news, and advertising intensity on agreement with the policy**

	Predictors (news and advertising, previous year)	Cross-sectional	Auto-regressive
Shell	Favorability of support & criticism print news	-.05*	ns.
	Agreement with policy, $_{t-1}$	-	.36***
	Adjusted R^2	.00	.13
	df	(1, 1427)	(1, 502)
	F	3.87*	75.20***
	n	1,429	504
Albert Heijn	Favorability of support & criticism print news	-.1***	ns.
	Advertising intensity newspapers	.07*	ns.
	Agreement with policy, $_{t-1}$	-	.49***
	Adjusted R^2	.01	.23
	df	(2, 1295)	(1, 427)
	F	9.85***	130.95***
	n	1,298	428
Super de Boer	Amount of TV news	-.14***	ns
	Amount of print news	.11**	ns
	Agreement with policy, $_{t-1}$	-	.47***
	Adjusted R^2	.01	.22
	df	(2, 887)	(1, 247)
	F	7.34**	71.23***
	n	890	249
NS	Favorability of support & criticism print news	-.09*	ns
	Amount of print news	-.14***	ns
	Agreement with policy, $_{t-1}$	-	.42***
	Adjusted R^2	.01	.18
	df	(2, 1375)	(1, 482)
	F	6.86***	104.87***
	n	1,378	484
ABN AMRO	Favorability of support & criticism TV news	-.07**	-.11*
	Advertising intensity newspapers	.07*	ns
	Agreement with policy, $_{t-1}$	-	.48***
	Adjusted R^2	.01	.23
	df	(2, 1259)	(2, 415)
	F	6.93***	64.52***
	n	1,262	418

TABLE 5.10.   (Continued)

	Predictors (news and advertising, previous year)	Cross-sectional	Auto-regressive
Rabobank	Favorability of support & criticism print news	-.07*	ns
	Advertising intensity newspapers	.08**	ns
	Agreement with policy, $_{t-1}$	-	.53***
	Adjusted R^2	.01	.28
	df	(2, 1239)	(1, 384)
	F	5.14**	153.56***
	n	1,242	386
Police	Favorability of support & criticism print news	.08**	.09*
	Advertising intensity TV	.07**	ns
	Agreement with policy, $_{t-1}$	-	.43***
	Adjusted R^2	.01	.19
	df	(2, 1383)	(2, 462)
	F	7.62***	56.08***
	n	1,386	465
Agricultural sector	Amount of print news	-.08**	ns
	Agreement with policy, $_{t-1}$	-	.29***
	Adjusted R^2	.01	.08
	df	(1, 1322)	(1, 454)
	F	8.99**	42.91***
	n	1,324	456

Note. Cell entries are Betas from Ordinary Least Squares regression.
For BP and Schiphol, no significant effects were found.
* $p \leq .05$; ** $p \leq .01$; *** $p \leq .001$

Table 5.10 will be discussed per hypothesis, starting with the non-autoregressive models. Hypothesis 1 (about the effects of the amount of media coverage on corporate reputation) will be discussed first.

In none of the models did the amount of TV news lead to more agreement with the policy of the organizations. In the case of Super de Boer, the amount of TV news even led to less agreement with the policy of the organization ($ß = -.14$, $p < .001$). It is not clear why this effect occurred. In agreement with the expectations, the amount of print news led to more agreement with the policy of the organizations not frequently in the news, such as Super de Boer ($ß = .11$, $p < .01$). The amount of print news had a negative impact on the agreement with the policy of the organizations in the news frequently, as is the case with the NS ($ß = -.14$, $p < .01$) and the agricultural sector ($ß = -.08$, $p < .001$). Apart from some changes in the regression coefficients, these results did not differ from the results presented earlier for the model with "success of the policy" as the dependent variable, Table 5.8. The effects of the amount of news were no longer significant in the autoregressive models due to the lower number of cases.

Table 5.10 shows that in the models of five organizations (Shell, Albert Heijn, NS, ABN AMRO, and Rabobank), support & criticism news had a negative impact on the agreement with the policy of an organization. This implies that the more an organization is criticized in the media by its competitors, the more respondents will agree with the policy of the organization. On the other hand, support and criticism news had a positive impact on the agreement with the policy of the police. This means that if the police are criticized in the media, the agreement with the policy of the police will decrease and vice versa.

As described in the previous subsections, a similar pattern was found in the models in which the impact of support & criticism news on reputation were estimated. The positive impact of support & criticism news on the police can be explained by the fact that the police do not have "real" competitors. Most support & criticism news came from the government, which is probably perceived by the respondents as impartial. Criticism of the police given by the government may decrease the agreement with the policy of the sector, while criticism by a competitor of a company led to an increase in the agreement with the policy of an organization.

In the autoregressive models, the effects of support & criticism news on the agreement with the policy of an organization were still significant in the models of ABN AMRO and the police. The effects of support & criticism news disappeared in the models of Shell, NS, and Rabobank due to the lower number of cases. The effect of support & criticism news disappeared in the model of Albert Heijn due to the addition of the lag dependent variable, the agreement with the policy of an organization in the previous year.

In the cross-sectional models of Albert Heijn, ABN AMRO, Rabobank, and the police, higher advertising expenditures led to more agreement with the policy of these organizations. However, these effects all disappeared in the autoregressive models due to the lower number of cases.

The low Adjusted coefficients of determination, the Adjusted R^2, in some of the non-autoregressive models will be dealt with in the discussion of the results in the next chapter.

Subsequently, a model was estimated in which the data of the ten organizations were combined. In Table 5.11, the pooled dataset is used to examine the impact of the favorability of support & criticism news on the agreement with the policy of an organization. The model was controlled for the impact of advertising.

Table 5.11 shows that in the non-autoregressive pooled model, the amount of print news had a negative impact on the agreement with the policy of the organizations ($ß = -.04$, $p < .001$). The amount of TV news was nearly significant ($ß = .03$, $p = .086$). In addition, more advertising expenditures in newspapers enhanced the agreement with the policy of the organizations ($ß = .06$, $p < .001$).

TABLE 5.11.   Confirmatory regression analysis of the effects of the amount of news, support & criticism news, and advertising on agreement with the policy

Predictors	Cross-sectional	Auto-regressive
Key variables		
Amount of TV news	.03	-.01
Amount of print news	-.04***	-.02
Favorability of support & criticism TV news	.02	-.03
Favorability of support & criticism print news	-.01	-.00
Advertising intensity TV	-.01	.01
Advertising intensity newspapers	.06***	.05*
Control variables		
Shell	-.04***	-.04*
BP	-.03*	-.01
Albert Heijn	.08***	.02
Super de Boer	.06***	.05**
NS	-.14***	-.09***
Schiphol	.04***	.03
ABN AMRO	.04**	.01
Rabobank	.04**	.03
Police	-.14***	-.07**
Agreement with policy, $_{t-1}$	-	.43***
Adjusted R^2	.07	.25
df	(15, 12903)	(14, 4294)
F	70.19***	92.58***
N	12, 919	4,309

Note. Cell entries are Betas from Ordinary Least Squares regression.
* $p \leq .05$; ** $p \leq .01$; *** $p \leq .001$

In the autoregressive pooled model, the effects of the advertising expenditures in newspapers remained significant. The effects of the amount of print news disappeared due to the lower number of cases.

5.8   THE IMPACT OF THE FREQUENCY OF ISSUE NEWS ON ISSUE SALIENCE (H. 4)

This section will model the impact of the frequency of issue news on the salience of an issue. In accordance with the agenda-setting hypothesis, it is expected that if respondents are asked what they think of when they think about an organization, they will mention associations (issues) that are mentioned in the media. At the same time, this hypothesis is important for testing the hypothesis on issue ownership. If the salience of an issue influences reputation, and the frequency of issue news influences the salience of an issue, the issue-ownership hypothesis is tested in a two-step approach. The model of this section is as follows:

Salience $_{i, a, t, o}$ = b$_o$ + b1 (Frequency of issues television news$_a$) + b2 (Frequency of issues print news$_a$) + e$_i$

Because the dependent variable is dichotomous, logistic regression was used. For the significant predictor variables, the unstandardized regression coefficients and the odds ratio were reported. The odds ratio for the independent variables was computed by exponentiating its regression coefficient that is, exp(b). The odds ratio represents the change in the odds of thinking about a certain issue for a one-unit change in the media coverage about that issue. Odds ratios greater than 1.0 show the increase in odds of the salience of an association for a one-unit increase in the amount of media coverage about a certain issue (a positive predictor-outcome relation). If the odds ratio is less than 1.0, it reflects a decrease in odds of the salience of an association with a one-unit change in the predictor (i.e., negative predictor-outcome relation). The models are presented in Table 5.12.

TABLE 5.12.   **Logistic regression analysis of the frequency of issue news on issue salience**

	Predictors (variable names)	Cross-sectional		Autoregressive	
		B	Odds Ratio	B	Odds Ratio
Shell	Frequency of issues television news	.01***	1.01	.01***	1.01
	Frequency of issues print news	.02**	1.02	ns	
	Salience issue, $_{t-1}$	-	-	1.91***	6.78
	Constant	-2.47***		-2.77***	
	Nagelkerke R^2	.01		.10	
	n	14,700		5,270	
Albert Heijn	Frequency of issues television news	-.05***	.96	ns	
	Frequency of issues print news	.23***	1.24	ns	
	Salience issue, t-1	-	-	2.96***	
	Constant	-2.00***		-2.66***	
	Nagelkerke R^2	.03		.31	
	n	14,190		4,910	
Super de Boer	Frequency of issues television news	.07***	1.07	ns	
	Frequency of issues print news	-.36**	.7	-1.08*	.34
	Salience issue, t-1	-	-	2.69***	14.73
	Constant	-2.42***		-2.91***	
	Nagelkerke R2	.01		.20	
	n	14,340		5,090	
NS	Frequency of issues television news	.06***	1.06	.06***	
	Salience issue, t-1	-	-	2.37***	10.74
	Constant	-2 15***		-2.72***	
	Nagelkerke R2	.04		.26	
	n	12,951		4,716	

TABLE 5.12.   **(Continued)**

	Predictors (variable names)	Cross-sectional		Autoregressive	
		B	Odds Ratio	B	Odds Ratio
Schiphol	Frequency of issues television news	.003***	1.003	.011***	1.01
	Frequency of issues print news	.04***	1.04	ns	
	Salience issue, $_{t-1}$	-	-	3.34***	28.20
	Constant	-1.92***		-2.86***	
	Nagelkerke R^2	.02		.40	
	n	11,552		4,056	
ABN	Frequency of issues television news	.02***	1.02	ns	
	Frequency of issues print news	.03***	1.03	.072***	1.07
	Salience issue, $_{t-1}$	-	-	1.83***	6.26
	Constant	-3.27***		-3.33***	
	Nagelkerke R^2	.03		.07	
	n	10,896		3,760	
Rabobank	Frequency of issues television news	.03***	1.04	ns	
	Frequency of issues print news	.05***	1.05	.03*	1.03
	Salience issue, $_{t-1}$	-	-	.22***	9.40
	Constant	-3.41***		-3.44***	
	Nagelkerke R^2	.01		.08	
	n	10,832		3,600	
Police	Frequency of issues television news	.01***	1.01	.01***	1.01
	Frequency of issues print news	.01**	1.01	ns	
	Salience issue, $_{t-1}$	-	-	1.59***	4.92
	Constant	-2.20***		-2.47***	
	Nagelkerke R^2	.14		.24	
	n	7,030		2,405	
Agricultural sector	Frequency of issues television news	.06**	1.06	.04***	1.04
	Frequency of issues print news	-.04*	.97	ns	
	Salience issue, $_{t-1}$	-	-	1.942	6.97
	Constant	-2.00***		-2.53***	
	Nagelkerke R^2	.05		.21	
	n	8,472		2,970	
Pooled	Frequency of issues television news	.01***	1.01	.01***	1.01
	Frequency of issues print news	.01***	1.01	.006	1.01
	Salience issue, $_{t-1}$	-	-	2.54***	12.61
	Constant	-2.28***		-2.81***	
	Nagelkerke R^2	.03		.22	
	N	118,893		41,697	

Note. Cell entries are unstandardized regression coefficients and odds ratios from binary logistic regression. For BP, no significant media effects were found. General issues were not included in the models.
NS not significant; * p ≤ .05; ** p ≤ .01; *** p ≤ .001

It appears from Table 5.12 that the odds ratio of issues television news was significant in the hypothesized direction for the models of eight organizations, while the odds ratio of issue news in the print media was significant in the predicted direction in the models of six organizations. In the case of Shell, this means if there is more television coverage about Shell and the environment, people will associate Shell more with the environment than if the television coverage addressed another topic. Most effects remained after controlling for the lagged salient association. Television and print news were both significant in the non-autoregressive pooled model (b television news = .01, odds ratio = 1.01; b print news = .01, odds ratio = 1.01).

Remarkably, in the case of the supermarkets and the agricultural sector, there also seemed to be a negative effect of the amount of news on the salience of an issue. In the case of Albert Heijn, the amount of television news about an issue seemed to decrease the salience of an issue (b = −.06, odds ratio = .96 in the model without the lagged dependent variable). This effect disappeared in the autoregressive model. In the case of Super de Boer (b = −.36, odds ratio = .7 in the non-autoregressive model) and the agricultural sector (b = −.036, odds ratio = .97 in the non-autoregressive model), the amount of print news about an issue seemed to decrease the salience of an issue. These effects remained in the autoregressive model.

As remarked in section 5.1, general associations were excluded from these analyses since they could not be matched with media coverage. In the case of the agricultural sector, it should be noted there was a relatively large amount of media coverage about BSE, but that this association was not provided to the respondents. As a result, this news was not taken into account in the analysis. Furthermore, the respondents were asked if they associated a certain company with the "Dutch economy." However, in the case of media coverage, foreign economic issues that may affect the focal company (such as the crisis in Asia) also belong to the category "economy." Appendix F displays the match between associations and media coverage. The associations and issues that belong to the category "other" were not taken into account in the analyses.

Overall, it appears from Table 5.12 that the amount of television news about an issue influences the salience of that issue. Television news about a certain issue influenced the salience of that issue in the models of eight organizations, and print news about a certain issue influenced the salience of that issue in the models of six organizations.

5.9 THE IMPACT OF ISSUE NEWS ON THE IMPORTANCE OF THAT ISSUE IN FORMING REPUTATION (H. 5)

The previous section tested whether the amount of news influences the salience of an issue. This section will test whether an issue becomes more *important* in determining report mark if there is more news about that issue in the media (priming); whether the reputation gets better or worse does not play a role in the test of the priming hypothesis. This results in the following equation:

$\text{Reputation}_{i,t,o} = b_0 + b1$ (Policy of the organization regarding a certain issue) + $b2$ (Amount of print news about an issue) + $b3$ (Amount of TV news about an issue) + $b4$ (Policy of the organization regarding a certain issue × Amount of print news about the issue) + $b5$ (Policy of the organization regarding a certain issue × Amount of TV news about the issue) + $b6$ (Advertising expenditures television) + $b7$ (Advertising expenditures newspapers) + e_i

The priming hypothesis is tested with $b4$ and $b5$. These parameters estimate the additional impact of the policy of an organization regarding a certain issue resulting from exposure to news about that issue. A statistically significant and positively signed $b4$ and/or $b5$ means that news coverage about a certain issue increases the weight that media users grant to the policy of an organization regarding that issue in their overall evaluation of the organization. The lower-order terms amount of print news and amount of TV news about an issue are included, since this is implied by the higher-order interaction terms.

Testing the priming hypothesis for each of the organizations separately rendered the following results (see Table 5.13).

In Table 5.13, only models that contain a significant interaction effect are included. This was the case for the models of three organizations. There was a positive interaction effect between the success of the policy of the organization regarding a certain issue and the amount of news about that issue in the models of the NS sector (ß = .06, $p < .05$) and in the model of the agricultural sector (ß = .09, $p < .05$). As is explained above, this means that news coverage about a certain issue increases the weight that media users granted to the policy of an organization regarding that issue in their overall evaluation of the organization. Remarkably, for Albert Heijn the opposite seemed to be the case. News coverage about a certain issue decreased the weight media users granted to the policy of Albert Heijn regarding that issue in their overall evaluation of the organization. Due to the lower number of cases, the interaction effects disappeared in the autoregressive models. If the non-autoregressive models are estimated with the same number of cases as the autoregressive models (respondents who participated in 2000 and in 1999), the interaction effects were no longer significant. If the data of all ten organizations were combined in one model, it appeared that the interaction effects were not significant.

TABLE 5.13. **Stepwise regression analyses of the amount of issue news on reputation**

	Predictors	Cross-sectional	Auto-regressive
Albert Heijn	Success of policy concerning association	.47***	.33***
	Success of policy amount of issues in print news	-.07**	ns
	Reputation, $_{1999}$	-	.59***
	Adjusted R^2	.26	.59
	df	(2, 1387)	(2, 300)
	F	187.69***	215.62***
	n	1,390	303
NS	Success of policy concerning association	.41***	.30***
	Success of policy amount of issues in print news	.06*	ns
	Reputation, $_{1999}$	-	.43***
	Adjusted R^2	.19	.35
	df	(2, 1185)	(2, 263)
	F	138.89***	72.90***
	n	1,188	266
Agricultural sector	Success of policy concerning association	.31***	.22***
	Success of policy amount of issues in TV news	.09**	ns
	Advertising expenditures on television	.07*	.20***
	Reputation, $_{1999}$	-	.53***
	Adjusted R^2	.15	.46
	df	(3, 1136)	(3, 257)
	F	65.32***	75.97***
	n	1,140	261

Note. Cell entries are Betas from Ordinary Least Squares regression.
For Shell, BP, ABN AMRO, Rabobank, Super de Boer, Schiphol, and the police, no significant interaction effects were found.
* $p \leq .05$; ** $p \leq .01$; *** $p \leq .001$

Furthermore, the model was also estimated with the agreement with the policy of an organization variable instead of the success of the policy of an organization variable. It appeared there was a positive interaction term in the models of two organizations, as well (Schiphol and the agricultural sector), but there was also a negative interaction term in the models of two organizations (Albert Heijn and Rabobank).

5.10 THE IMPACT OF THE NEWS ABOUT OWNED ISSUES ON CORPORATE REPUTATION (H. 6)

This section will estimate the effects of the frequency of news about owned issues on the reputation of the organizations (hypothesis 6). It is expected that the more news about owned issues that is published, the better the reputation of the organization will be. Two separate approaches were used in order to test this hypothesis. In the first approach, it was determined by the researcher whether an issue was an owned issue or not (see Appendix F). Nearly all the issues were included in the models. Regression analyses were used to test these models.

In the second approach, the hypothesis was tested in two steps. The first step overlapped with the agenda-setting hypothesis, which tested whether the impact of the amount of media coverage on a certain issue influenced the salience of that issue. The second step was to examine whether the reputation of a company would get better or worse if the respondents associated the organization with a different issue. If respondents give Shell a lower report mark when they associate Shell with the environment than when they associate Shell with gas stations, it can be concluded that the environment is not an owned issue for Shell. This will be tested by using Repeated Measures ANOVA. Only two issues will be examined at the same time.

5.10.1 No direct signs of issue ownership

This subsection will model the impact of the frequency of news about owned issues on reputation. As shown in Appendix F, not all the issues were taken into account, because the associations of respondents did not always correspond with the news. The models were controlled for the impact of the amount of news and advertising expenditures. This results in the following equation:

$$\text{Reputation}_{i,t,o} = b_0 + b1 \text{ (Frequency of owned issues in TV news}_o) + b2 \text{ (Frequency of owned issues in print news}_o) + b3 \text{ (Amount of TV news}_o) + b4 \text{ (Amount of print news}_o) + b5 \text{ (Advertising intensity TV}_o) + b6 \text{ (Advertising expenditures newspapers}_o) + e_i$$

When this model was tested for all the organizations, it turned out that the key variable "frequency of owned issues" was not significant in any of the models. It can be questioned whether the criteria used to determine whether or not an issue was an owned issue were correct. In his study on issue ownership in presidential elections, Petrocik (1996) asked the voters which party would be more successful in resolving a problem. In this study, the criteria used to determine whether or not an issue is an owned issue were created by the researcher. It may be possible that respondents would have defined the "owned issues" differently. Therefore, the hypothesis about issue ownership was also tested with a second approach. In the previous section (section 5.8) it was tested whether the frequency of issue news influences the salience of an issue. The second step – whether respondents give a different report mark to the organization if they have a different association with the organization – will be tested in this subsection.

5.10.2 Different salient associations lead to different reputations

In this subsection, specific comparisons will be made per organization between an issue that is not expected to be an owned issue of the organization (such as the issue "environment" in the case of Shell and BP) and a more general issue (such as the issue "gas stations" in the case of the two oil companies).

It is expected that the respondents will assign a lower report mark to the organization if they associate the organization with an issue that is not an owned issue for the organization than they would if they associate the organization with a more general issue. No difference in report mark is expected if respondents associate the organization with the same issue twice. Repeated Measures ANOVA with between-subjects factors will be used to test the significance of the difference.[11] "Report mark" is the repeated measures variable (measured in 1999 and in 2000). The respondents will be categorized in a maximum of four groups, according to their association with the organization in a certain year. However, respondents can occur in two different groups if they gave two associations per organization. The results will be discussed for each organization separately: between the different groups at one point in time, and within the groups over the two years.

Association "environmental pollution" leads to a lower report mark for Shell than association "gas stations"
In the case of Shell, respondents were divided into the following three groups: a group of respondents who associated Shell with gas stations in 1999 and with environmental pollution in 2000 ($n = 23$); a group of respondents who associated Shell with environmental pollution in 1999 and with gas stations in 2000 ($n = 46$); and a control group of respondents who associated Shell in 1999 and in 2000 with gas stations ($n = 137$). There were no respondents who associated Shell twice with environmental pollution.

There was a significant main effect of the associations with Shell on the report mark of Shell $F (2, 203) = 9.34$, $p = .000$. This means that respondents will assign a different report mark if they associate Shell with a different issue. Figure 5.1 shows the respondents grouped by their associations with Shell.

The line at the top of the figure represents respondents who associated Shell with gas stations in 1999 and in 2000. The line is nearly straight. This indicates that, as expected, the respondents assigned Shell nearly the same reputation in 1999 as in 2000 ($M = 6.5$, $SD = 1.37$ in 1999, $M = 6.4$, $SD = 1.37$ in 2000). The simple main effects method shows that this difference in report mark is not significant ($p > .05$). This is in accordance with the expectation that respondents who do not change their association with Shell will not give a different report mark. The middle line represents the group of respondents who associated Shell with gas stations in 1999 and with environmental pollution in 2000. In accordance with the expectation, the respondents seemed to be more positive in 1999

FIGURE 5.1.   **Respondents gave Shell a lower report mark when they associated Shell with "environmental pollution" than when they associated Shell with "gas stations"**

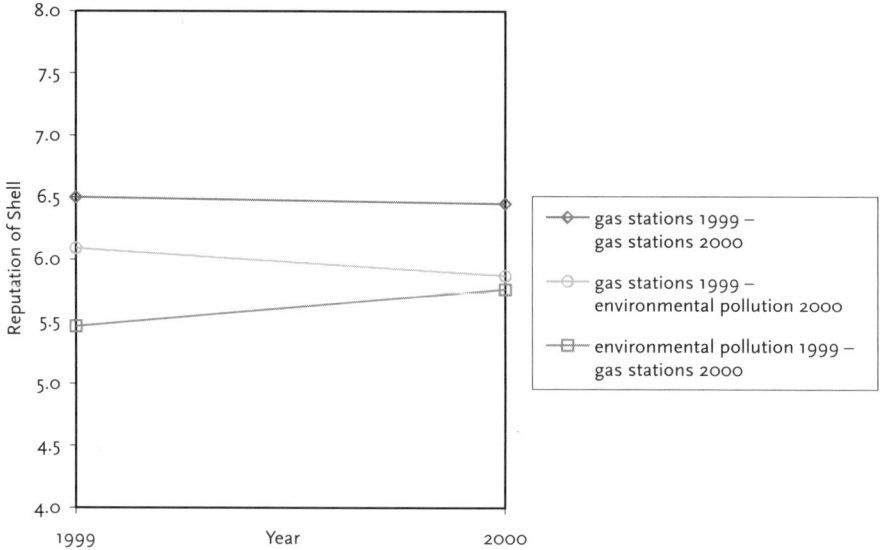

$(M = 6.1, SD = 1.16)$ when they associated Shell with gas stations than in 2000 when they associated Shell with environmental pollution $(M = 5.9, SD = 1.18)$. This is a non-significant difference, however $(p > .05)$. The bottom line represents the group of respondents who associated Shell in 1999 with environmental pollution and in 2000 with gas stations. This group seemed to be more negative in 1999 $(M = 5.5, SD = 1.71)$ than in 2000, when they thought about gas stations $(M = 5.8, SD = 1.32)$. Again, these results are in line with the hypothesis, but the difference is not significant $(p > .05)$.

As may become clear from the above section, the significant main effect that was found was not caused by differences in the report marks of 1999 and 2000 within a certain group of respondents. This effect was caused by differences *between the groups* within a certain year. Figure 5.1 shows there is a relatively large distance between the top line of the figure (the group of respondents who associated Shell twice with gas stations) and the bottom line of the figure (the group of respondents who associated Shell with environmental pollution in 1999 and with gas stations in 2000). The group of respondents who associated Shell twice with gas stations assigned a significantly $(p < .000)$ higher report mark $(M = 6.5, SD = 1.37)$ than respondents who associated Shell with environmental pollution in 1999 $(M = 5.5, SD = 1.71)$.

*For BP, the association "environmental pollution" does not lead to different report
mark than for association "gas stations"*
In the case of BP, the same categorization of groups was maintained: a group of
respondents who associated BP twice with gas stations (n = 142), a group of respondents who associated BP with gas stations in 1999 and with environmental
pollution in 2000 (n = 18), and a group of respondents who associated BP with
environmental pollution in 1999 and with gas stations in 2000 (n = 27). The report marks for BP given by the three groups are displayed in Figure 5.2.

FIGURE 5.2.   **The different associations did not have a significant influence on report mark of BP**

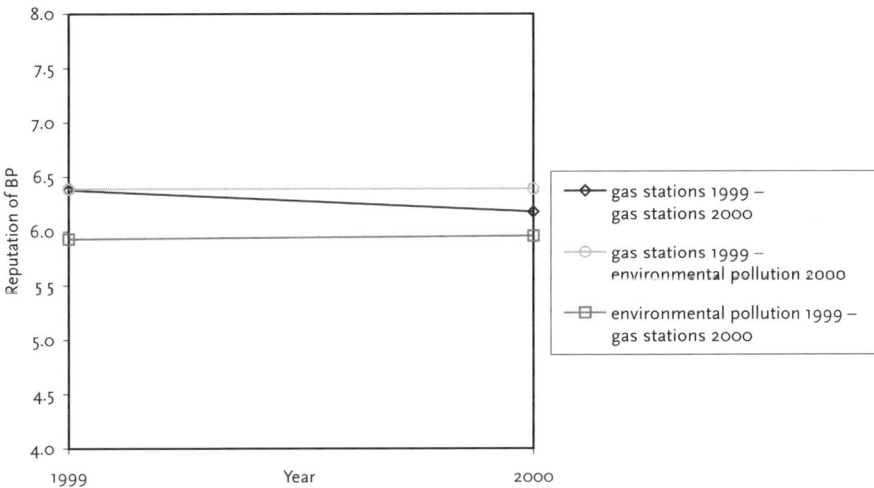

For BP, no significant main effect of the associations on corporate reputation was
found F (2, 184) = 1.70, p >.05. This could be due to the fact there was relatively
little issue news in the case of BP.

*Association "profit" leads to better report mark for ABN AMRO than association
"bank accounts and bank cards"*
In the case of ABN AMRO and Rabobank, there was no reason to assume that the
banks would have a bad reputation on any of the listed issues. Therefore, a comparison was made between an issue perceived to be an owned issue for the banks
("profit") and a more general issue ("bank accounts and bank cards", hereafter
"bank cards"). In the case of ABN AMRO, the following categorization of groups
was made: a group of respondents who associated ABN twice with profit (n = 2), a
group of respondents who associated ABN AMRO twice with bank cards (n = 119),
a group of respondents who associated ABN AMRO with profit in 1999 and with
bank cards in 2000 (n = 8), and a group of respondents who associated ABN
AMRO with bank cards in 1999 and with profit in 2000 (n = 12). Due to the low

FIGURE 5.3. **Respondents gave ABN AMRO a higher report mark if they associated the bank with profit rather than with bank cards**

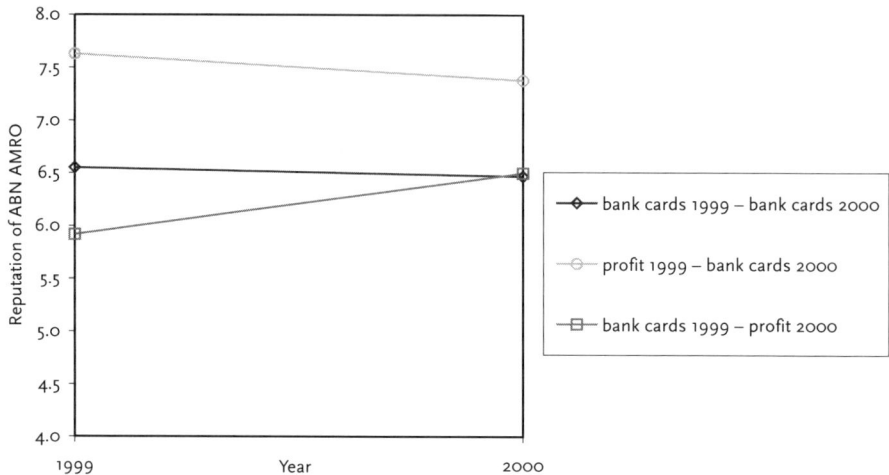

number of respondents in the first group, this group was excluded from the analysis. The report marks for ABN AMRO, which are given by the three groups, are displayed in Figure 5.3.

For ABN AMRO, the main effect of the associations on corporate reputation was significant $F (2, 136) = 3.51$, $p = .03$. In 1999 the effects between several groups were significant and in the expected direction. Respondents who associated ABN AMRO with profit in 1999 were more positive than respondents who associated ABN with bank cards. Respondents of the "profit-bank cards" group were more positive than respondents of the "bank cards-bank cards" group ($p = .03$) and respondents of the "bank cards-profit" group ($p = .01$). An unexpected effect occurred between two groups of respondents who both associated ABN AMRO with bank cards in 2000. The "profit-bank cards" group gave a higher report mark than the "bank cards-bank cards" group ($p = .04$).

Low number of cases for Rabobank
In the case of Rabobank, the same categorization of groups was maintained: a group of respondents who associated Rabobank twice with profit ($n = 0$), a group of respondents who associated Rabobank twice with bank cards ($n = 105$), a group of respondents who associated Rabobank with profit in 1999 and with bank cards in 2000 ($n = 9$), and a group of respondents who associated Rabobank with bank cards in 1999 and with profit in 2000 ($n = 5$). For Rabobank, the main effect of the associations on corporate reputation was not significant $F (2, 116) = 1.80$, $p = .43$. This may be due to the number of cases per group, which was relatively low, except for the group who associated Rabobank twice with bank cards.

Association "fresh products" leads to better report mark for Albert Heijn than association "price"
In the case of Albert Heijn, the respondents were divided into the following four groups: a group of respondents who associated Albert Heijn twice with the wide selection of products ($n = 57$), a group of respondents who associated Albert Heijn twice with the price of products ($n = 15$), a group of respondents who thought about the wide selection of products in 1999 and about the price of products in 2000 ($n = 26$), and a group respondents who thought about the price of products in 1999 and about the wide selection of products in 2000 ($n = 42$).

There was a significant main effect of the associations of respondents with Albert Heijn on the report mark of Albert Heijn, F (3, 136) = 16.75, p = .000. These differences were caused not only by differences between the groups, but also by a significant difference within a group of respondents. As shown in Figure 5.4, respondents who associated Albert Heijn with the wide selection of products in 1999 were more positive in 1999 (M = 7.0, SD = 1.08) than in 2000, when they associated Albert Heijn with the price (M = 6.4, SD = 1.33). This difference was significant (p < .001). Respondents who thought about the price in 1999 were more negative (M = 6.9, SD = 1.08) about Albert Heijn than in 2000, when they thought about the wide selection of products (M = 7.2, SD = .96). This difference was nearly significant (p = .059). Both results are in line with the hypotheses. Compared to the wide selection of products, the price of products is not an owned issue for Albert Heijn.

FIGURE 5.4.   **Respondents assigned Albert Heijn a lower report mark when they associated the supermarket with "the price" than when they associated it with "wide selection"**

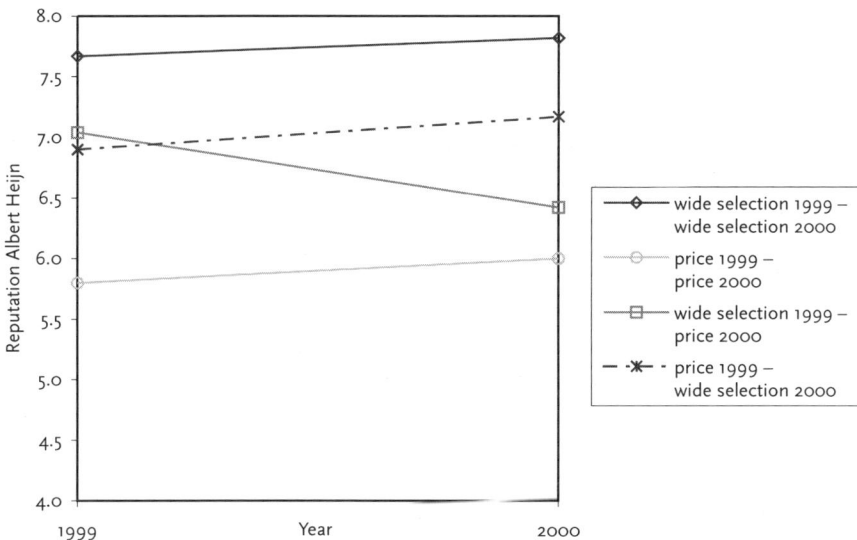

Figure 5.4 shows there are also differences between the groups of respondents. Respondents who thought about the wide selection of products twice when they thought about Albert Heijn assigned a higher report mark to Albert Heijn in 1999 ($M = 7.7$ in 1999) than respondents who thought about the price of products twice when they thought about Albert Heijn ($M = 5.8$ in 1999). This difference was significant ($p < .001$). The same applied for the year 2000. Respondents who thought about the wide selection of products twice when they thought about Albert Heijn assigned a higher report mark to Albert Heijn in 2000 ($M = 7.8$ in 2000) than respondents who thought about the price of products twice when they thought about Albert Heijn ($M = 6.0$ in 2000, $p < .001$). When the two middle lines were examined (the group of respondents who thought about the selection of products in 1999 and price in 2000, and the other way around for the other group of respondents), it appeared there was no significant difference in the report marks of 1999 ($p = .63$). However, there was a significant difference between the report marks of these two groups in 2000 ($p = .01$).

Association "fresh products" leads to better report mark for Super de Boer than association "price"
In the case of Super de Boer, the respondents were divided into the same groups as in the case of Albert Heijn: a group of respondents who associated Super de Boer twice with the wide selection of products ($n = 21$), a group of respondents who associated Super de Boer twice with the price of products ($n = 12$), a group of respondents who thought about the wide selection of products in 1999 and about the price of products in 2000 ($n = 18$), and a group respondents who thought

FIGURE 5.5. **Respondents assigned Super de Boer a lower report mark when they associated the supermarket with "the price" than when they associated it with "wide selection"**

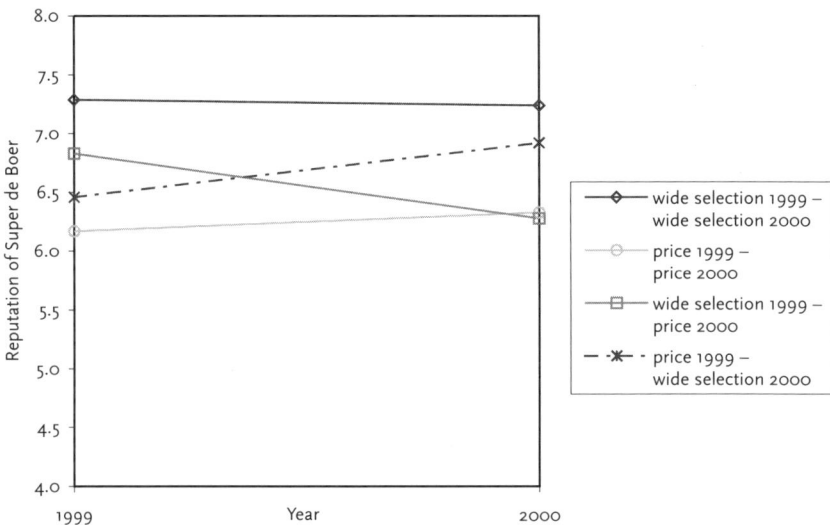

about the price of products in 1999 and about the wide selection of products in 2000 ($n = 26$).

The main effect of the associations with Super de Boer on the reputation of Super de Boer was significant, F (3, 73) = 4.29, $p < .01$. There were two different significant effects within a group. Respondents who associated Super de Boer with the wide selection of products in 1999 were more positive in 1999 ($M = 6.8$) than in 2000, when they associated Super de Boer with the price ($M = 6.3$). Respondents who thought about the price in 1999 were more negative ($M = 6.5$) about Super de Boer than in 2000, when they thought about the wide selection of products ($M = 6.9$). These differences were significant at $p < .05$. Both results were in line with the hypotheses. Compared to the wide selection of products, the price of products is not an owned issue for Super de Boer.

Not surprisingly, respondents who associated Super de Boer twice with the wide selection of products assigned a higher report mark to Super de Boer ($M = 7.3$ in 1999, $M = 7.2$ in 2000, $n = 21$) than respondents who associated Super de Boer twice with the price of products ($M = 6.2$ in 1999, $M = 6.3$ in 2000, $n = 12$). For 1999, these differences were significant at $p = .004$ and for 2000 at $p = .013$. When the two middle lines were examined (the group of respondents who associated Super de Boer with the selection of products in 1999 and with price in 2000, and the reverse for the other group of respondents), it appeared there was no significant difference in the report marks of 1999 ($p = .24$). There was a significant difference, however, between the report marks of these two groups in 2000 ($p = .035$).

Association "KLM" leads to better report mark for Schiphol than association "environment"
In the case of Schiphol, respondents were divided into the following four groups: a group of respondents who associated Schiphol twice with KLM ($n = 84$), a group of respondents who associated Schiphol twice with environmental pollution ($n = 6$), a group of respondents who thought about the airline KLM in 1999 and about environmental pollution in 2000 ($n = 22$), and a group respondents who thought about the environment pollution in 1999 and about KLM in 2000 ($n = 31$).

There was a significant main effect of the associations with Schiphol on the report mark of Schiphol, F (3, 139) = 13.09, $p = .000$. There was one significant and one nearly significant difference within a group. Respondents who associated Schiphol with KLM in 1999 were more positive in 1999 ($M = 6.8$) than in 2000, when they associated Schiphol with environmental pollution ($M = 6.3$). This difference was nearly significant ($p = .06$). Respondents who thought about the environment in 1999 were more negative ($M = 5.4$) about Schiphol than in 2000, when they thought about KLM ($M = 5.9$). This was a significant difference ($p = .02$). Both results are in line with the hypothesis that respondents will assign a higher report mark to

FIGURE 5.6.   **Respondents assigned a lower report mark to Schiphol when they associated the airport with "environmental pollution" than when they associated it with "KLM"**

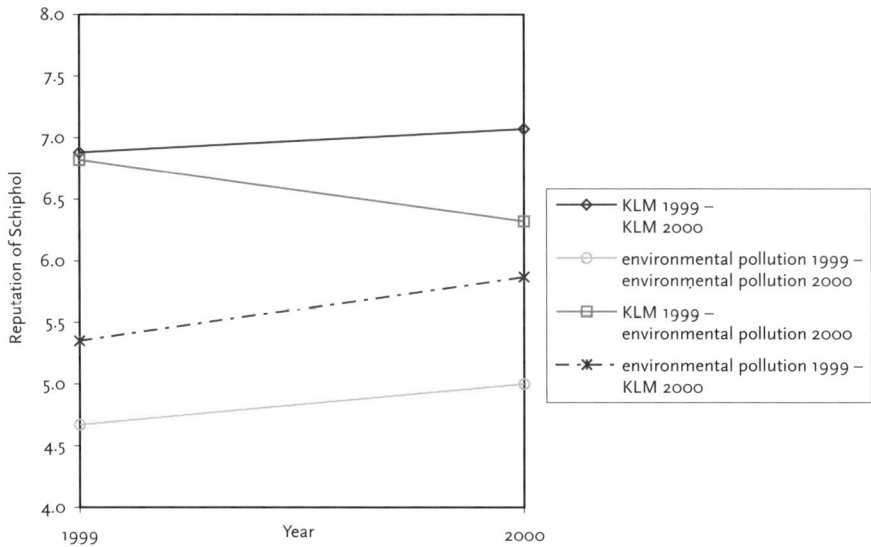

Schiphol when they associate the airport with KLM than when they associate the airport with environmental pollution, not an owned issue for Schiphol.

When the two middle lines were examined (the group of respondents who thought about the KLM in 1999 and the environment in 2000 and the reverse for the other group of respondents), it appeared there was a significant difference in the report marks of 1999 (p = .000). The difference between the report marks of these two groups was not significant (p = .24) in 2000.

Association "transport" leads to better report mark for the NS than association "delays"
In the case of the NS, respondents were divided into the following four groups: a group of respondents who associated the NS twice with transport (n = 28), a group of respondents who associated the NS twice with delays (n = 45), a group respondents who associated the NS with transport in 1999 and with delays in 2000 (n = 22), and a group respondents who associated the NS with delays in 1999 and with the environment in 2000 (n = 72).

The main effect of the associations with the NS on the report mark of the NS was significant, F (3, 163) = 18.35, p = .000. Remarkably, the respondents who associated the NS twice with delays were more positive in 2000 than in 1999. This difference was nearly significant (p = .07). This may suggest that the NS simply performed better, i.e., that there were fewer delays in 2000 than in 1999. There were no significant differences, however, within the group of respondents (see Figure 5.7). This means that differences in report mark were mainly due to differences in the several groups.

FIGURE 5.7.   **The respondents assigned the NS a lower report mark when they associated the railway company with "delays" than when they associated it with transport**

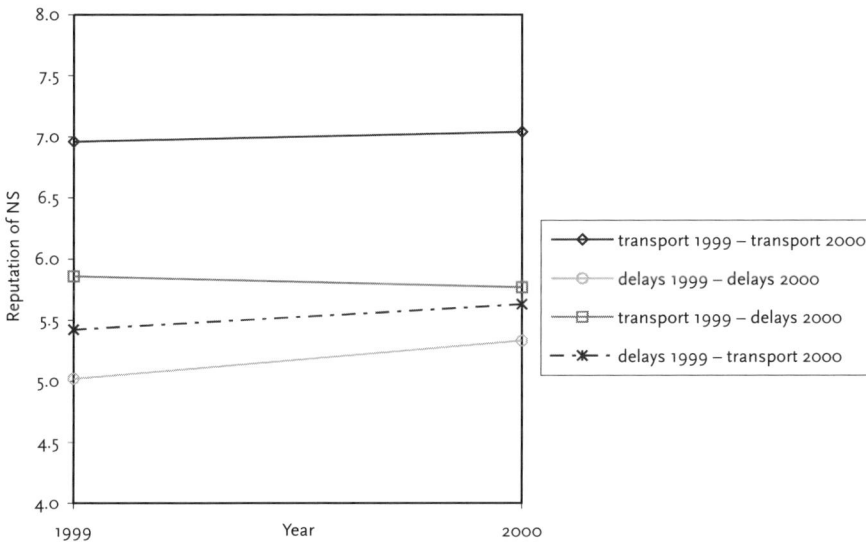

If the group who associates the NS twice with transport is compared to the other three groups, it appears that the "twice-transport" group gave a significantly higher report mark in 1999 and in 2000 compared to the other three groups. For five comparisons $p = .000$, the difference in report mark in 1999 between the "twice-transport group" and the "transport in 1999 and the delays in 2000" group was significant at $p = .003$. There was also a significant difference in 1999 between the "transport in 1999 and delays in 2000" and the "delays in 1999 and delays in 2000" group ($p = .01$). In agreement with the expectations, respondents who associated the NS with transport give a higher report mark than respondents who associated the NS with delays.

Association "maintaining authority" leads to better report mark for police than association "weak actions"
In the case of the police, respondents were divided into the following four groups: a group of respondents who associated the police twice with maintaining authority ($n = 74$), a group of respondents who associated the police twice with their weak actions ($n = 11$), a group of respondents who associated the police with maintaining authority in 1999 and with weak actions in 2000 ($n = 10$), and a group of respondents who associated the police with weak actions in 1999 and with maintaining authority in 2000 ($n = 17$).

The main effect of the associations with the police on the report mark of the police was significant, $F(3, 108) = 11.21$, $p = .000$. In addition, there were also significant differences within certain groups. Respondents who associated the po-

FIGURE 5.8.   **Respondents assigned the police a lower report mark when they associated the police with "weak actions" than when they associated the police with "maintaining authority"**

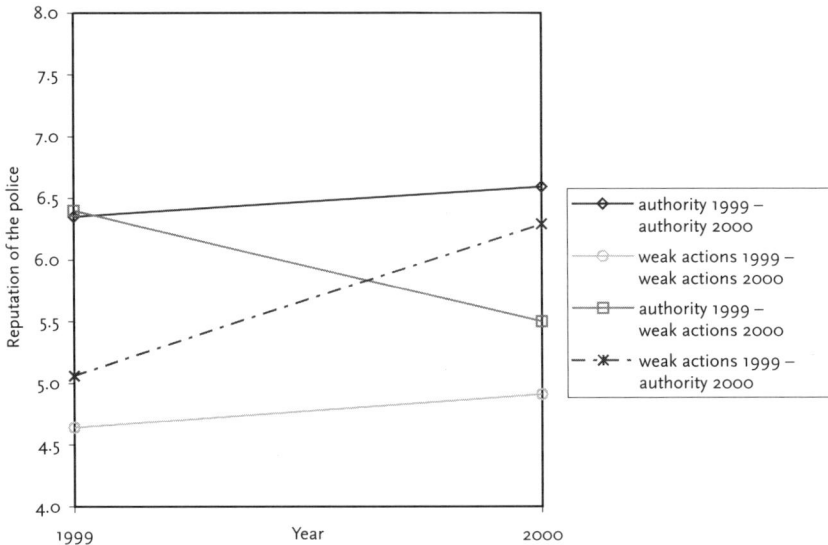

lice with authority in 1999 were more positive in 1999 ($M = 6.4$) than in 2000 when they associated the police with their weak actions ($M = 5.5$), $p = .02$. Respondents who associated the police with weak actions in 1999 were more negative ($M = 5.06$) about the police than in 2000, when they associated the police with maintaining authority ($M = 6.29$), $p = .000$. Both results are in line with the hypothesis that respondents will assign a higher report mark to the police when they associate the police with maintaining authority than when they associate the police with weak actions. Furthermore, respondents who associated the police twice with maintaining authority were more positive in 2000 ($M = 6.6$, $SD = .98$) than in 1999 ($M = 6.4$, $SD = 1.18$). This may suggest that the police simply performed better in 2000 than in 1999. Nevertheless, as reported in the subsection above, respondents who associated the police in 2000 with weak actions were more negative than the year before, when they associated the police with maintaining authority.

There were also differences between the several groups. In 1999, respondents who associated the police twice with maintaining authority were more positive than respondents who associated the police with weak actions ($p = .000$ for both the "weak-weak group" and the "weak-authority group"). This was also the case in 2000 ($p = .000$ for both the "weak-weak group" and the "authority-weak group)." Furthermore, the respondents who associated the police twice with weak actions were more negative in 1999 than the group of respondents who associated the police with maintaining authority in 1999 and with weak actions in 2000, $p = .002$, and more negative in 2000 than respondents who associated the police with weak actions in 1999 and with maintaining authority in 2000, $p = .000$. Respondents who associated the police with maintaining authority were more positive about the police than respondents who associated the police with weak actions. This was in agreement with the expectations.

Association "fresh products" leads to better reputation for agricultural sector than association "environment"
In the case of the agricultural sector, the respondents were divided into the following four groups: a group of respondents who associated the agricultural sector twice with fresh products ($n = 81$), a group of respondents who associated the agricultural sector twice with the environment ($n = 18$), a group of respondents who associated the agricultural sector with fresh products in 1999 and with the environment in 2000 ($n = 24$), and a group of respondents who associated the agricultural sector with the environment in 1999 and with fresh products in 2000 ($n = 31$).

The main effect of the associations with the agricultural sector on the report mark of the agricultural sector was significant, $F (3, 150) = 7.54$, $p = .000$. In addition, there was also a significant difference within a group. Respondents who associated the agricultural sector with the environment in 1999 were more negative in 1999 ($M = 6.1$) than in 2000, when they associated the agricultural sector with fresh products ($M = 6.7$), $p = .004$. There was no significant difference in report mark in the "product-environment" group.

FIGURE 5.9.   **Respondents gave the agricultural sector a lower report mark when they associated the sector with "the environment" than when they associated it with "fresh products"**

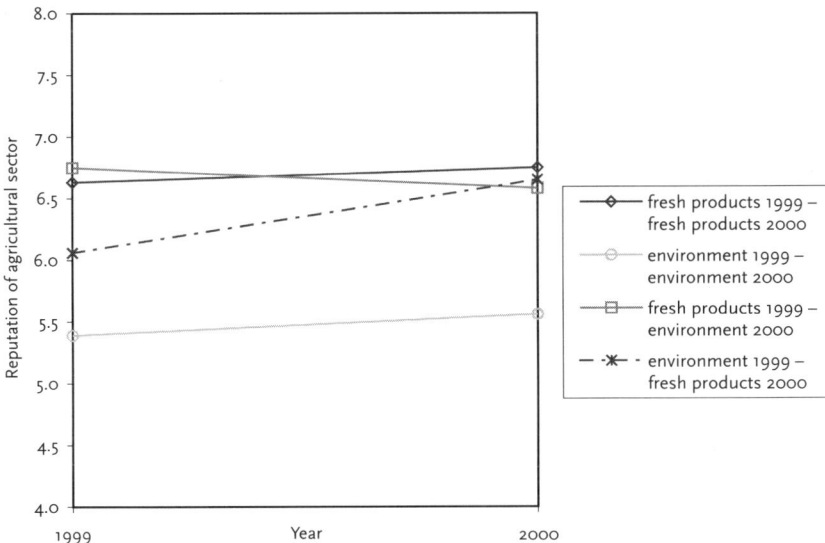

There were also differences between the several groups. In 1999, respondents who associated the agricultural sector twice with fresh products were more positive than respondents who associated the agricultural sector twice with the environment ($p = .000$ for both the "environment-environment group" and $p = .03$ for the "environment-fresh products group").

For 2000, there was a significant difference between the group of respondents who associated the agricultural sector twice with fresh products and the group of respondents who associated the agricultural sector twice with the environment $p = .000$. There was no significant difference in 2000 between the group of respondents who associated the agricultural sector twice with fresh products and the group of respondents who associated the agricultural sector with fresh products in 1999 and with the environment in 2000. Similarly, respondents who associated the agricultural sector twice with environment were more negative in 1999 than respondents who associated the agricultural sector with fresh products in 1999 and with the environment in 2000 $p = .000$. This was also the case in 2000 between the "twice-environment group" and the "product-environment group" ($p = .001$). Remarkably, there was also a significant difference in 2000 between the "twice-environment group" and the "product-environment group" ($p = .003$). Although both groups associated the agricultural sector with the environment in 2000, the group that associated the agricultural sector with fresh products in 1999 gave a higher report mark in 2000 than respondents who associated the agricultural sector in 1999 with the environment.

The section above demonstrates that respondents who associate the organization with a poorly owned issue will give a lower report mark to an organization than if they associate the organization with a more general issue.

5.11 OVERVIEW OF THE RESULTS OF THE HYPOTHESES

The hypotheses were tested in the previous sections in various types of models (among them non-autoregressive models, auto-regressive models, models on the level of the organization, pooled models, models that are controlled for education, and models that are not controlled for education). The use of different models provides insight into the stability of the effects of the media coverage and advertising on reputation. This section will present a more general overview of the results per hypothesis.

5.11.1 The effects of the amount of television coverage on reputation (Hypothesis 1a)

Hypothesis 1a predicted that the more an organization is visible in the television the news, the better its reputation. In Table 5.14, an overview is presented of the results. The left column displays the tables of the previous sections, in which models are presented that estimate the effects of the amount of television coverage. The second column describes the number and the direction of the significant Beta coefficients. The notation "3+" means that the amount of television coverage was significant in the models of three organizations, in a positive direction. The control variables are presented in the last column. All the models were controlled for advertising intensity.

TABLE 5.14.   **Overview of the results of the effects of the amount of TV news on reputation (hypothesis 1a)**

	Direction β	Control variables other than advertising
Table 5.2: Shell, NS, Schiphol, police	4+	
Table 5.3: pooled model	+	favorability and direction of success
Table 5.2 and 5.3: autoregressive model per organization and pooled	ns	& failure news
Table App. K1: NS, Schiphol, police	3+	favorability and direction of success
Table App. K2: pooled model	+	& failure news and educational level
Table 5.4: Shell, NS, police	3+	
Table 5.5: pooled model	+	favorability and direction of support
Table 5.4 and 5.5: autoregressive models per organization and pooled	ns	& criticism news
Table App. L1: police, agricultural sector	2+	favorability and direction of support
Table App. L2: pooled model	+	& criticism news and educational level
Table 5.6: Shell, NS, police, agricultural sector	4+	
Table 5.7: pooled model	+	favorability and direction of success
Table 5.6 and 5.7: autoregressive models per organization and pooled	ns	& failure and support & criticism news
Table 5.8: Super de Boer	–	favorability of success & failure news,
Table 5.9: pooled (non-)autoregressive models	ns	success policy as dependent variable
Table 5.10: Super de Boer	–	Favorability of support & criticism
Table 5.11: pooled (non-)autoregressive models	ns	news, agreement with policy as dependent variable

Table 5.14 shows that if the amount of television news has a significant effect on reputation, it is a positive effect. This means that the more television coverage about an organization, the better the reputation of the organization will be. The effect of the amount of television news was significant in the predicted direction in the models of three or four organizations, depending on the control variables. Note that the amount of television news was significant in the models of the organizations that were *most frequently* in the news (police, Schiphol, agricultural

sector, Shell, NS). Two explanations are suggested as to why the effect of the amount of television news on reputation could not be demonstrated in the models of the organizations that were less frequently in the news. The first explanation is that there was simply not enough television coverage about these organizations to influence the reputation. Assuming that television viewers possess a relatively low level of organizational awareness (see Chapter 2), there may just not have been enough television coverage to reach the television viewers. A second explanation is that the impact of relatively little television news fades away more easily than if there is more television news. This can be resolved by applying a shorter time lag (see Chapter 6).

Surprisingly, the amount of television news had a negative effect on the perceived success of the policy of Super de Boer and the agreement with the policy of Super de Boer. It can be wondered why this effect occurs, since Super de Boer received hardly any television coverage (see Table 4.1). Furthermore, the effects of the amount of television coverage disappeared in the autoregressive models. In the autoregressive models on the organizational level, this was mainly due to the lower number of cases in such models. In the autoregressive pooled models, this was due to the addition of the lagged dependent variable. However, the amount of television news was nearly significant ($p = .079$) in the autoregressive pooled model if the model was controlled for the tenor of success & failure news.

In sum, hypothesis 1a is largely confirmed. This means that the more television coverage about an organization, the better the reputation of the organization will be.

5.11.2 The effects of the amount of print news on reputation (Hypothesis 1b)

Hypothesis 1b proposed that if the amount of print news about an organization is low, the reputation of the organization will improve. If the amount of print news about an organization goes beyond a certain point and becomes relatively high, the reputation of the organization will worsen.

In Table 5.15, an overview is presented of the effects of the amount of print news on corporate reputation. The left column displays the tables from the previous sections in which models are presented that estimate the effects of the amount of print news. The second column describes the number and the direction of the significant Beta coefficients. The notation "-" means that the amount of print news was significant in the model of an organization in a negative direction. The control variables are presented in the last column. All the models were controlled for advertising intensity.

Table 5.15 shows that in most models, the amount of print news has a negative impact on corporate reputation, in contrast to the amount of television news. This means that the more print coverage about an organization, the worse the reputation of the organization will be. This was the case for the models of the following four organizations: Schiphol (controlled for success & failure news), the police (controlled for support & criticism news), the agricultural sector (controlled for success & failure news and for support & criticism news), and the NS (with success of the policy or agreement with policy as dependent variables). In addition, the amount of print news had a negative impact on reputation in various pooled models. The amount of print news remained significant in the autoregressive pooled model with the success of the policy as the dependent variable. Contrary to the models with "reputation" as the dependent variable, this model is based on three points in time, and thus on more cases. In the case of organizations that were rarely in the news, such as Super de Boer and BP, more print coverage improved their reputations.

TABLE 5.15. **Overview of the results of the effects of the amount of print news on reputation (hypothesis 1b)**

	Direction ß	Control variables other than advertising
Table 5.2: Super de Boer	+	
Table 5.2: Schiphol	−	favorability and direction of success
Table 5.3: pooled	−	& failure news
Table 5.2 and 5.3: (pooled) autoregressive model	ns	
Table App. K1: Super de Boer and Schiphol	ns	favorability and direction of success
Table App. K2: pooled model	−	& failure news and educational level
Table 5.4: police	−	
Table 5.5: pooled	−	favorability and direction of support
Table 5.4: autoregressive model of BP	+	& criticism news
Table 5.5: autoregressive pooled model	ns	
Table App. L1: model per organization	ns	favorability and direction of support &
Table App. L2: pooled model	ns	criticism news and educational level
Table 5.6: Super de Boer	+	
Table 5.6: Agricultural sector	−	favorability and direction of success &
Table 5.7: pooled model	−	failure and of support & criticism news
Table 5.6 and 5.7: (pooled) autoregressive model	ns	
Table 5.8: Super de Boer	+	
Table 5.8: NS, Schiphol, and agricultural sector	3−	favorability of success & failure news,
Table 5.9: non-autoregressive pooled model, autoregressive pooled model	2−	success policy as dependent variable
Table 5.10: Super de Boer	+	
Table 5.10: NS and the agricultural sector	2−	Favorability of support & criticism
Table 5.11: non-autoregressive pooled model	−	news, agreement with policy as dependent var.
Table 5.11: autoregressive pooled model	ns	

The effect of the amount of print news disappeared in some of the models when the models were controlled for the educational level of the respondents. This was the case in the model of Schiphol. However, in the models of Super de Boer and the police, the amount of print news was no longer significant due to a slight decrease[12] in the number of cases. In the pooled model of support & criticism news, the amount of print news was nearly significant ($ß = -.03$, $p .086$), because the education variable was added to the model. In the pooled model of success & failure news, the amount of print news remained significant after controlling for education, educational level was added to the model.

These results provide support for hypothesis 2b. In accordance with the expectations, the reputation of organizations that were rarely in the news (such as Super de Boer and BP) improved according to the amount of print news they received, while the reputation of organizations that were frequently in the print news (such as Schiphol, the agricultural sector, and the police) worsened the more print news they received.

5.11.3 The effects of success & failure news on reputation (Hypothesis 2a)

Hypothesis 2a postulates that the tenor of success & failure news has a positive impact on reputation. In Table 5.16, an overview is presented of the effects of the favorability or the direction of success & failure news (the tenor) on corporate reputation. The first half of the table presents the effects of television news about the success & failure of the organizations. In the second half of the table, the estimated impact of print news about the success & failure of the organizations is displayed. The left column shows the tables of the previous sections, in which models are presented that estimate the effects of success & failure news. The second column describes the number and the direction of the significant Beta coefficients. The notation "3+" means that the favorability or the direction of success & failure news was significant in the models of three organizations in a positive direction. The control variables are presented in the last column. All the models were controlled for advertising intensity.

The first half of Table 5.16 shows that if the favorability or the direction of success & failure television news influenced corporate reputation it was a positive influence, except for the agricultural sector. This means that the more success is attributed to the organizations in the television news, the better their reputations will be. This was the case in the models of Shell, Schiphol, and the police (controlled for success & failure print news and the amount of news), Schiphol (controlled for success & failure print news, educational level, and the amount of news), and the model of Shell and the police (controlled for success & failure print news, support & criticism news, and the amount of news). In the three different pooled models, the favorability or direction of success & failure news also had a positive influence on the reputation of the organizations.

TABLE 5.16. **Overview of the results of the effects of the success & failure news on reputation (hypothesis 2a)**

	Direction ß	Control variables other than advertising
Favorability or direction of success & failure TV news		
Table 5.2: Shell, Schiphol, police	3+	
Table 5.2: agricultural sector	−	success & failure print news,
Table 5.3: pooled model	+	mount of news
Table 5.2: autoregressive model Schiphol	+	
Table 5.3: autoregressive pooled model	ns	
Table App. K1: Schiphol	+	success & failure print news,
Table App. K1: agricultural sector	−	amount of news and educational
Table App. K2: pooled model	+	level success & failure print news,
Table 5.6: Shell, police	2+	amount of news, support &
Table 5.7: pooled model	+	criticism news
Table 5.6 and 5.7: autoregressive models per organization and pooled	ns	
Favorability or direction of success & failure print news		
Table 5.2: Super de Boer	+	
Table 5.2: Schiphol	−	success & failure TV news,
Table 5.3: pooled model	ns	amount of news
Table 5.2: autoregressive model BP	+	
Table 5.3: autoregressive pooled model	ns	
Table App. K1: Schiphol	−	success & failure TV news, amount
Table App. K2: pooled model	ns	of news and educational level
Table 5.6: Super de Boer	+	
Table 5.6: Schiphol, agricultural sector	2−	
Table 5.7: pooled model	ns	success & failure TV news, support
Table 5.6: autoregressive model BP	+	& criticism news, amount of news
Table 5.6: autoregressive model Schiphol	−	
Table 5.7: autoregressive pooled model	ns	

For the agricultural sector, the opposite seems to be the case. The more television news about their struggle to keep farming (their "failure"), the better their reputation will be. This suggests that in the case of the agricultural sector, an underdog effect appeared. As has been described in section 5.3, the influence of the direction or favorability of success & failure television news on reputation was no longer significant in the autoregressive models. This was mainly due to the lower number of cases in the autoregressive models.

The second half of Table 5.16 summarizes the effects of success & failure news in the print media on the reputation of the organization. Although Super de Boer and BP were not frequently in the print news, the reputations of these organizations improved the more success was attributed to them in the media. In contrast to Super de Boer and BP, the reputation of the farmers not only improved if there was more television news about their struggle to keep farming, but also if they were more in the print media with their "failure." Furthermore, newspaper readers reacted in the opposite way if they read about the successes and failures of Schiphol. They did not seem to appreciate Schiphol's success, such as the increase in the number of passengers. As with the effect of the total amount of news, newspaper readers may be more skeptical than television viewers concerning the media favorability of success & failure print news.

Overall, these results provide support for hypothesis 2a. The tenor of success & failure television news had a positive effect on reputation, except in the case of the agricultural sector where the underdog effect showed up. The tenor of success & failure print news also had a positive effect on reputation, except for the agricultural sector and Schiphol.

5.11.4 The effects of success & failure news on success of policy (Hypothesis 2b)

Hypothesis 2b assumed that the favorability of success & failure news has a positive impact on the perceived success of the policy of the organization. An overview of the effects of the tenor of success & failure news on the perceived success of the policy of the organization is presented in Table 5.17.

TABLE 5.17.   Overview of the results of the effects of favorability of success & failure news, controlled for the amount of news and advertising on success of policy (hypothesis 2b)

	Direction β
Favorability of success & failure TV news	
Table 5.8: Albert Heijn	−
Table 5.8: ABN AMRO, police	2+
Table 5.9: non-autoregressive pooled model	+
Table 5.8 and 5.9: autoregressive models per organization and pooled model	ns
Favorability of print news	
Table 5.8: Schiphol, agricultural sector	2−
Table 5.9: non-autoregressive pooled model	ns
Table 5.8 and 5.9: autoregressive models per organization and pooled model	ns

These results show that the respondents not only assigned a higher report mark to the organization the more an organization was attributed success in the television news, they also perceived the policy of the organization to be more successful. This was the case in the model of ABN AMRO, the police, and in the model that included all ten organizations. Remarkably, the opposite effect occurred in the case of Albert Heijn, and the more success was attributed to the supermarket, the less successful its policy was judged by the respondents.

The favorability of success & failure print news had a negative impact on the perceived success of the policy of Schiphol and of the agricultural sector. Newspaper readers judged the policy of Schiphol to be less successful the more the airport was in the news with its business. The more news there was about the hard times for the agricultural sector, the more successful the policy of the sector was perceived. The tenor of success & failure print news was significant in neither the models of the rest of the organizations nor in the pooled model. The effects of the tenor of success & failure news disappeared in the autoregressive models due to the lower number of cases.

These results provide moderate support for hypothesis 2b. The more success is attributed in the news to an organization, the more successful the policy of the organization will be perceived by the respondents.

5.11.5   The effects of support & criticism news on reputation (Hypothesis 3a)

Hypothesis 3a predicts that the tone of support & criticism news is negatively related to reputation if the organization is criticized by its competitors. In Table 5.18, an overview is presented of the effects of the favorability or the direction of support & criticism news (the tenor) on corporate reputation. The first half of the table presents the effects of the support & criticism received by the organizations in the television news. In the second half of the table, the estimated impact of print news about the support & criticism of the organizations are displayed. The left column shows the tables of the previous sections, in which models are presented that estimate the effects of support & criticism news. The second column describes the number and the direction of the significant Beta coefficients. The notation "2–" means that the favorability or the direction of support & criticism news was significant in the models of two organizations in a negative direction. This implies that the more the organization was criticized in the media, the better the reputation of the organization and vice versa. The control variables are presented in the last column. All the models were controlled for advertising intensity.

The effects of support & criticism television news and support & criticism print news do not seem to be different. Therefore the table will be discussed as a whole. It appears from Table 5.18 that if support & criticism news had a significant impact on reputation, for most of the organizations it was a negative impact.

This means that the more an organization is criticized in the media, the better its reputation will be. This was the case for models of Shell, Schiphol, Rabobank, Albert Heijn, and Schiphol. However, this influence was not found in any of the pooled models. Moreover, support & criticism news had a positive impact on the reputation of the sectors. The reputation of the agricultural sector and the police improved if they received support and vice versa.

TABLE 5.18.   Overview of the results of the effects of success & failure news on reputation (hypothesis 3a)

	Direction ß	Control variables other than advertising
Favorability or direction of support & criticism TV news		
Table 5.4: Shell	−	support & criticism print
Table 5.4: autoregressive model Schiphol, Rabobank	2−	news, amount of news
Table 5.5: pooled model, autoregressive pooled model	ns	
Table App. L1: Shell, NS	2−	support & criticism print
Table App. L2: pooled model	ns	news, amount of news and educational level
Table 5.6: Shell, Schiphol	2−	support & criticism print
Table 5.6: autoregressive models Schiphol and Rabobank	2−	news, amount of news,
Table 5.7: pooled model, autoregressive pooled model	ns	success & failure news
Favorability or direction of support & criticism print news		
Table 5.4: Albert Heijn and Schiphol	2−	
Table 5.4: agricultural sector	+	support & criticism TV news,
Table 5.4: autoregressive model Schiphol	−	amount of news
Table 5.4: autoregressive model ABN AMRO	+	
Table 5.5: pooled model, autoregressive pooled model	ns	
Table App. L1: autoregressive models per organization	ns	support & criticism TV news,
Table App. L2: pooled model	ns	amount of news and educational level
Table 5.6: Albert Heijn, Schiphol	2−	support & criticism TV news,
Table 5.6: police	+	success & failure news,
Table 5.6: autoregressive model Schiphol	−	amount of news
Table 5.6: autoregressive model ABN AMRO	+	
Table 5.7: pooled model, autoregressive pooled model	ns	

The negative effect of support & criticism news on the reputations of Shell, Albert Heijn, Schiphol, and Rabobank may be explained by the fact that the sources of criticism were usually from competitors, who are more or less "traditional opponents." Criticism of opponents may have the opposite reaction, which means that the more they criticize the organization, the better the reputation of the besieged organization will be.

The positive influence of support & criticism news on the reputation of the agricultural sector and the police may be explained by the fact that most criticism came from the government. The government may be perceived by the respondents to be an impartial source or even an ally. The respondents concluded that if such a credible source criticizes the agricultural sector, there must indeed be something wrong. Therefore, the reputation of the agricultural sector and the police was influenced in the same direction as the support & criticism given by the government. The positive influence of support & criticism news on the reputation of the ABN AMRO is remarkable, however, and cannot be accounted for by this explanation.

These results provide some support for hypothesis 3a. The more traditional opponents criticized the focal organizations in the news, the better the reputation of these besieged organizations. The opposite was the case for the police and the agricultural sector. The more criticism the sectors received from an impartial actor like the government, the worse their reputation became.

5.11.6 The effects of support & criticism news on agreement with policy (Hypothesis 3b)

Hypothesis 3b predicts that the tenor of support & criticism news has a negative impact on the agreement with the policy of the organization. An overview of the effects of the tenor (which consists of the favorability and the direction) of support & criticism news on the agreement of respondents with the policy of the organization is presented in Table 5.19.

TABLE 5.19. **Overview of the results of the effects of the tenor of support & criticism news on agreement with policy controlled for advertising (hypothesis 3b)**

	Direction
Favorability of support & criticism TV news	
Table 5.10: ABN AMRO	−
Table 5.10: autoregressive model ABN AMRO	−
Table 5.11: pooled model, autoregressive pooled model	ns
Favorability of support & criticism print news	
Table 5.10: Shell, Albert Heijn, NS, Rabobank	4−
Table 5.10: police	+
Table 5.10: autoregressive model police	+
Table 5.11: pooled model, autoregressive pooled model	ns

Table 5.19 shows that the tenor of support & criticism news had a negative impact on the agreement with the policy of an organization in the non-autoregressive models of five organizations (ABN AMRO, Shell, Albert Heijn, NS, and Rabobank). This implies that the more any of these organizations is criticized in the media by its competitors, the more respondents will agree with the policy of that organization. For the police, the opposite was the case, and support and criticism news had a positive impact on the agreement with the policy of the police. This means that if the police are criticized in the media, the agreement with the policy of the police will decrease and vice versa. In the autoregressive models, the effects of support & criticism news on the agreement with the policy of an organization was still significant in the models of ABN AMRO and the police. The effects of support & criticism news disappeared in the models of Shell, NS, and Rabobank, due to the lower number of cases.

These results provide moderate support for hypothesis 3b. The results were in line with the results that were found in the models of the effects of support & criticism news on corporate reputation. The tenor of support & criticism news had a negative effect on agreement with the policy of the organizations and a positive influence on agreement with the policy of the police sector.

5.11.7 The impact of issue news on the salience of issues (Hypothesis 4)

Hypothesis 4 poses that the frequency of news about a certain issue influences the salience of that issue. This was tested by means of logistic regression analysis. Table 5.20 gives an overview of the results. The left column shows the table (in this case only Table 5.12) of the previous section, in which the models are presented that estimate the effects of issue news on the salience of an issue. The second column describes the number and the direction of the significant unstandardized regression coefficients. The notation "+" means that the frequency of news about a certain issue stimulated the salience of that issue. This implies that the more news there is about a certain issue in relation to an organization, the sooner respondents will associate the organization with that issue. No control variables were entered in the models.

In the models of eight organizations, more television news about a certain issue in relation to the organization stimulated the salience of that issue. These effects remained in five autoregressive models on the organizational level. The results found in the pooled model and in the autoregressive pooled model also showed that television news about a certain issue enhances the salience of that issue.

The frequency of print news about a certain issue in relation to the organization stimulated the salience of that issue in the models of six organizations and in the pooled model. The effects remained in two autoregressive models of the organizations and disappeared in the autoregressive pooled model. Surprisingly, a negative impact of the amount of news about an issue on the salience of an issue was found for the two supermarkets and for the agricultural sector.

TABLE 5.20. **Overview of the results of the effects of issue news on the salience of issues (hypothesis 4)**

	Direction ß
Frequency of issues television news	
Table 5.12: Shell, Super de Boer, NS, Schiphol, ABN AMRO, Rabobank, police, agricultural sector	8+
Table 5.12: Albert Heijn	−
Table 5.12: pooled model	+
Table 5.12: autoregressive model Shell, NS, Schiphol, police, agricultural sector	5+
Table 5.12: autoregressive pooled model	+
Frequency of issues print news	
Table 5.12: Shell, Albert Heijn, Schiphol, ABN AMRO, Rabobank, police	6+
Table 5.12: Super de Boer, agricultural sector	2−
Table 5.12: pooled model	+
Table 5.12: autoregressive model ABN AMRO, Rabobank	2+
Table 5.12: autoregressive model Super de Boer	−

All in all, most results in Table 5.20 confirm hypothesis 4 (news about an issue influences the salience of that issue). For example, the more the media reported about the delays of the NS trains, the more people associated the NS with delays.

5.11.8 The impact of issue news on the importance of issues in forming reputation (Hypothesis 5)

The previous subsection presented an overview of the results in which the effects of the amount of news about an issue on the salience of that issue (hypothesis 4) were tested.

This subsection will present an overview of the results on the testing of whether an issue becomes more *important* in determining reputation if there is more news about that issue in the media (hypothesis 5). Whether the reputation gets better or worse does not play a role in the testing of the priming hypothesis.

Table 5.21 gives an overview of the results. The left column shows the table (in this case only Table 5.13) of the previous subsection, in which the models are presented that estimate the effects of priming. The priming hypothesis is tested by the regression coefficients of the interaction effect between the "success of the policy of an organization and the news about a certain issue." A statistically significant and positively signed Beta coefficient means that in their overall evaluation of the organization, news coverage about a certain issue increases the weight media users grant to the policy of an organization regarding that issue.

TABLE 5.21.   Overview of the results of the effects of issue news on the importance of issues in forming reputation (hypothesis 5)

	Direction ß
Success policy issue television news	
Table 5.13: agricultural sector	+
Table 5.13: autoregressive models	*ns*
Pooled model, autoregressive pooled model	*ns*
Success policy issue print news	
Table 5.13: NS	+
Table 5.13: Albert Heijn	−
Table 5.13: autoregressive models	*ns*
Pooled model, autoregressive pooled model	*ns*

In the models of the agricultural sector and the NS, a positive interaction effect was found between the success of the policy of the organization regarding a certain issue and the amount of news about that issue. Surprisingly, for Albert Heijn a negative interaction effect was found. Due to the lower number of cases, the interaction effects disappeared in the autoregressive models, and no interaction effect was found in either the pooled model or in the autoregressive pooled model.

This means that only weak support is found for hypothesis 5. In the models of two organizations, the issue became more important in determining reputation if there was more news about that issue in the media. However, in one model a negative effect was found.

5.11.9 The effects on reputation of news about owned issues (Hypothesis 6)

This subsection will discuss the effects of news about owned issues on reputation. It is assumed in hypothesis 6 that news about owned issues will lead to a better reputation. The effect of news about owned issues on corporate reputation has been tested by means of two approaches and statistical techniques: regression analysis and Repeated Measures ANOVA. If it was determined *a priori* whether or not an issue was an owned issue, and if the effects of the frequency of news about owned issues were tested subsequently by means of regression analyses, it turned out that these effects were not significant. It can be questioned whether the criteria used to determine whether an issue was an owned issue or not were correct. In the study by Petrocik (1996), owned issues were not determined a priori by the researcher but by the respondents.

Therefore, the hypotheses about issue ownership were also tested in a two-step approach. The first step was to test whether the amount of news about issues influences the salience of an issue (this step overlaps hypothesis 4). The second step examined whether respondents gave different report marks to the organization if they had a different association with the organization.

In order to test whether a difference in the association with the organization leads to a difference in report mark, respondents were categorized in several groups according to their association with the organization. It was expected that respondents would assign a lower report mark to the organization if they associated the organization with an issue the organization was perceived to be incapable of handling (such as "environment" in the case of Shell, a poorly owned issue), than when they associated the organization with a more neutral issue (such as "gas station" in the case of Shell). Repeated Measures ANOVA with between-subject factors was used to test the differences between and within the groups of respondents.

TABLE 5.22.   **Overview of the results of the effects of the salience of an association on reputation**

	Difference between groups	*Difference within group*
Shell: associations "gas stations" and "environment"	Yes	No
BP: associations "gas stations" and "environment"	No	No
ABN AMRO: associations "bank cards" and "profit"	Yes	No
Rabobank: associations "bank cards" and "profit"	No	No
Albert Heijn: associations "price" and "fresh products"	Yes	Yes, 1 group + 1 group ($p = .059$)
Super de Boer: associations "price" and "fresh products"	Yes	Yes, 2 groups
Schiphol: associations "KLM" and "environment"	Yes	Yes, 1 group + 1 group ($p = .063$)
NS: associations "transport" and "delays"	Yes	No
Police: associations "order" and "weak action"	Yes	Yes, 2 groups
Agricultural sector: associations "fresh products" and "environment"	Yes	Yes, 1 group

Note. Only "relevant" differences are mentioned. In the case of the NS, a significant difference was found within the group of respondents that associated the NS twice with delays. This suggests that the performance of the NS could have been improved.

In eight of the focal organizations/sectors, a difference *between groups* was found. This means that the respondents who associated the organization twice with a neutral issue (e.g., "gas stations") gave a higher report mark than the respondents who associated the organization twice with a poorly owned issue (e.g., "environment" in the case of Shell). Or in the case of ABN AMRO, respondents who associated the bank with "profit" assigned a higher report mark to ABN AMRO than respondents who associated the bank with the more neutral issue "bank cards."

Moreover, for five organizations, significant differences *within groups* were found. This means that the report mark of respondents differed per year if the respondents associated the organization with another issue in each of the years. For example, respondents who associated Super de Boer in 1999 with "price" and in 2000 with "fresh products" gave a higher report mark in 2000 than in 1999. In the case of Super de Boer, two differences within groups were found. This indicates that a significant difference was not only found within the "price 1999-fresh product 2000" group, but also within the "fresh product 1999-price 2000" group. The latter means that the respondents who in 1999 associated Super de Boer with "fresh products" and in 2000 with "price" gave a lower report mark in 2000 than in 1999.

In sum, in the previous subsection it has been demonstrated that news about a certain issue determines the salience of that issue (hypothesis 4). For example, the more Shell is in the news with its environmental policy, the more respondents will associate Shell with the environment. The tables in the present subsection reveal that the salience of an issue determines corporate reputation. This means that the respondents will assign a worse reputation to Shell if they think Shell is incapable of handling its environmental policy (a poorly owned issue for Shell).

5.11.10   The impact of advertising intensity on reputation (Hypothesis 7)

Hypothesis 7 poses that the higher the advertising intensity of a company or sector, the better the reputation of the organization will be. In Table 5.23, an overview is presented of the effects of television advertising expenditures on corporate reputation. The left column shows the tables of the previous sections, in which models are presented that estimate the effects of advertising. The second column describes the number and the direction of significant Beta coefficients. The notation "2+" means that advertising was significant in the models of two organizations, both in a positive direction. The control variables are presented in the last column.

Table 5.23 shows that if advertising expenditures on television have a significant impact on reputation, it is a positive influence. This means that the higher the advertising intensity on television, the better the reputation of the organization. Significant effects were found in models of Super de Boer, Shell,[13] and the agricultural sector.[14] BP and Schiphol did not advertise on television during the research period. No significant effects of television advertising expenditures on reputation were found in the pooled models.

In the models of two organizations (the NS and the police), higher advertising expenditures on television led to a more successfully perceived policy of the organization (Table 5.8). Moreover, the higher the advertising expenditures on television, the more respondents agreed with the policy of the police (see Table 5.10).

TABLE 5.23.   **Overview of the results of the effects of advertising expenditures (TV) on reputation (hypothesis 7)**

	Direction ß	*Control variables*
Table 5.2: Super de Boer	+	
Table 5.3: pooled model	ns	success & failure news,
Table 5.2: autoregressive model agricultural sector	+	amount of news
Table 5.3: autoregressive pooled model	ns	
Table App. K1: model per organization	ns	success & failure news, amount
Table App. K2: pooled model	ns	of news and educational level
Table 5.4: Super de Boer, agricultural sector	2+	support & criticism news,
Table 5.4: autoregressive model agricultural sector	+	amount of news
Table 5.5: non-autoregressive and autoregressive model	ns	
Table App. L1: model per organization	ns	support & criticism news,
Table App. L2: pooled model	ns	amount of news, educational level
Table 5.6: Super de Boer	+	success & failure print news,
Table 5.6: autoregressive model agricultural sector	+	amount of news, support &
Table 5.7: pooled model, autoregressive pooled model	ns	criticism news
Table 5.8: non-autoregressive model NS	+	favorability of success & failure
Table 5.8: autoregressive model police	+	news, success policy as
Table 5.9: (non)-autoregressive pooled model	ns	dependent var.
Table 5.10: non-autoregressive model police	+	favorability of success & failure
Table 5.11: (non)-autoregressive pooled model	ns	news, agreement policy as dependent var.

Table 5.24 on the next page presents an overview of the effects of advertising expenditures in newspapers on corporate reputation.

Table 5.24 shows that if advertising expenditures in newspapers have an impact on corporate reputation, it is a positive one. This impact was quite consistent, and occurred in models of several organizations (BP, Albert Heijn, Rabobank, and the agricultural sector) and in the pooled models.

Moreover, it was found that the higher the advertising expenditures in newspapers, the more successful an organization was judged by the respondents (Albert Heijn in Table 5.8) and the more respondents agreed with the policy of the organization (see the models of Albert Heijn, ABN AMRO, and Rabobank in Table 5.10). This effect was also found in the pooled models (Table 5.9 and Table 5.11). In the pooled models, the advertising intensity in newspapers influenced both the perceived success of the policy of the organization as well as the agreement with the policy of the organization.

TABLE 5.24. **Overview of the results of the effects of advertising intensity newspapers on reputation (hypothesis 7)**

	Direction β	Control variables
Table 5.2: BP, Albert Heijn, Rabobank	3+	
Table 5.3: pooled model	+	success & failure news,
Table 5.2: autoregressive model police	+	amount of news
Table 5.3: autoregressive pooled model	ns	
Table App. K1: BP, Albert Heijn, Rabobank	3+	success & failure news,
Table App. K2: pooled model	+	amount of news and educational level
Table 5.4: BP, Albert Heijn, Rabobank, agricultural sector	4+	support & criticism news,
Table 5.5: pooled model	+	amount of news
Table 5.4: autoregressive model police	+	
Table 5.5: autoregressive pooled model	ns	
Table App. L1: BP, Albert Heijn, Rabobank	3+	support & criticism news,
Table App. L2: pooled model	+	amount of news, educational level
Table 5.6: BP, Albert Heijn, Rabobank, agricultural sector	4+	success & failure print news,
Table 5.7: pooled model	+	amount of news, support &
Table 5.6: autoregressive model police	+	criticism news
Table 5.7: autoregressive pooled model	ns	
Table 5.8: non-autoregressive and autoregressive model Albert Heijn	2+	favorability of success & failure news, success policy
Table 5.9: (non)-autoregressive pooled model	2+	as dependent variable
Table 5.10: non-autoregressive models Albert Heijn, ABN AMRO, Rabobank	3+	favorability of success & failure news, agreement
Table 5.11: (non)-autoregressive pooled model	2+	policy as dependent variable

Last but not least, the effects of advertising expenditures in newspapers remained significant in the autoregressive pooled models if they were based on three points in time (see Table 5.9 and Table 5.11): in other words on more cases.

Overall, the results provide support for the hypothesis that the higher the advertising intensity of an organization, the better the reputation of the organization will be. These results are in line with the results found by Fombrun and Shanley (1990). Stronger evidence was found in the case of advertising expenditures in newspapers than in the case of advertising expenditures on television news.

6
CONCLUSIONS AND DISCUSSION

Although media coverage can have a large impact on the reputation and market value of an organization, as a field of research this topic is relatively young. The few pioneering empirical studies about the effects of media coverage on reputation (Fombrun & Shanley, 1990; Verčič, 2000; Wartick, 1992) showed mixed results. Because the study of the effects of media coverage on attitudes has a long tradition in the field of political communication, theories from this field were used to try to explain these mixed results from the earlier studies and to serve as a theoretical framework of the present study. In order to apply these theories, different types of news were distinguished. Issue news, for example, is a type of news often focused upon in political communication studies. Moreover, the effects of paid publicity – advertising intensity – were taken into account. The present study contributes to studies on advertising effects because of the research method used and the real-life situations of respondents.

The project was designed to investigate the content of the news and reputations of eight large companies and two sectors. The analysis of three years of media coverage by five daily newspapers and television news from two TV channels resulted into 15,883 assertions. Corporate reputation was measured at three points in time by conducting a survey in which 1,289 different respondents participated. A longitudinal panel design was used, because this design is more appropriate for showing that cause precedes effect than a cross-sectional design. In contrast to previous empirical studies on the effects of media coverage on reputation, the media use of the individual respondents was taken into account; the respondents were the units of analysis. The present study was conducted in the Netherlands. This country is a suitable one for testing the effects of media coverage on reputation in a real-life situation. One of the reasons for this is that the Netherlands has excellent research agencies in the field of public opinion polling. Moreover, subscriptions to daily newspapers are responsible for 85% of the total circulation (Bakker & Scholten, 2003). This means that the respondent's subscription(s) to a certain newspaper is a good predictor of the print news consumption of the respondent, because the impact of single issues is relatively small.

This final chapter is structured as follows. First, in section 6.1 the key findings of the project are summarized. Then the limitations and shortcomings of this study will be discussed in section 6.2. In the subsequent section, suggestions will be given for future research and theoretical issues will be touched upon. Finally, in 6.4 the implications of the findings of this study for communication managers and the possibility of developing a media monitor will be discussed.

6.1 FINDINGS

This section will summarize and discuss the findings of the study. The first sub-section will focus upon the media profiles of the focal companies and sectors, followed by a short description of their reputation. The next subsection will present the conclusions with regard to the hypotheses.

6.1.1 The media profiles, advertising intensity, and the reputation of the focal companies

The police received most media coverage, and issue news occurred most
Of the focal organizations and sectors, the Dutch police received most media coverage (32%). Schiphol ranked second, with 19% of the news about the focal organizations and sectors about Schiphol. Shell received a relatively large amount of news (11%) compared to its British competitor BP (1%). Supermarkets Albert Heijn (2%) and Super de Boer (0.5%) were rarely in the news compared to the other companies. ABN AMRO received more news (9%) than Rabobank (4%), which is not listed on the stock exchange. The NS (11%) and the agricultural sector (12%) both received an average amount of news.

"Issue news" was the most common type of news with regard to the ten organizations. It made up 38% of the total amount of the news about the ten focal companies. This was due to the fact that support & criticism and success & failure was taken into account only if the focal company was in the "object position" of the assertion. Next to issue news, support & criticism news occurred most (19%) in the media coverage of the focal companies. At 17%, the total amount of success & failure news came close to support & criticism news. A big part of media coverage (26%) about the ten organizations consisted of "other news" (such as support & criticism given by the focal organization).

The agricultural sector struggled to keep farming
The agricultural sector was attributed most failure (−.41 on a scale ranging from −1 to +1) of the ten focal organizations. The message was loud and clear: the agricultural sector was suffering from the slump in their sector. Next to the agricultural sector, Shell was attributed most failure (−.26). This was due to several topics: anticipation of a hard year, riots at Shell Nigeria, and the fine Shell had to pay for violating the oil embargo against Iraq. BP took a middle position (−.15). Supermarket Super de Boer received scarcely any success & failure news ($n − 6$), but was

attributed most failure (–.25) after the agricultural sector and Shell. Competitor Albert Heijn had to pay a fine, but received more success (–.10) than Super de Boer, because Albert Heijn was still number one with regard to turnover.

The NS (–.16) and Schiphol (–.15) both received more failure than success, but were in the middle group compared to the other companies. ABN AMRO (–.08) was attributed a bit more failure than success, whereas this was the other way around for Rabobank (.12). Of the focal companies, the police received the most success (.18) due to its success in fighting crime.

Albert Heijn was criticized most
Because of the criticism by diet guru Montignac, the supermarket Albert Heijn received relatively the largest amount of criticism (–.29 on a scale ranging from –1 to +1). Competitor Super de Boer was doing considerably better, and had the most friends compared to the other focal companies (.75).

Next to Albert Heijn, the NS (–.22) and the police (–.22) received most criticism. The NS was criticized by the government, and quarreled with the unions. The cooperation with other organizations such as Schiphol and KLM led to support for the NS. The police had a cool relationship with the government, but in the eyes of the media they were crazy. The latter revealed itself in headlines such as "Police have gone crazy" (*De Telegraaf*, December 1, 1997, p. 3), "Police powerless at illegal house party" (*de Volkskrant*, July 19, 1999, p. 4). Schiphol (–.15), the agricultural sector (–.12), ABN AMRO (–.12) and Shell (–.10) formed the middle group when it came to support & criticism news. Like the police, the agricultural sector received a lot of support & criticism from the government (46%) and from the media (23%). Rabobank (.31) and BP (.38) both received more support than criticism. The support for Rabobank came from its cooperation with other companies.

The issues in the media broadly matched the associations of respondents
Shell was in the news most often with the issues profit, the Dutch economy, and environmental pollution. Although the associations of respondents did not match these issues in the media exactly, a relatively large number of the respondents associated Shell with environmental pollution. BP was associated with the environment by the respondents considerably less than Shell. Even though there was a relatively large amount of issue news about BP (*n* = 53), only in 6% of the issue news was BP related to the environment. However, many of the respondents (18%) did not have an association with BP. ABN AMRO was frequently in the news with the issues mergers and takeovers, the Dutch economy, and profit. Although respondents did associate the bank with these issues, most respondents associated ABN AMRO with general associations such as bank accounts and financial services. Rabobank received a relatively large amount of news about its racing team. After the general associations, the respondents associated Rabobank most with advertising.

Of all the issues, in the first and third research period the media wrote most about Albert Heijn's new products. This fit with the association "wide selection of products," which was mentioned most often by the respondents when asked about their association with Albert Heijn. In the case of Super de Boer, most respondents associated the supermarket with its price level. Except for the issue "mergers," there was very little issue news about Super de Boer.

Schiphol was associated by the respondents mainly with KLM and in the news with its expansion. Next to the expansion of Schiphol, the media paid most attention to environmental pollution, and this association was mentioned by the respondents, as well. The other transport company, NS, was associated by the respondents mainly with delays. There was some media coverage about the delays at NS. The police were in the news mainly with the issue "crime." This issue was the second most frequently mentioned association by the respondents. In the case of the agricultural sector, in the media the issue "environment" came second, and was the second most frequently mentioned association by the respondents.

Albert Heijn had the highest advertising expenditures, followed by Rabobank
The advertising expenditure patterns of Shell and BP were rather different. Over the three research periods, Shell spent the highest amount of advertising expenditures in the third research period, namely 1.2 million Euros. BP on contrary spends the highest amount of advertising expenditures in the first research period, namely 1.1 million euros. Next to supermarket Albert Heijn, the two banks had the highest total amount of advertising expenditures over the three research periods. Rabobank had the highest amount of advertising expenditures per research period, namely 11.9 million euros in the second research period. Both banks spent more on advertising on public television channels than on RTL 4.

Compared to Schiphol, the advertising expenditures of the NS were relatively high. The advertising expenditures of the airport came to 1.9 million euros in the total research period, while the advertising expenditures of the NS came to 13.1 million euros within the total research period. At 27.6 million euros, Albert Heijn had the highest advertising expenditures of the ten focal organizations and sectors within the total research period. Albert Heijn spent more money on advertising in newspapers (mainly in *De Telegraaf*) than on advertising on television. At 10.3 million euros within the total research period, Super de Boer did not even have half the advertising budget of Albert Heijn.

The advertising expenditures of the agricultural sector (1.6 million euros) and the police (1 million euros) were relatively low compared to the other organizations. The police did not spend money on television advertising in the first two research periods.

Albert Heijn had the best and the NS the worst reputation
The overall reputation was measured by asking the respondents to give a report mark to the focal organization or sector, on a scale from 1 to 10. Albert Heijn had

the best reputation in 1999 ($M = 6.9$, $SD = 1.52$) and in 2000 ($M = 6.8$, $SD = 1.39$). In both 1999 and in 2000, Schiphol had the second best reputation ($M = 6.6$, $SD = 1.48$ in 1999; $M = 6.7$, $SD = 1.46$ in 2000). Of the ten focal companies, the NS had the worst reputation in 1999 ($M = 5.5$, $SD = 1.39$) and in 2000 ($M = 5.7$, $SD = 1.44$). In 1999, the police had the second worst reputation ($M = 6.0$, $SD = 1.54$). In the year 2000, the police were doing better ($M = 6.2$, $SD = 1.32$).

6.1.2   The effects

The more television news about the organization, the better its corporate reputation
The findings indicate that for most companies, more television news about the focal company (Shell, NS, Schiphol, the police, or the agricultural sector) improved corporate reputation. This effect was controlled for the impact of the tenor of success & failure news, the tenor of support & criticism news and for advertising intensity.

These findings support the "two-stage familiarity-then-something wrong model" based on Cacioppo and Petty (1979), and on the reception axiom of Zaller's model. Assuming that the organizational awareness of television viewers is relatively low, it follows from this model that more television news leads to a greater chance of receiving the message and to a more favorable attitude towards the organization.

At a certain point, more print news worsens corporate reputation
In accordance with the expectations, the reputation of organizations rarely in the news (such as Super de Boer and BP) improved according to the amount of print news they received, while the reputation of organizations frequently in the print news (such as Schiphol, the agricultural sector, and the police) worsened the more print news they received.

These results support the "two-stage familiarity-then-something wrong model" also mentioned in the previous subsection. Assuming that the organizational awareness of newspaper readers is relatively high, it follows from this model that once a certain point has been passed, more print news evokes counterargumentation and leads to a more negative attitude towards the organization.

The effects of success & failure news: TV viewers jump on the bandwagon
The reputation of Shell, Schiphol, and the police improved the more they were in the television news with their successes. The results indicated that in the case of television news the bandwagon effect, which can be expressed in the slogan "success breeds success," applied for several organizations and for the pooled model. The agricultural sector was an exception, and the favorability of success & failure news negatively influenced corporate reputation. The farmers seemed to gain the sympathy of the public due to the frequent media coverage about the slump in their sector.

As with the effect of the total amount of news, newspaper readers may be more skeptical than television viewers concerning the media favorability of success & failure print news. Newspaper readers did not seem to appreciate Schiphol's success, like the increase in the number of passengers. The reputation of organizations not frequently in the news, such as Super de Boer and BP, improved the more success was attributed to them in the newspapers.

The respondents not only assigned a higher report mark to the organization the more an organization was attributed success in the news, they also perceived the policy of the organization to be more successful. This was the case in the models of ABN AMRO, the police, and in the model that includes all ten organizations. Surprisingly, the opposite effect occurred in the case of television news about Albert Heijn: the more success was attributed to the supermarket, the less successful its policy was judged by the respondents. The latter may be due to a ceiling effect. This would mean that the reputation of the supermarket is so good, news about it evokes counterargumentation and leads to a policy that is perceived as less successful.

The effect of support & criticism news: criticism by competitors improves reputation
In agreement with research in the field of political communication, it is expected that newspaper readers and television *viewers* will react suspiciously if an organization is criticized by its competitors or supported by its friends. In accordance with the expectation, the direction of support & criticism news (in newspapers and on television) had a *negative* effect on the corporate reputation of Shell, Schiphol, Rabobank, Albert Heijn, and Schiphol. These organizations were criticized mainly by their competitors, who are more or less "traditional opponents." Criticism by opponents results in a boomerang effect, which means that the more the competitors criticize an organization, the better the reputation of the besieged organization will be.

Support & criticism news had a *positive* impact on the reputation of the sectors. The reputations of the agricultural sector and the police improved the more they received support in the media, and worsened the more they were criticized. The positive influence of support & criticism news on the reputations of the agricultural sector and the police can be explained by the fact that most criticism came from the government. In addition, both sectors received a relatively large amount of severe criticism from the media. The government and the media may be perceived by the respondents as impartial sources, or in the case of the government, even as an ally. Therefore the reputations of the agricultural sector and the police was influenced in the same direction as the support & criticism given by the government and the media.

The results of the effects of the tenor of support & criticism news on the agreement with the policy of the organization were in line with the effects of the tenor of support & criticism news on corporate reputation. The tenor of support

& criticism news had a negative effect on agreement with the policy of the organizations and a positive influence on agreement with the policy of the sectors.

The frequency of news about a certain issue influences the salience of that issue
The present study found strong support for the hypothesis that news about a certain issue in relation to the organization stimulated the salience of that issue. These effects were demonstrated in the models of eight organizations and in five autoregressive models.

The results indicate that the second level of the agenda-setting effect not only applies to news about political parties but to companies and sectors as well. This means that if there was a lot of television news about Shell and the environment, the respondents were more likely to think about the environment when they thought about Shell. Surprisingly, for the two supermarkets and for the agricultural sector, a negative impact was found to result from the amount of news about an issue on the salience of that issue.

The impact of issue news on the importance of issues in forming corporate reputation
In the models of the agricultural sector and the NS, evidence was found that supports priming. This means that the more news there was about a certain issue, the more important the issue became in the formation of corporate reputation. Surprisingly, a negative effect was found for Albert Heijn.

In sum, weak support was found for priming. The results indicate that the theory about priming applies not only to news about political parties, but may be applicable to companies and sectors as well. This means that Iyengar and Kinder's (1987, p. 63) definition of priming ("By calling attention to some matters while ignoring others, television news influences the standards by which governments, presidents, policies, and candidates for public office are judged.") may also be applicable for organizations ("By calling attention to some matters while ignoring others, news influences the standards by which organizations are judged.").

The salience of an issue influences corporate reputation
Issue-ownership theory was tested in a two-step approach. The first step was to test whether the amount of news about issues influences the salience of an issue. This step overlaps with the testing of the second level of agenda-setting theory, which was discussed in the previous subsection. The second step was to examine whether respondents gave different report marks to the organization if they had a different association with the organization.

For five organizations (Albert Heijn, Super de Boer, Schiphol, the police, and the agricultural sector) significant differences were found within groups. This means that the report mark of respondents differed per year if the respondents associated the organization with another issue in each of the years. For example, respondents who associated Super de Boer in 1999 with "price" and in 2000 with "fresh products" gave higher report marks in 2000 than in 1999.

In sum, the second level of agenda-setting hypothesis demonstrates that the amount of news about issues determines the salience of an issue. In testing issue-ownership theory, it is shown that the salience of an issue determines corporate reputation. This means that issue-ownership theory is proved using a "two-step approach."

The higher the advertising intensity of a company, the better its reputation
In the models of four organizations (BP, Albert Heijn, Rabobank, and the agricultural sector) higher advertising intensity in the print media resulted in a better reputation. In two models (Super de Boer, agricultural sector) higher advertising intensity on the television news resulted in a better reputation. In none of the models did higher advertising intensity worsen corporate reputation.

These findings support the mere repeated exposure theory, which postulates that the mere exposure of the individual to a stimulus object enhances his or her attitude towards it.

These findings contrast with the results of the studies in which an inverted U-shaped relationship between ad exposure and ad effectiveness was found (Anand & Sternthal, 1990; Calder & Sternthal, 1980; Craig et al., 1976; Nordhelm, 2002; Schumann et al., 1990). This may be due to the use of different research methods.

6.2 LIMITATIONS OF THE STUDY

A number of limitations pertain to the present study. The findings must therefore be interpreted as explorative, and as an attempt to enrich the study of the effects of media coverage on reputation by applying and empirically testing theories from the field of political communication. In the subsection below, limitations concerning the media segment will be discussed first, followed by the limitations of the reputation segment, and limitations with respect to the research design.

6.2.1 Limitations concerning the media segment

Actual media use and interests of respondents
In this study, media coverage was linked to the respondents according to their individual media use. It was known whether the respondents had a subscription to a particular newspaper, and if they watched a certain television program or not. In his reaction on the media monitor, Cramwinckel (2000) posed that a limitation of the media monitor is that it remains unclear whether the respondents actually saw the news articles about the focal organizations and whether they read them.

Although it is practically impossible to find out if each of the respondents has read a certain newspaper article or watched a certain news program, I would suggest that in future research, one could take into account respondents' interests in business news. This variable could also be used to study the "organizational awareness" of the respondents.

Different forms of communication, media sample, sampling units, and images
In this study, the five largest daily newspapers and television news of one public and one commercial television channel were analyzed, as is also done in election research (Kleinnijenhuis et al., 1998; Kleinnijenhuis et al., 1995). However, Schoenbach and Lauf (2002) found in their study on the factors that influence voter turnout that personal conversations worked better to persuade the uninterested to turn out and vote than television (and newspapers) did. Although the political communication setting differs from a business news and reputation context, it will be interesting to examine the joint effects of personal conversations and media coverage on reputation. Moreover, future research may try to include different media types such as current affairs television programs and weekly magazines.

In the present study, only headlines of newspaper articles were coded. In subsequent studies it could be considered to code the summary of the article as well. As in other studies on the effects of media coverage on reputation (Fombrun & Shanley 1990; Verčič 2000; Wartick, 1992), images (photos in newspapers and image on the television news) were not taken into account.

6.2.2 Limitations concerning the corporate reputation part

Measurement of the overall reputation measure
In contrast to the variables of "perceived success of the policy of the organization" and "agreement with the policy of the organization," the overall reputation measure is measured on a single-item scale. Although in another study on the effects of media coverage on reputation (Verčič, 2000) reputation was also measured on a single-item scale, a multiple-item scale for measuring reputation is preferred. An advantage of using multiple items (and hence, a disadvantage when using only one item) is that measurement errors cancel each other out as the number of items increases (Fishbein & Ajzen, 1975; Himmelfarb, 1993).

Nevertheless, in several political communication studies that focused on the effects of media coverage on public opinion, single-item scales were also used (Dalton, Beck & Huckfeldt, 1998; Hertog & Fan, 1995; Krosnick & Kinder, 1990; Miller & Krosnick, 2000; Shah et al., 2002).

Associations with a company
Van Riel et al. (1998) examined six different image-measurement techniques. They concluded that open methods can be used as input for subsequent quantitative analysis. In this study the advantage of the use of open methods was recognized as well, and they provide respondents the opportunity to mention the beliefs that are salient to them. Therefore, respondents were asked to give their opinions

about the organization with regard to their two most salient beliefs. The respondents did not have to give their opinions about an exhaustive list of items that meant nothing to them.

It can be remarked that the pre-listed items were predominantly of a cognitive nature, while marketing communications seems to be taking more notice of the "softer" elements such as atmosphere and emotions (Maathuis, 1999; Pruyn, 2002). Berens and Van Riel (2004) distinguished three different types of associations people may have with regard to companies. These three different types of associations are based on three different conceptual streams. The first stream is based on the concept of social expectations, and examples of these types of associations are the financial performance of a company and the quality of the products. The second stream is based on the concept of corporate personality, and in this case the company is conceptualized as a person. Examples of these types of associations are "friendly," "pleasant," and "empathetic." The third main stream distinguished by Berens and Van Riel (2004) is the stream that uses the concept of *trust*. Examples of these types of associations are "truthful" and "honest."

This study focused upon associations with regard to social expectations, since they were most closely linked to the content of the news and therefore found to be the most suitable for testing the hypotheses.

Success and agreement with policy: the number of salient beliefs and linearity
Attitude models are criticized by several authors (Bouwman, 1998; Reynolds & Gutman, 1984) as measurement methods of reputation, because they make use of a predetermined set of items which are not guaranteed as being important to the respondents. Therefore, in the present study only the two most salient beliefs of respondents were taken into account when they were asked to evaluate the success of the policy of the organization and whether they agreed with the policy of the organization.

However, the attitudes of respondents are based on the total set of a person's salient beliefs (Ajzen & Fishbein, 1980), which may amount to more or less than the two most salient beliefs that were taken into account in the present study. McGuire (1985) argued that he found it very implausible that people form their attitudes by a tedious process. He proposed a "construction by aspect" process in which the person starts with just one salient characteristic, presumably the most important attribute. Van der Pligt and Eiser (1984) suggested that three to five beliefs determine attitude. Eagly and Chaiken (1993) stated that functional considerations should be taken into account. In important areas of consumer decision-making such as buying a house, people are probably willing to spend more time and energy to investigate the attributes of attitude objects than when buying tissues or butter.

In addition to the criticism about the number of attributes people use in forming an attitude, the linear compensatory character of the expectancy value model is called into question (Van der Pligt & De Vries, 1995; Pras & Summers, 1975). The expectancy value model assumes that attitude is additive and based on different attributes towards an object, which can be weighted equally. Researchers in the marketing area and political sciences have also described and tested nonlinear models (Heeler, Kearney, & Mehaffey, 1973; Pras & Summers, 1975).

Success and agreement with policy: difficulties with the word policy
The respondents were asked to evaluate the success of the policy of the organization and to state whether they agreed with the policy of the organization. Several respondents remarked that it was sometimes hard for them to answer these questions, because they had difficulties with the word "policy" and they were not aware of the policy of a certain organization. In 1998 for example, 88% of the respondents answered that they thought of gas stations when they thought of Shell. The next question for these respondents was, "Do you agree with the policy of Shell concerning gas stations?" In 1998, a total of 50% of the respondents said they did not know the policy of Shell when they were asked if they agreed with Shell's policy.

These operationalizations were chosen due to the theoretical premises that the agreement with the policy of an organization is influenced by support & criticism news, while the perceived success of the policy of an organization is influenced by success & failure news.

6.2.3 Limitations of research design, scope of the effects, and causality

Research design: decay of influence of media coverage on reputation over time
As is mentioned in Chapter 3, the datasets that were used in this study were constructed by assigning to the respondents the media coverage of the year before the public opinion polling took place. A limitation in the present study is that the time interval between the date of the article or news item and the measurement of corporate reputation was not taken into account. This means that if corporate reputation was measured on, for example, August 24, 1998, an article from August 20, 1998 was weighted the same as an article from November 15, 1997. It is likely however, that the more recent article from August 20, 1998 will have a larger impact on the formation of reputation than the article from November 15, 1997.

Kiousis and McCombs (2004) pointed out that effects were found in agenda-setting studies that were based on different time lags. For example, effects were found in both studies that included the media coverage from the week before the public opinion survey took place (Wanta & Foote, 1994), and in studies that included media coverage from several months previous (Stone & McCombs, 1981) or nine months previous (Atwood, Sohn, & Sohn, 1978) to when the public opinion survey took place.

Future research on the effects of media coverage on reputation could aim to take into account the decay of the influence of media messages on reputation over time (for examples, see Fan & Tims, 1990; Hertog & Fan, 1995; Trumbo, 1995).

The scope of the effects

The results show that in addition to the media effects, there are other factors that influence reputation. To put it more precisely, the adjusted coefficient of determination (Adjusted R^2) is relatively low in the non-autoregressive models. For example, the adjusted R^2 was 0.5% in the non-autoregressive models of NS and Rabobank and 1% in the cross-sectional models of Shell, BP, and Albert Heijn (see Chapter 5, Table 5.2).

It is not possible to compare these findings with the findings of the "media effects on reputation studies" described earlier, however. In the case of the article by Fombrun and Shanley (1990), Adjusted R^2 was based on 12 different independent variables (such as profitability, risk, size, and diversification), and only three of the independent variables consisted of media coverage (media visibility, media favorability, and the interaction between the two variables). Therefore, it remains unclear what part of the variance in reputation is explained by the variance in media coverage. Wartick (1992) reported correlation coefficients only. Verčič (2000) did not find any significant media effects. Deephouse (2000) studied the effects of media coverage on relative return on average assets (ROA); he found that media reputation increased the pseudo[1] R^2 of ROA by 2%.

Many studies in the field of political communication that focus on the effects of media coverage on public opinion report an Adjusted R^2 that is based on several independent variables (or control variables), of which media coverage is one (Dalton, Beck, & Huckfeldt, 1998; Miller, Goldenberg, & Erbring, 1979; Miller & Krosnick, 2000; Page, Shapiro, & Dempsey, 1987; Peter, 2003; Waterman, Jenkins-Smith, & Silva, 1999). These studies did not explicitly address which part of the variance in the dependent variable is explained by the variance in media coverage only. Hertog and Fan (1995) did find high coefficients of determination, however. They examined the effects of media content on public beliefs concerning the likelihood of HIV transmission by way of sneezing, insects, and toilets. Hertog and Fan (1995) found an R^2 of 13% for sneezing, 36% for insects, and 88% for toilets.

In the present study, two possible explanations are suggested for the differences in the explained variance between the present study, and the study of Deephouse (2000) on one hand and that of Hertog and Fan (1995) on the other. An important difference is the influence of people's direct experience. In addition to news and advertising, companies' reputations are based on the companies' products and services (Brown, 1998). This means that most of the consumers have their own experiences with the company. Direct experience did not play a role in the study of Hertog and Fan (1995), however. Hertog and Fan stated that the HIV/AIDS case was a unique test case, since most of the population does *not* have contact with someone infected with HIV, although this proportion is of course increasing. Another explanation is the "decay of the influence of media messages on reputation over time," which was taken into account by Hertog and Fan (1995), but not by either the present study or by Deephouse (2000).

Another aspect that should receive attention is the amount of news during the research period. Several media hypes took place just before and just after the research period. For example, the outbreak of swine fever took place in 1997, before the research period started. The intention by holding company Laurus to close down supermarket Super de Boer (November 2000), the resignation of top management at the NS (January 2002), and last but not least, the scandal at Ahold, Albert Heijn's parent company (February 2003) all took place after the research period ended. It could be questioned as to whether this was a relatively "quiet" period with regard to the news. This explanation would fit in with the findings that the reputation of the companies in the present study was relatively stable during the research period, as was demonstrated in Chapter 4. In order to detect media effects, it is an advantage to study organizations with different trends in their reputations, as in the study by Verčič (2000). Verčič chose three organizations: one organization whose reputation got better over time (Post Office), one organization with a stable reputation (British Airways), and one organization whose reputation got worse over time (Shell). However, it might be difficult to choose organizations from the same industry with different trends.

The consistent pattern of media and advertising effects found in the various models (see the "overview of the results of the hypothesis" in Chapter 5) indicates that although in some of the models the media effects involved may be weak, they are by no means random. This was confirmed by the autoregressive models. Although in some tables media effects are not significant due to the lower number of cases in the autoregressive models, the models of more cases show that some of the media effects remain significant in the autoregressive models. See the pooled model of Table 5.9. In the case of testing the agenda hypothesis, the frequency of issue news remained significant in the autoregressive models of eight organizations. Moreover, the effects of issue television news remained significant in the autoregressive pooled model.

The causality question
The present study reveals that media coverage influences reputation. It can be questioned, however, whether there is also a reverse effect, for example whether the reputation of an organization influences the amount of news. Feedback or bilateral causality is said to exist if media coverage influences corporate reputation and corporate reputation influences media coverage.

A frequently used test to detect bilateral causality is the Granger causality test (Granger, 1980). Hertog and Fan (1995) used the Granger causality test in their study, which was briefly described in the previous subsection. The application of the Granger causality test in the study by Hertog and Fan showed that the prediction was unidirectional – from media coverage to public beliefs, and not from public beliefs to media coverage.

Essentially, the Granger test says that if predictions of corporate reputation based on past reputation can be improved by past media coverage, then the press can be considered to be a cause of reputation. Note that the autoregressive models tested in the previous subsections belong to this part of the Granger causality test. The media coverage was analyzed in the period before corporate reputation was measured, and past reputation was included in the models. The Granger causality test is also run in the reverse to see if the prediction of media influence based on past media influence can be improved by past reputation. These analyses were performed for several of the pooled models in this study (see Appendix M). In the present study, various media variables (such as the amount of television news, the amount of print news, the direction of success & failure television news, and the direction of success & failure print news) were entered in the models as the dependent variables, while the rest of the media variables and corporate reputation were entered in the models as independent variables. The results show that for the amount of news and the tenor of the news, the causality is unidirectional – from media coverage to corporate reputation, and not from corporate reputation to media coverage.

In the case of issue news, a bidirectional effect was found and the salience of an issue influenced the amount of news about that issue in the next year (see Appendix M). This is in agreement with the results by Trumbo (1995), who also found a simultaneous relationship. Trumbo argued that if public concern is shaped by media coverage, the effects take place in a shorter time span than the one measured in his study, the month. This may explain why a bi-directional effect was found in the present study, which used a time span of a year. Hertog and Fan (1995) assumed that the persuasive impact of news declines over time in an exponential manner and employed a half-life of one day, and indeed found a unidirectional relationship from media coverage to public opinion. This confirms Trumbo's suggestion that news shapes public concern in a time span shorter than one month.

It should be remarked that the causality test is very sensitive to the number of lags used in the analysis. It is therefore recommended to use more rather than fewer lags (Gujarati, 1995). Since the overall reputation variable was measured at two points in time, it was possible to specify one media lag only for the amount of news and the tenor of the news. In the case of issue news, it was possible to specify two media lags, since the salience of an issue was measured at three points in time.

6.3 DID WE MAKE PROGRESS AND WHAT LIES AHEAD?

In this book, several theoretical perspectives are taken to study the effects of media coverage and advertising intensity on reputation. In the subsections below, these perspectives are discussed briefly and suggestions are made for further research.

Studying the effects of the amount of television news and the amount of print news
In this study Zaller's (1992) model of attitude change and Cacioppo and Petty's (1979) two-stage attitude-modification process were used as a theoretical framework to study the effects of the amount of media coverage. The thought has been raised that the level of "organizational awareness" of newspaper readers and television viewers may be different. Reasoned from Zaller's model, it follows that repeating a message if the level of awareness is low (as is the case for television viewers) leads to a larger probability of attitude change. In the case of a high level of organizational awareness (newspaper readers), repeating the message frequently leads first to a more positive and then to a more negative attitude towards the organization, since after a while repeated exposure leads to counterargumentation.

Although these ideas are supported empirically, future research is needed with regard to the following three issues. The first issue is whether it is the *medium user* or the *medium format* that determines the effect of the amount of television and the amount of print news. It could be empirically tested to see if television viewers really do possess a lower level of organizational awareness than newspaper readers. Many of the respondents subscribe to a newspaper and they watch the NOS or RTL 4 news. This raises the question as to whether it is not the medium *user* but the medium *format* that determines the effects of the amount of media coverage. De Vreese (2003) mentioned that newspapers provide more political than economic news, and that studies suggest newspapers are processed more intensively. Moreover, DeFleur, Davenport, Cronin, and DeFleur (1992) demonstrated that news stories presented by newspaper or computer screen were recalled better than were facts from the same stories when presented via television or radio. This fits with the idea that print news may evoke counterargumentation sooner than television news.

A second issue that needs more theoretical underpinnings is the application of a theory from social psychology (the two-stage attitude-modification process) to a study on media effects. In the studies by Cacioppo and Petty (1979; 1985), persuasive messages – recommendations – were used. For example, Cacioppo and Petty (1979) rated the agreement with the recommendation that university expenditures be increased. Cacioppo and Petty (1985) also rated the agreement with the recommendation that all seniors should be required to pass a comprehensive exam in their major area of study prior to being allowed to graduate. However, Chapter 4 of this book shows that companies are not always "recommended" in the news, and they receive failure and criticism in the news as well. As demonstrated by Oegema, Meijer, and Kleinnijenhuis (2000), even slightly negative news can improve reputation. The explanation we gave was that readers and viewers get used to the critical tone of the news and compensate for this. This is only the case up to a certain point – if the news is very negative, more media coverage will worsen the reputation.

The distinction in different types of news

Valence models of media effects do not always adequately explain the change in public opinion (Shah et al., 2002). Therefore, an attempt was made in the present study to move beyond a study of the media effects on reputation, which is based solely on a rough division of positive, neutral, and negative news. An attempt was made to preserve the information in the texts and to build a theoretical framework for the effects of various types of news instead. Three different types of news were distinguished in the present study: success & failure news, support & criticism news, and issue news. The study of the effects of issue news on corporate reputation may contribute to one of the directions of research advocated by Wartick and Heugens (2003). They suggested a better integration of issues management research with the growing literature on the management of corporate reputation.

In the field of political communication, however, the emphasis has been on issue news in the past 20 years of empirical research on the agenda-setting process (McCombs, 1992). This was surprising to McCombs (1992), while in the widely cited article by McCombs and Shaw (1972) it has already been mentioned that a considerable amount of campaign and other news was *not* devoted to discussion of the major political issues but rather to analysis of the campaign itself. Although this study focused on business news, it was likely that news about issues would not be the only type of news. Therefore, the effects of two other types of news were focused upon as well.

Predicting the effects of success & failure news

The present study postulated that success & failure news will lead mainly to a bandwagon effect. Only a small minority will change to the side of the looser (Kleinnijenhuis, 2004). Future research may try to formulate and predict a priori under what conditions success & failure news will lead to an underdog effect, and under what conditions success & failure news will lead to a bandwagon effect. Or to put it in the context of the stock exchange, under which conditions will stockbrokers follow the news (buy shares when there is positive news) and when will they oppose it (buy shares when there is failure news). It should be mentioned that this type of research is very complicated, and until now we have only succeeded in using the theoretical framework to interpret the results, not to predict them a priori (Kleinnijenhuis, 2004).

Predicting the effects of support & criticism news

It was expected in the present study that support & criticism news would lead to a boomerang effect if the source of criticism was perceived as partial. For future studies on support & criticism news, the same would apply with regard to what has been said about success & failure news in the previous section. Future research could try to formulate and predict a priori under which conditions support

& criticism news will lead to an effect regardless of the source, and under which conditions the reputation of the source of criticism will be taken into account.

The congruity principle of Osgood and Tannenbaum (1955) may be helpful in this regard, as is addressed in subsection 2.5. Future research on the effects of support & criticism news could be focused more explicitly on the role of the source (or in NET-concepts, of the "subject") that criticizes or supports the organization. For example, it can be empirically tested whether criticism of competitors has a different effect on corporate reputation than criticism by the employees. One difficulty may be the operationalization of the reputation of the source. The concept of support & criticism news is used most often in the analysis of news about election campaigns (Kleinnijenhuis et al., 2003; Kleinnijenhuis et al., 1998; Kleinnijenhuis et al., 1995). During the elections it may be less complicated to determine who is the enemy (the other political parties) and who is supposed to be a friend (politicians of the same party). A headline in which the political leader of the conservative party calls the leader of the labor party a liar is relatively easy to code. The leader of the conservative party is in this example the "enemy" of the labor party, and the labor party is criticized by the enemy. These types of headlines hardly ever occur in business news. The reputation of the source may be context-dependent. For example, BP may be a "traditional friend" of Shell when the government wants to introduce a higher tax on gasoline, but BP might be seen as Shell's "enemy" when they are competing for the right to drill in a certain oil field.

Standardization of the calculation of intercoder reliability

In the period 1991 to 1995, nearly three quarters (72%) of content analysis studies reported intercoder reliability assessment (Riffe & Freitag, 1997). In a more recent study, Lombard, Snyder-Duch, and Campanella Bracken (2002) found that even fewer content analysis studies reported intercoder reliability. Lombard et al. (2002) analyzed all articles indexed in communication abstracts for the years 1994 through 1998. Only 69% of the studies contained any report of intercoder reliability.

Although intercoder reliability was assessed in the present study, its calculation was complicated. This was mainly due to the fact that in the case of the NET-method, the coders rather than the researcher determined the number of assertions per headline. As a consequence, the number of assertions per headline may differ per coder. In order to assess reliability, the subject-object combinations distinguished by different coders have to be compared with each other, although these coders may not have distinguished the same number of assertions. Despite the fact that missing data may be a fairly common problem, no solution to this is provided by any of the computer programs available for the assessment of the reliability in content analysis studies, such as PRAM 0.4.5 (Skymeg Software, 2002; Neuendorf, 2002).

Elaboration on the issue of the calculation of intercoder reliability in future content analysis studies is strongly recommended. This point corresponds with the recent call for standardization in content analysis reliability by Lombard, Snyder-Duch, and Campanella Bracken (2004). At the present time, a new software program with the preliminary title of *iNET* (Van Atteveldt, in press) is being developed for coding according to the NET-method, as mentioned in Chapter 3. It would be advisable to make the calculation of intercoder reliability a standard component of the program, especially since there are few automated tools for calculating intercoder reliability, and this is usually calculated by hand (Lombard et al., 2002). This intercoder reliability measure should be controlled for chance agreement (Krippendorff, 1986; Neuendorf, 2002; Riffe, Lacy, & Fico, 1998).

Elaboration and application of the concepts of first- and second level agenda-setting
At the first level of agenda-setting, more news about a certain company leads to a higher public awareness of that company. At the second level of agenda-setting, news about particular attributes of a firm is positively related to the proportion of the public that defines the firm by those attributes (Carroll & McCombs, 2003). As suggested by Carroll and McCombs (2003), these theoretical ideas that stem from political communication studies fit equally well into the study of the effects of business news on reputation. The results of the present study show that the media influence the salience of the associations with a certain company if the daily issues (such as gas stations in the case of oil companies) are excluded from the analyses.

A theoretical question for future studies could be whether there are also intermediate forms of the first and second levels of agenda-setting. For example, if there is news about an issue that is *not* related to a particular company (such as the high salaries of CEOs) and as a result the public defines a certain firm by those issues, would this be considered to be a first level or a second level of the agenda-setting effect? Moreover, future studies on the second level of agenda-setting could take into account the effects of advertising on the salience of an issue. It can be questioned whether "The proportion of the public that defines a firm by a certain set of attributes depends on advertising devoted to those attributes." The experimental study of MacKenzie (1986) that was conducted in this respect may provide some starting points.

Framing business news
A scholarly discussion is going on about the difference between the concepts "second level agenda-setting" and "framing" (Kiousis, Bantimaroudis, & Hyun, 1999; McCombs, Lopez-Escobar & Llamas, 2000; McCombs, Shaw, & Weaver, 1997; Scheufele, 1999; 2000; De Vreese, 2003). The framing concept has been defined in two ways, as "central theme" versus "aspects" (McCombs et al., 2000). In the present study, the perspective of the second level of agenda-setting approach was used, because this study did not fit with the central theme definition

of framing. As demonstrated in Chapter 3 about the NET-method, it is important to recognize that an article can address several types of news. Therefore, the sentence in said article should be regarded as the coding unit rather than the whole article. This contrasts with the theory of framing, when framing is perceived as a central organizing idea.

Another reason why the concept of agenda-setting (and the second level in particular) was used in the present study is because it stresses the effects of media coverage on public opinion (McCombs et al., 1997; McCombs et al., 2000). In some publications about framing, framing refers only to the way information is presented and not to its effects on public opinion (Entman, 1991; Knight, 1999; Ruth & York, 2004).

Future framing research may question whether it is possible for a media message to consist of two or more different frames. Is an article with a headline like "Mamma's boy leads BP Amoco" an economic consequences frame, a conflict frame, or a human interest frame (see Valkenburg, Semetko, & De Vreese 1999). Or should different frames be applied to analyze business news as is done when analyzing political news?

Priming in advertising and in media coverage

The relationship between priming and agenda-setting is also a topic of discussion (McCombs et al., 2000; Scheufele, 2000; De Vreese, 2003). In this book, priming is seen as a consequence of the first level of agenda-setting (Iyengar & Simon, 1997; McCombs et al., 2000). Priming theory postulates that a certain issue will be *more important* in the formation of the overall evaluation, the more that issue is in the news.

This means that in agenda-setting the amount of news about issues is the independent variable and issue salience is the dependent variable. In the case of priming, the independent variables are "the news about an issue" and "the performance on that issue." The overall evaluation (in this case, of a company) is the dependent variable. The coefficient of the interaction term between the news about an issue and the performance on that issue tests the priming effect. This is in agreement with the conceptualization of priming by Iyengar and Kinder (1987).

As mentioned in Chapter 2, the application of feature priming in consumer research comes close to the priming concept as used by Iyengar and Kinder (1987). It could be interesting to study the joint effects of priming in advertising and media coverage on the reputation of an organization. In order to study priming effects in advertising, the content of the ad or the commercial should be taken into account. This will be elaborated upon in the next subsection.

Applying issue-ownership theory

In this study, issue-ownership theory (Budge & Farlie, 1983; Petrocik, 1996) was used to examine the effects of issue news. As mentioned in the subsection about the theoretical framework of the present study, issue-ownership theory postulates that the reputation of an organization will be improved if the organization is in the news with "owned issues." This theory was tested in two steps, since it was not measured whether or not certain issues were owned issues of the organizations or sectors. Consequently, future research needs to be tested in one step. In order to do so, a public opinion poll should be conducted to determine a company's owned issues.

In order to determine the list of owned issues, one should consider if the same procedure can be used as in political communication. In the study by Petrocik (1996), respondents were asked if the Democrats or the Republicans would be more successful in handling a certain issue. Imagine that in this study respondents were asked if Shell or BP was doing better concerning the environment, and that BP came out as handling the issue of the environment better than Shell. That would bypass the possibility that a third party like Greenpeace is the "real issue owner" of the environment. If subsequently BP and Greenpeace were in the news with the issue environment, this could damage the reputation of BP, since Greenpeace was the "real issue owner" of the environment and not BP.

In other words, when it comes to "owned issues," in political communication, the main competitor of a political party is probably the other political party. Since the "product" of political parties is their ideology, they are expected to profile themselves in different areas. Aside from other political parties, they probably have no competitors that can steal their issue. In the case of companies, the main competitor of a company is not always another company, but can also be a non-governmental organization (NGO) such as Greenpeace. This is rather likely to occur, since the aim of commercial companies is primarily to maximize profit for companies, and therefore companies may be looked at more suspiciously if they profile themselves on issues other than economic ones. If the respondents are asked only if company A or company B is better at handling a certain issue, the possibility of another actor being the "real issue owner" is ignored. This could be solved by simply posing the question differently. Instead of asking if company A or B is better at handling a certain issue, the respondent could be asked what he or she thinks of company A's ability to handle a certain issue.

Studying the effects of advertising in a longitudinal design

Studies that examine the effects of advertising on corporate reputation in a real-life situation are rare. In the study of Fombrun and Shanley (1990) and in the present study, advertising expenditures were used to operationalize advertising intensity. It would be interesting to elaborate on studies in this field and, for example, to use the perspective of classical conditioning described in Chapter 2 as a starting point. As remarked in the previous subsection, it is important to take the content of the ad or the commercial into account. Priming and classical conditioning can only serve as a framework to study the effects of advertising on reputation if the content of the ads and the commercials are unraveled.

Moreover, this study showed that for roughly half of the organizations, television commercials positively influenced the reputation of the organization, while for three organizations the television commercials did not have an impact on reputation. Content analysis of the commercials could provide insights into why this difference occurs.

6.4 A REFLECTION ON THE MEDIA MONITOR[2]

The present study reveals that if communication managers want to gain insights into how media coverage influences the reputation of their companies, it is not enough to focus only on the amount and the valence of the news (see also Shah et al., 2002). Instead, they should also focus on the other actors who support or criticize their company in the news, on the issues with which their company is in the news, and whether their company is attributed success or failure in the news. Each organization has a unique media profile that requires a tailor-made press policy. This will be illustrated by focusing briefly on the media profiles of some of the focal companies and the effects of these profiles on reputation.

Although the Brent Spar affair (see Appendix A) took place four years before Shell's reputation was measured for the present study, the aftermath of this still seemed to be present. Respondents were more negative about Shell when they associated the oil company with the environment than when they associated Shell with the general association "gas stations." This was not the case for Shell's competitor BP, where the type of association (either gas station or environment) did not influence BP's reputation. It follows from issue-ownership theory that Shell should not seek to profile itself in the media with the environment, since this will increase the salience of this issue and hence also the respondents' associations with the bad reputation of the company on this issue.

Supermarket leader Albert Heijn received the most severe criticism (although still moderate) of all of the focal companies because of its battle with French diet guru Michel Montignac. The publisher of the Montignac books tried to protect its interest by threatening to initiate proceedings against Albert Heijn if the supermarket continued to use labels for its products stating they were "suitable for the Montignac method." People chose the side of the supermarket, and increased criticism of the diet guru enhanced the reputation of Albert Heijn. This illustrates that communication teams do not always have to panic if their company is criticized in the media. It is important not only to study the valence of the news, but also to take into account which actor is giving the criticism or support.

The NS was in the news with the disruption of train traffic and the issues "delays" and "strikes." Some headlines no longer even mentioned *what* was wrong but simply *that* there was something wrong: "Chaos in the NS." The financial success of the NS received considerably less media attention. It may not come as a surprise that the NS has the worst reputation of the ten focal companies studied. The first association respondents had with the NS was not the general association "transport," but the association "delays." This shows that the most important aspect of communication policy is the actual performance of the organization in its core business, one that is hard for a communication team to manage. Nevertheless, it can be wondered whether people react differently to delays when they travel by car and when they travel by train.

Amsterdam Airport Schiphol had to contend with the opposite situation. The increase in the number of passengers was the basis of the success news about Schiphol, but also led to news about the claims against the airport as a result of noise nuisance. Television viewers appreciated the economic success of the airport, while the claims about noise nuisance worsened the airport's reputation. Nevertheless, Schiphol has a good reputation compared with the rest of the focal companies. Most respondents associated the airport with Royal Dutch airline KLM and with holiday trips; environmental pollution ranked third and fourth together with the association "Dutch economy." Like Shell, Schiphol could take measures to improve its environmental policy, but would do better to profile itself in the media with another issue, such as its contribution to the Dutch economy.

The media profile of the Dutch police shows they are successful at solving crimes. Of all the sectors and companies, the police were most in the news with their successes, which improved the reputation of the police. Remarkably, the police also received a great deal of criticism by the media that harmed their reputation, since the media are perceived by the public as reliable sources. The reputation of the police is the second worst of all the focal companies and sectors. This raises the following questions about the press policy of the police: Why are the media skeptical about the police in spite of their successes? Do the journalists know whom to call when the police are having difficulties, or is their press policy mainly oriented towards the distribution of press releases in the case of a success? These questions lead to the remaining challenges for the implementation of a media monitor.

The challenges
A media monitor based on the insights of theories from communication science can contribute to the professionalization the communication policy of organizations. Although executive managers recognize the strategic importance of communication, the daily routine of communication employees consists of practical manual work (Van Ruler, 2000). Several communication professionals remarked that some challenges remain with regard to the implementation of the media monitor.

In her reaction to the usefulness of the media monitor in daily communica tion practice, Smits (2000) argued that although the production of the news is an important factor in determining an organization's media profile, it is hard for communication managers to control. The production of the news content depends not only on the actual event, but also on coincidence, such as the mood of the individual journalist and the presence of pictures (Smits, 2000). The present study postulates that news selection factors, the socio-demographic background of journalists, and the possibilities of journalists to communicate with the news source do indeed influence the production of news (Galtung & Ruge, 1965; Harcup & O'Neill, 2001; Kleinnijenhuis, 1990; De Vreese, 2003; Zoch & VanSlyke Turk, 1998). Whereas in the present study the focus was on the effects of news on reputation, future studies may aim to take into account the effects of business press policy on the news and news production.

Tiemeijer (2000) raised another challenge, stating that the complexity of the media monitor limits its practical usefulness. He stresses the motto that "time is money" in industry and the government. Documents that are longer than two pages should be accompanied by a summary, and presentations should take half an hour at the most. However, the present study shows that sophisticated theories and methods are necessary in order to examine the effects of news on reputation. Each of the various types of news can have a different impact on the receivers. Therefore, a thorough press analysis will never be so simple that two or three key figures will do. Moreover, new research techniques require time and effort from organizations as well.

For example, consider the development of public opinion research, which has been extensively described by Price (1992). Price indicated that the concept of public opinion is closely linked to liberal political philosophies of the late 17th and 18th centuries. During the 1930s, public opinion research got a boost due to two phenomena: the growth of psychological measurement and the application of scientific sampling research. It took ten years before survey research centers had been established in universities, governement bureaus, and private industry (Price, 1992). From then until the present day, knowledge about public opinion research has been exchanged, and public opinion polls have been widely applied in both academic and applied settings. At the moment, opportunities for monitoring media coverage are optimal due to three developments: the professionalization of content analysis, the increasing knowledge about the effects of media coverage on public opinion, and the opportunity to combine organizations' media profiles with panel surveys.

It has been stressed by other communication professionals that corporate reputation is determined by more factors than media coverage alone (for example, paid publicity) (Cramwinckel, 2000; Heijting & Scheeringa, 2000). Because of this, there are events that will always be news (including such events as an airplane crash, or mass lay-offs). Heijting and Scheeringa (2002) suggested that no media monitor is resistant to this type of news.

It is acknowledged that corporate reputation depends on more factors than media coverage alone, as mentioned earlier. The present study controlled for paid publicity, and it is recommended that product usage be taken into account in future studies. The fact that future research will improve the media monitor does not make the present media monitor useless. On the contrary: in our opinion, a media monitor can fulfill the same role in shaping corporate communication policy as a financial annual report does in shaping financial policy. In the same way a financial report gives an overview of the financial situation, the media monitor gives an overview of the media coverage about the organization. A financial annual report cannot always prevent losses, and a media monitor cannot always prevent "bad news."

A financial annual report is indispensable, however, because it gives the management the opportunity to take specific measures with regard to the strong parts – or perhaps even more so, the weak parts – of a company. For communications professionals, a media monitor provides insights into which issues are crucial for the organization, which stakeholders support or criticize the organization in the media, and how successfully the organization is depicted in the media. This information enables communication professionals to evaluate, and if necessary adjust, corporate communication policy.

Appendices

Shell and BP (oil companies)

Shell (Royal Dutch/Shell Group of Companies) grew out of an alliance made in 1907 between Royal Dutch Petroleum Company and the "Shell" Transport and Trading Company, plc, in which the two companies agreed to merge their interests on a 60:40 basis, while keeping their separate identities. On October 28 of this year Shell announced that the two parent companies would merge into a single company, to be known as "Royal Dutch Shell plc," with headquarters in the Netherlands. Shell is mainly known as a supplier of fuel and lubrication products for local filling stations, and as a company that carries out exploration on land and at sea for the discovery and production of oil. However, the organization is also engaged in the production of chemical products, and worldwide exploration for and production of natural gas. Shell operates in over 145 countries and employs more than 119,000 people (Shell, 2004). In 2002, the Royal Dutch/Shell group was ranked fourth on *Fortune* magazine's list of world's largest corporations, with revenues of 179.4 billion dollars and profits of 9.4 billion dollars ("Global 500: The world's largest corporations," 2003, F1).

A few years before the research period of this study, Shell received scientific attention in the communication field due to the two crises it faced in 1995: the "Brent Spar case," which will be described first, and "Nigeria." In 1995, Shell received a lot of media attention because of its decision to sink the Brent Spar – an oil storage and loading buoy – in the sea. Greenpeace campaigned against this idea. After a consumer boycott in Germany, Shell decided to dismantle the oil buoy instead of sinking it. Oegema, de Haan, and van Leur (1998) investigated the media coverage during the "Brent Spar period" from May 1995 to the end of June 1995. They found that media about Shell during this period was extremely negative (the quality was −.76, on a scale ranging from −1 to +1), and that the wording of the newspapers was suggestive: *dumping* the Brent Spar instead of *sinking* the Brent Spar. Greenpeace is clearly the issue owner of the issue "environment," while Shell has a bad reputation on this issue (Oegema et al.,

1998). This suggests that the more Shell is in the news with the issue environment, the worse its reputation will be. In the aftermath of the news about the Brent Spar, it appeared from independent research that the Brent Spar was not as big a source of pollution as Greenpeace had claimed (containing 75 tons of oil and sludge, rather than 5,500 tons). Although this sheds another light on the issue, it barely got any media attention, since the media were busy covering Greenpeace's next battle, this time against French nuclear testing in Mururoa (Oegema et al. 1998).

Van den Bosch and Van Riel (1998) examined the Brent Spar case from the perspective of two different strategies: a buffering and a bridging strategy.[1] Livesey (2001) studied the role of wording (language games) during the "Brent Spar crisis." Shell was talking about the "Best practicable environmental option" (the rational approach), while Greenpeace referred to the Brent Spar as a "Toxic Time Bomb" in which the Brent Spar was turned into a symbol of man's misuse of the oceans, irrespective of the reality.

Livesey (2001) also focused on the role of wording of another crisis of Shell in 1995, namely Shell's operations in Ogoniland, Nigeria. Shell was accused of profiteering with its dealings with the ruling military junta in Nigeria that had taken power by force, and of forsaking the rights of the Ogoni people to a fair share of the oil profits that came from their lands (Fombrun & Rindova, 2000). In November 1995, the Nigerian government executed nine Ogoni environmentalists, among them Ken Saro-Wiwa, an internationally acclaimed journalist and the outspoken representative of the Ogoni people. Shell was blamed for doing too little to intervene. Fombrun and Rindova (2000) described reputation management at Royal Dutch/Shell after the two crises. They concluded that practicing reputation management demands that stakeholders' expectations be monitored routinely to ensure that performance results are not jeopardized by shifting expectations. In addition, they stress that the management of organizational reputation is inextricably linked to the management of organizational identity (Fombrun & Rindova, 2000).

BP emanates from the merger of British Petroleum, Amoco, Arco, and Castrol. In the summer of 2000, it announced it would operate under one worldwide brand. Like Shell, BP supplies fuel and lubrication products to local filling stations, and is also engaged in the production of chemicals and worldwide exploration for and production of natural gas. BP operates in 100 countries and employs approximately 103,700 people (BP, 2004). In 2002, BP was fifth on *Fortune* magazine's Global 500 list of the largest corporations, with revenues of 178.7 billion dollars and profits of 6.8 billion dollars ("Global 500: The world's largest corporations," 2003, F1). BP does not seem to get much attention from scholars in the field of communication science compared to Shell, despite the fact that BP was only one rank lower than Shell[2] in 2002. Nevertheless, BP was the object of case studies because of its alliance with Mobil[3] (Robson & Dunk, 1999) and its merger with Amoco (Salama, Holland, & Vinten, 2003).

ABN AMRO and Rabobank (banks)

ABN AMRO emanates from a merger between Algemene Bank Nederland (ABN) and the Amsterdam-Rotterdam (AMRO) Bank. In 1991, these two banks merged as ABN AMRO Bank. The bank has over 3,000 branches in more than 60 countries, and a staff of over 107,000 full-time employees. ABN AMRO is listed on the Euronext, London, and New York stock exchanges. There are three main business units: consumer & commercial clients,[4] private clients & asset management,[5] and wholesale clients.[6] The consumer & commercial clients business units serve approximately 15 million clients, mainly through their presence in the Midwestern United States, the Netherlands and Brazil (ABN AMRO, 2004).

With revenues of 34.6 billion American dollars and profits of 2.1 billion dollars in 2002, ABN AMRO ranked one-hundred and ninth on *Fortune* magazine's Global 500 list of the largest corporations ("Global 500: The world's largest corporations," 2003, F3) and third on *Fortune*'s list of the largest Dutch corporations ("Global 500: The world's largest corporations," 2003, F35).

The Rabobank group emanates from a merger in 1972 between two central banks: the "Coöperatieve Centrale Raiffeisen-Bank" and the "Coöperatieve Centrale Boerenleenbank."[7] The Rabobank group consists of 321 independent local cooperative "Rabobanks" in the Netherlands, together with specialized subsidiary organizations (such as Interpolis, Robeco Groep, De Lage Landen) and "cooperative" Rabobank Nederland.[8] Due to its cooperative structure, the Rabobank Group is not listed on the stock exchange. Despite the fact that outside the Netherlands the Rabobank Group is represented in 35 countries, the Netherlands is its main market. Therefore, the Rabobank Group is market leader in the Netherlands in most of the areas of financial services: mortgages (26%), savings (38%), small and medium-sized businesses (39%), and the agricultural sector (85%). The Rabobank Group serves 9 million clients (Rabobank, 2004).

In *Fortune*'s list of the largest Dutch corporations, Rabobank occupies the seventh place, with revenues in 2002 of 20.5 billion dollars ("Global 500: The world's largest corporations," 2003, F35).

Albert Heijn and Super de Boer (supermarkets)

Albert Heijn Senior started his first supermarket in 1887 in the small Dutch town of Oostzaan. Nowadays there are approximately 700 Albert Heijn supermarkets; Albert Heijn is market leader in the supermarket industry. According to the ranking list of research agency GFK (2002), Albert Heijn has the most customers. Approximately 75% of Dutch households bought something at Albert Heijn at least once in 2002.

In 1973, parent company Ahold was established, with Albert Heijn as the most important subsidiary in the Netherlands. Dutch drugstore chain ETOS was one of Ahold's first specialty shops. In the 1990s, Ahold gained a presence in the US by the acquisition of supermarket chain Stop & Shop, Red Food Stores, and Mayfair (Ahold, 2004). As mentioned in the introductory chapter, in 2003 Ahold was in the news as a result of accounting scandals.

Super de Boer has 374 supermarkets in the Netherlands. Super de Boer, Konmar (134 supermarkets), and Edah (263 supermarkets) are part of parent company Laurus. Laurus emanates from the merger between De Boer Unigro (Super de Boer) and Vendex Food (Edah, Konmar) in 1998. Super de Boer ranks fourth in GFK's ranking list (2002). Approximately 39% of the Dutch households bought something at least once at Super de Boer in 2002.[9] In April 2001, after the research period of this study, Laurus was in the news frequently with its plan to rename and restyle supermarkets Super de Boer and Edah into more luxurious Konmar supermarkets. The plan failed, and an interim manager had to be appointed to restore the damage (de Volkskrant, September 25, 2001, E3).

The Dutch Railways (NS) and Schiphol airport (transport companies)

The N.V. Nederlandsche Spoorwegen was founded in 1937 by the clustering of the two large rail companies "SS" and "HSM" (NS, 2004). The NS has a monopoly position in rail transport of passengers in the Netherlands. It gained its independence 1995, but the shares are owned by the state. The government is responsible for the maintenance of the railway network, while the NS takes care of the transportation operation. Lovers Rail tried to compete with the NS in the transport of passengers[10] in the period from 1996 to 1999, but gave up in the end (see for example, "Lovers Rail gives up" in Trouw, September 15, 1999).

Every day the NS transports more than 1 million passengers and provides work to its approximately 23,000 employees (based on FTEs). Two business units take care of the domestic passenger transport: NS Travellers and supply company NedTrain. The latter takes care of the management and maintenance of the trains. The other business units are NS Commerce (development of the commercial policy of the NS), NS International (international passenger transport), NS Stations (management and operation of the stations), and NS Property (development around the stations). The revenues of the NS came to 2.3 billion euros in 2002, while the profits of the NS amounted to 47 million euros in 2002 (NS, 2002).

Amsterdam Airport Schiphol is the biggest airport in the Netherlands and has the ambition of becoming one of Europe's main airports (NRC Handelsblad,

November 7, 2003, p. 2). Amsterdam Airport Schiphol is part of the Schiphol Group. The state owns most of the shares (75.8%) of the Schiphol Group, followed by the city of Amsterdam (21.8%), and the city of Rotterdam (2.4%). In addition to Amsterdam Airport Schiphol, the Schiphol Group also owns Rotterdam Airport and Lelystad Airport, and owns 51% of the shares in Eindhoven Airport. The revenues for 2002 came to 774 million euros, and the net profit amounted to 141 million euros. Schiphol Group employed 2,100 employees in 2002 on a full-time basis (Schiphol Group, 2003).

Chief Executive Officer Cerfontaine wants to privatize Schiphol and list it on the stock exchange as soon as possible (see for example NRC *Handelsblad*, October 2, 2003, p. 11, and de Volkskrant, April 4, 2002, p. E3).

Police

The Dutch police are made up of 25 regional police corps and the Netherlands National Police Agency. The 25 police corps contribute to safety, a livable climate, and combating crime in their respective parts of the Netherlands. The Netherlands National Police Agency organizes the national police tasks carried out by services such as the traffic police, the river and harbor police, the national criminal investigation department, and the police on horseback (the mounted police). The Minister of the Interior is responsible for the Dutch police as a whole. Because the Dutch police are made up of 25 police corps, one of the mayors in the region (most often the mayor of the largest city) takes the role of corps commander (Police, 2003). In 2002, the police employed 52,452 employees including executive staff and administrative and technical staff. Police costs in 2002 were approximately 3.2 billion euros (Police, 2004).

The regional structure of the police is organized in the "Police Act" of 1993. The coalition agreement states that the structure of the Dutch police will be evaluated in 2005. A discussion is going on as to whether the police should be organized nationally rather than in regions (see for example "National police should not be established", NRC *Handelsblad*, November 14, 2003, p. 7; "National police are not being discussed", de *Volkskrant*, November 8, 2003, p. 2; "Dutch police should be one organization", NRC *Handelsblad*, September 2, 1997, p. 7).

Agricultural sector

The agricultural sector employs approximately 258,169 persons (Centraal Bureau voor de Statistiek, 2003). Market gardening is the most important agricultural sector with regard to revenue. In 2001, the total production value of the market gardening sector increased to 6.8 billion euros, with 4.4 billion euros coming from greenhouse market gardening. Approximately 10,800 farms with greenhouses supplied work directly to 72,500 people (LTO, 2003). There were 41,266 cattle farms and 11,851 hog farms in the Netherlands in 2002 (Centraal Bureau voor de Statistiek, 2002). The cattle farms accommodated 1.49 million cows while the hog farms accommodated 11.65 million hogs (Centraal Bureau voor de Statistiek, 2002).

The agricultural sector is united in LTO Nederland (Land- en Tuinbouworgani-satie Nederland – the Dutch Organization for Agriculture and Horticulture). As an umbrella organization for five regional and sixteen sectoral organizations in agriculture and horticulture, the LTO has a particular focus on political activities (LTO, 2003).

APPENDIX B – KEY TERMS ENTERED

Box B1 lists the key terms entered to retrieve the articles from the FactLANE data-base. For reasons of efficiency, articles that contained the names of two of the focal companies or sectors (for example, in the headline "Shell and BP sell Texas subsidiary") were only selected once. In order to do so, Boolean operators were used. After coding headlines that contained the names of two of the focal organi-zations, they were taken into account for both of the focal organizations.

Box B1.   Key terms entered to retrieve the media coverage about the focal companies

Shell:	"Shell or BP"
BP:	"BP not Shell"
ABN AMRO:	"ABN AMRO or Rabobank"
Rabobank:	"Rabobank not ABN AMRO"
Albert Heijn:	"Albert Heijn or AH or Super de Boer"
Super de Boer:	"Super de Boer not (Albert Heijn or AH)"
Schiphol:	"Schiphol"
NS:	"Nederlandse Spoorwegen or NS"
Agricultural sector:	"farmer or farmers or market gardener or market gardeners or cattle breeder or cattle breeders or 'hog farmer' or hog farmers or BSE or swine fever not (Albanian farmer or Argentian farmers or American farmer or American farmer or Den Boer et cetera)"
in Dutch:	"boer or boeren or tuinder or tuinders or veehouder or veehouders or varkenshouder or varkenshouders or agrariër or agrariërs or BSE or varkens-pest not (Albanese boer or Argentijnse boeren or Amerikaanse boer or Amerikaanse boeren or Den Boer et cetera)."
Police:	"police not (Basque police or Irish police or Indonesian police or North Irish police or Palestinian police or Turkish police or Spanish police or police Spain or Algeria et cetera)"

Appendix C — Percentage of assertions (unweighted and weighted) per organization per medium

The media coverage was weighted in order to take into account the length of the article (or news item) and the page number of the article (or the time of broadcasting). The headlines of very short newspaper articles in the last pages of the newspaper were weighted less heavily than large articles on the front page of the newspaper. In the table below, the consequences of the weighting of the news are illustrated per organization per medium.

Before weighting the news, 66% came from the print media while 34% came from television news. After a weight factor was applied, nearly half of the news (45%) comes from the print media while a bit more than half (55%) of the news comes from television news. This is due to the fact that there were relatively few front-page newspaper articles about the focal organizations and there was a relatively large amount of 8:00 o'clock evening news. Since television news and print news are treated as separate variables, this difference in the ratio of television news to print news does not influence the testing of the hypotheses, however.

TABLE C1.　Weighted and unweighted media coverage per medium

	Unweighted							Total		Weighted							Total
	AD	NRC	Tel	Trouw	VK	NOS	RTL			AD	NRC	Tel	Trouw	VK	NOS	RTL	
Shell	15%	22%	9%	14%	14%	19%	8%	100%	Shell	10%	15%	6%	10%	12%	36%	11%	100%
BP	18%	38%	5%	16%	13%	7%	4%	100%	BP	15%	30%	4%	12%	13%	18%	8%	100%
Albert Heijn	19%	14%	8%	19%	20%	13%	6%	100%	Albert Heijn	16%	10%	8%	15%	17%	24%	9%	100%
Super de B	7%		29%	2%	2%	43%	17%	100%	Super de B	3%		17%	2%	2%	63%	14%	100%
NS	18%	13%	8%	13%	15%	22%	10%	100%	NS	13%	10%	6%	9%	11%	41%	10%	100%
Schiphol	16%	13%	11%	15%	13%	21%	10%	100%	Schiphol	11%	10%	8%	11%	10%	38%	11%	100%
ABN AMRO	13%	16%	15%	11%	17%	22%	6%	100%	ABN AMRO	9%	11%	13%	7%	13%	40%	7%	100%
Rabobank	11%	8%	22%	11%	9%	29%	10%	100%	Rabobank	7%	4%	19%	6%	7%	47%	11%	100%
Police	15%	10%	11%	13%	12%	28%	10%	100%	Police	9%	7%	8%	9%	7%	47%	13%	100%
Agriculture	13%	12%	9%	16%	14%	22%	14%	100%	Agr culture	8%	9%	7%	10%	10%	40%	15%	100%
Row %	15%	13%	11%	14%	14%	24%	10%	100%	Row %	10%	10%	8%	10%	10%	41%	12%	100%
N	2,410	2,116	1,722	2,197	2,147	3,746	1,545	15,883	N	1,597	1,521	1,317	1,511	1,586	6,501	1,851	15,884

Note. Due to rounding off, percentages may not add up to 100%.

Appendix D – Coding instructions[11]

This appendix will briefly discuss the coding instructions. Bearing in mind the length of this appendix, only the general part of the coding instructions will be focused upon. This means that the general rules for the use of meaning objects will be described first, followed by the general classification of the meaning objects. The general part of the coding instructions was drawn up to give coders an idea of how the meaning object lists per two companies were set up, and it was not used by the coders to do the actual coding. In the original coding instructions, this general part was followed by specific coding instructions per two organizations and per sector: Shell/BP, ABN AMRO/Rabobank, Albert Heijn/Super de Boer, Schiphol/NS, and the police and the agricultural sector.

The classification of the meaning objects is based on a "tree structure." A distinction can be made between *actors* and *issues*, and these are the two "main branches." The actors can be classified into different types of actors, such as companies, the government, the third/fourth and fifth powers, and other "people." Each of these types of actors is divided once again. For example, the third/fourth and fifth powers are divided into pressure groups (to which all the NGOs belong) and "other powers" (like a judge or consultancy firms). Meaning objects that belong to the same category were coded with the same first letter. For example, if an NGO was mentioned in the headline, the letter **g** (from the category "group") was coded in front of the NGO so that in the analysis, the meaning objects could be classified into the same group if necessary.

Issues can be divided into corporate issues, economic issues, mergers, societal issues, and issues like "image." These groups of issues were each divided into several meaning objects. In the analysis, these meaning objects were aggregated, if necessary, into larger or different groups in order to match the public opinion associations. See the column "media-issues" in the tables of Appendix B. Two fictive meaning objects were created, "reality" and "ideal." This was done in order to code success & failure news and assertions about the "media evaluation of an actor or issue," which was elaborated upon in Chapter 3.

General rules for the use of meaning objects

1. Do not use capital letters when formulating meaning objects.
2. With regard to the configuration, make sure the computer is adjusted to "England (United States)." This is important for a correct reflection of the value of the predicate (in other words, 0.5 instead of 0,5) and for a correct reflection of the data. Install the computer as follows: in the start menu press "settings," then click "configuration screen" and select England and the United States in "landinstellingen."

3. Do not add new meaning objects, or at least try to limit the addition of new meaning objects. The aim of this research is not to summarize the full content of a specific article but to reproduce the *evaluative meaning* of the article in terms of positive and negative media coverage concerning the main actors (Shell, BP, etc). Different groups may be grouped into one meaning object. For example four different ecology groups may be coded as "ecology group" instead of using a different core object for each of the different groups. The different groups may be treated as one, as long as we pay attention to the evaluative meaning of the text.

4. Try to code as specifically as possible. For example: Code "**g**ai" if Amnesty International is mentioned instead of ghumanrightorganisation or ggroup.

5. If a specific actor or issue is mentioned in an article and it is not mentioned in the list of meaning objects, please code the closest category at the higher level. If a mink-breeding organization is against supermarket Albert Heijn, then the closest core object for the mink-breeding organization is ggroup. If it is an organization against child abuse, the core object "**g**humanrightorganization" is closer than "ggroup."

6. In the case of very detailed headlines, it might be tempting to add new meaning objects. For example, in a headline that makes clear that Albert Heijn's profit is small for products with a reduced price, but large for the other products, it might be tempting to create two meaning objects "profit reduced price products" and "profit not reduced price products." The correct solution is to code the general meaning object, in this case cahprofit, and assign the value 1 for the question mark. So:
Reality / quality=+0.5; question mark=0.5 / cahprofit
This means that on average Albert Heijn makes a small amount of profit. The quality = +0.5 on a scale of [−1..+1], there is a certain variation however, question mark = 0.5 on a scale of [0..1].

7. When should new meaning objects be added? Unfortunately, the answer is abstract: if it becomes clear that the evaluative meaning of the text would otherwise be missed. For example, when there is a huge conflict developing between Shell in Germany and Shell in England, or between CEO X and CEO Y at supermarket Super de Boer. In that case, a subcategory in the main organization category should be created to distinguish the competing parties. If you do not know what to do, please do not hesitate to send an e-mail or call!

8. When a new meaning object is added, it must be reported at the next team meeting and be approved. All the coded text before the new meaning object was introduced must be recoded.

9. Always make sure the first letter of the meaning object is correct!

10. Prevent typing errors when naming the meaning objects; use suggestions by Excel.

Coding agreements

- In order to keep articles from being coded twice, when coding Shell please code the headlines about BP as well. If you are coding a BP file you do not have to code the headlines in which Shell is mentioned, because these headlines are coded by the "Shell coders." If you are coding an ABN AMRO file, please also code the headlines about Rabobank (when working in a Rabobank file, do not code headlines in which ABN AMRO is mentioned). The same applies to Albert Heijn (please also code the headlines about Super de Boer). When coding the headlines about Super de Boer, do not code the headlines about Albert Heijn. If you are coding Schiphol, code the NS as well, and when coding the NS, please do not code headlines in which Schiphol is mentioned.
- Please note that all meaning objects have a direction. In the case of the meaning object "cmaterial price," it is assumed that the material price is high. This means that if material prices have gone up, you code: Reality + cmaterial price.
- Change a passive headline into an active one in order to make sure the subject (the actor or issue where the "energy" comes from) and the object are proper placed. The headline: "Pete gets a cookie from John" is changed to "John gives a cookie to Pete." John fulfills the role of the subject, while Pete is the object.
- If a headline contains "will have to," assume that it is not happening yet.

Type of predicate:

FUS: MERGER OR TAKEover (that is or is not taking place). A relationship between two actors (this means an actor in the subject position and an actor in the object position), e.g., "Merger BP and Arco;" "BP no longer wants Rosneft."

AFF: AFFEctive relationship between two actors, e.g., "Schuitema-BP pact."

WIL: *actor* in the subject position and issue in the object position, e.g., "BP introduces a new smart pump."

CAU: *causal re*lationship, an issue in the subject and an issue in the object position, e.g., "Higher oil price increases Shell's profits."

TCH: *touch*, issue, or actor in the subject and actor in object position, e.g., "Crisis in South Asia hits Shell."

REA: *there* is no subject, and an actor or issue in the object position, e.g., "Profits of BP doubled."

EVA: *eval*uation of an actor or issue, e.g., "Shell policy praised."

RES: *used when* articles about the police are coded. Actor-issue or actor-actor news and at the same time also a success, e.g., the sentence "Police find 85 kilos of cocaine" is coded as follows:

Police (subject)/find (predicate) /RES (type)/-1/hcrime (object). And in the case of "police catch drug dealer": Police (subject) / catch (predicate) /RES (type)/-1/ hcriminals (object).

General classification of meaning objects

Actors
Companies
   Main actors:
- <u>sh</u>ell, <u>bp</u> ; <u>ab</u> (ABN/AMRO), <u>rb</u> (rabobank); <u>ah</u>, <u>sb</u> (Super de Boer) **ns, sc**
  (Schiphol) <u>pl</u> (police); <u>lt</u> (agriculture industry), ltmarketgardeners
  derivations of main actors:
- shshareholder etc ; shgermany etc.

<u>z</u> : zsamebranche
<u>u</u> : udifferentbranche

Politics
- <u>o</u>governement, ominpronk (Dutch minister Pronk), ominwijers (Dutch minister Wijers), ocouncil, oeu, oep, otk (members of parliament), opolparty,
  opvda, ocda, ovvd, od66
- **lcountryr**nl (government of the Netherlands), lcountryreu (government of
  one of the EU members), countryrvs, countryrnigeria, countryrrusland,
  countryriran, countryrchina, countryrother
- **lcountryb**nl (Dutch inhabitants/industry), lcountrybeu (inhabitants of one of
  the EU countries), lcountrybvs, lcountrybnigeria, lcountrybrusland,
  lcountrybiran, lcountrybchina, lcountrybother
- **lcountrym**nl (commercial interests in the Netherlands), lcountrymeu (pro-
  ject/future interests in one of the EU countries), lcountrymvs,
  lcountrymnigeria, lcountrymrusland, lcountrymiran, lcountrymchina,
  lcountrymother

Third, fourth, and fifth powers
<u>d</u> : djudge, dmedia, dzd (business-to-business), dconsultancyagency,
dadvertisingagency
g : ggroup (pressure groups, NGOs), ggreenpeace, gai, gecologygroup,
ghumanrightsorganization, gconsumersorganization

People
w : wemployees; *derivations* wshemployees, wbpemployees etc.
k : kcustomers (consumers, train passengers, etc); *derivations* kahcustomers etc.

Issues
<u>c</u> : *Corporate issues*
- cprofit, cmarketvalue, cturnover, cproductioncapacity etc ;
- *derivations:* cshprofit, cshmarketvalue ... cbpprofit, cbpmarketvalue, etc.

- cnewproduct, cnewproductmv (new socially responsible product such as solar energy or cleaner cars)
- cefficiency, creorganization, cdismissal
- ccodeofconduct
- cwagesemployees, cwagesmanagers (incl.options), cshlwagesemployees, cshwagesmanagers
- cadvertising
- csponsoring
- cinvestments

_e_ : _Economic issues_
- eworldeconomy, enleconomy, edollarexchangerate, eeuroexchangerate, epoundexchangerate, ebexhangerate (AEX, DowJones Index)
- etaxes
- eemployment, einflation, einterest rate
- ecapitalism, eprivatizing, eformationoftrustcartel
- econsumerloyalty, epurchasingpower

_f_: _Mergers_
fmerger

_m_ : _Societal issues_
menvironment, mhumanrights, mfraud, mcrime, mdiscrimination, mwomen, mpublic

_x_ : _Image, namedropping_
xnamedropping, ximage

_r_ : _Reality; i_ : _Ideal_

APPENDIX E – INTERCODER RELIABILITY

Separate training sessions were conducted for each pair of organizations, because each pair had its own coding instructions. Special training sessions were
held for the agricultural sector and the police, because separate coding instructions were drawn up for each sector. This resulted in six different types of training. Coders coded the news about at least two different organizations, although
the number of organizations coded per coder differed. The first training sessions
were held in November 2000, and the last training session took place in June
2001. The coding took place between January 2001 and August 2001.

For each pair of organizations and for each sector, a random selection of 10% of
the news was selected for reliability testing. This appendix will examine whether
the coders agreed on the following four topics: the number of assertions, the "subject" and "object" combination, the quality of the predicate, and the frequency of
the issues. The computations for arriving at the reliability figures were performed
by writing syntax files in SPSS.

The reliability of the number of assertions
The coders parsed each headline from the print news and each sentence from the
news broadcasts into as many assertions as they found necessary. Theoretically,
the number of assertions per headline could differ from 1 to infinity, although in
practice the minimum number of assertions was 1 and the maximum number of
assertions was 11. The following table presents the percentage of agreement between pairs of coders. The units of analysis to compute the agreement between
pairs of coders are the numbers of headlines coded by a specific pair of coders.
For each pair of coders, the number of units of analysis is enclosed in brackets.
There was a total of 18 pairs of coders.

TABLE E1.   **Agreement between the coders with regard to the number of assertions**

	Coder 1	Coder 2	Coder 3	Coder 4	Coder 5	Coder 6
Coder 2	100% (n = 9)					
Coder 3	81% (n = 153)	a				
Coder 4	83% (n = 536)	72% (n = 29)	83% (n = 150)			
Coder 5	86% (n = 439)	a	83% (n = 42)	83% (n = 63)		
Coder 6	85% (n = 692)	75% (n = 20)	75% (n = 4)	86% (n = 436)	88% (n = 257)	
Coder 7	82% (n = 203)	a	100% (n = 2)	81% (n = 200)	81% (n = 32)	89% (n = 203)

a Cannot be computed because the coders did not code the same content.

Table E1 demonstrates that the agreement between the coders about the number of assertions is high. If the number of assertions is taken into account with re- gard to the calculation of the overall percentage of agreement, this overall per- centage is 84%.

The reliability of the subject-object combination
After the sentences were parsed into assertions, the coders had to determine the subject and the object of each assertion. The assertions used to calculate inter- coder reliability with regard to the subject-object combination were selected in agreement with the analyses in order to test hypotheses 2a, 2b, 3a, and 3b. This meant that the assertions were selected only if they belonged to success & failure news or support & criticism news, and if they contained the focal organization in the object position.

The units of analysis for determining whether two coders agreed on the ques- tion of whether a specific subject-object-combination was mentioned in the head- line of an article are the headlines of the articles coded by these two coders. The percentage of agreement between coders was computed for each pair of coders and for each subject-object combination. Next, the overall agreement between all coders on all subject-object pairs was computed as the weighted average of these percentages of agreement, whereby each coefficient of agreement was weighted by the number of units of analysis on which a specific percentage of agreement was based, i.e., on the number of headlines coded by a given pair of coders.

The overall agreement of the subject-object combinations is 98%, $n_{\text{subject-object combina-tions}} = 54$, $n_{\text{sum of the number of headlines coded by the different pairs of coders}} = 1,107$. This overall agreement is largely based on the fairly trivial agreement that a subject-object combination is not present in most sentences. If most subject-object combinations are not present in most sentences, then one would expect a high agreement by chance that coders will indeed agree that a specific subject-object-combination will not be present in a given sentence. Therefore, a reliability coefficient that corrects for agreement by chance is called for. Since the presence of a subject-object-combination is a nomi- nal variable, Scott's Pi has been applied. The value of Scott's Pi $= .68$, $n_{\text{subject-object combina-tions}} = 54$, $n_{\text{sum of the number of headlines coded by the different pairs of coders}} = 1,107$. Lombard et al. (2002) indi- cated that coefficients of .70 are appropriate in exploratory studies, but that lower criteria can be used for indices known to be more conservative[12], such as Scott's Pi, Cohen's Kappa, and Krippendorff's Alpha. This means that the reliability of the subject-object combination is satisfactory.

The reliability of the quality
In order to assess the reliability of the quality (H. 2a, 2b, 3a, and 3b), assertions were selected that contained the same subject-object combination. The overall correlation is high, $r = .81$, $n = 648$ assertions. This means that if coders agree on the subject-object combination of an assertion, they will also agree on the direc- tion of the quality of the assertion.

The reliability of the frequency of issues

In order to test the reliability of the coding of issue news (H. 4, 5 and 6), the position of the issue in the assertion (subject or object) and the direction of issue news did not have to be taken into account, because the hypotheses focused on the effects of the frequency of issue news. Except for this difference, the procedure used to calculate the reliability of the frequency of issues was similar to the calculation of the reliability of the subject-object combinations. The headlines of the articles coded by these two coders were the units of analysis for determining whether two coders agreed on the question of whether a specific subject-object combination was mentioned in the headline of an article.

The overall agreement of the subject-object combinations is 98%, $n_{issues} = 23$, $n_{\text{sum of the number of headlines coded by the different pairs of coders}} = 1,376$. This overall agreement is largely based on the agreement that in most sentences an issue is not present. Because the presence of an issue is a nominal variable, Scott's Pi was applied to correct for chance agreement. The value of Scott's Pi $= .80$, $n_{issues} = 23$, $n_{\text{sum of the number of headlines coded by the different pairs of coders}} = 1,376$ which is a good level of reliability.

APPENDIX F — MATCHING MEDIA AND PUBLIC OPINION ISSUES, PLUS OWNED ISSUES

The tables below show how issues mentioned in the media in relation to the organizations (the first column of the tables) are matched with the associations of the respondents with a certain company (the second column of the tables). In contrast to other "issue-ownership studies," it is determined by the researcher whether an issue is an owned issue or not, as remarked in Chapter 3. The value "1" in the column "owned issue" means that an issue is thought to be successfully handled by the organization, the value "0" means that an issue is not thought to be handled either successfully or unsuccessfully by the organization, and the value "-1" means that an issue is thought to be handled unsuccessfully by the organization.

It should be mentioned that the media issue "reorganization" is categorized under the category "efficiency," while it might also have been categorized under the category "employment." In the case of the agricultural sector, "swine fever" and other animal diseases are not pre-listed items.

TABLE F1.   **Owned issues of Shell and BP**

Media issues	Associations of the respondents	Owned issue
Economic issues (such as investments, production capacity, and world economy)	Dutch economy	1
Material price	Materials and lack of materials	0
Advertising (+ issue image)	Advertising	0
BP/Shell return	Profitability	1
Efficiency (including reorganization)	Efficiency (and inefficiency)	0
Employment (and dismissal)	Employment	1
Mergers	Mergers and takeovers	0
Societal issues	Standards and values	0
Environment (new sustainable products)	Environment (pollution)	-1
Gas station shops	Mini-supermarkets in gas stations	1
Other: corporate issues, new product	Other: gas stations	

TABLE F2.   **Owned issues of ABN AMRO and Rabobank**

Media issues	Associations of the respondents	Owned issue
Economic issues (such as investments, production capacity, and world economy)	Dutch economy	1
Advertising, (in the case of Rabobank, the bicycle racing team which is sponsored by Rabobank)	Advertising	0
ABN AMRO/Rabo return	Profitability	1
Efficiency (including reorganization)	Efficiency (and inefficiency)	0
Employment (and dismissal)	Employment	1
Mergers	Mergers and takeovers	0
Societal issues	Standards and values	0
Environment, (new sustainable products)	Environment (pollution)	0
Other: Corporate issues, new product	Other: bank accounts, financial services	

TABLE F3.   **Owned issues of Albert Heijn and Super de Boer**

Media issues	Associations of the respondents	Owned issue
Economic issues (such as investments, production capacity, and world economy)	Dutch economy	1
Advertising (and image)	Advertising	0
Return	Profitability	1
Efficiency (including reorganization)	Efficiency (and inefficiency)	0
Employment (and dismissal)	Employment	1
Bonus Card	Price	1
Mergers	Mergers and takeovers	0
Societal issues	Standards and values	0
Environment ($n = 1$), new product (d)	Environment (pollution)	0
New product	Wide selection	1
Other: corporate issues	Other: opening hours, service, material	

TABLE F4.   **Owned issues of NS**

Media issues	Associations of the respondents	Owned issue
Economic issues (such as investments, production capacity, and world economy)	Dutch economy	0
Efficiency (including reorganization)	Efficiency (and inefficiency)	0
Advertising (and image)	Advertising	0
Employment	Employment	1
Delays	Delays	0
Strike	Strike	0
Return	Profitability	0
Service	Service in trains and stations	1
Environment	Environment (noise, air)	1
Other: corporate issues, new product, high-speed railway	Other: transport	

Table F5. **Owned issues of Schiphol**

Media issues	Associations of the respondents	Owned issue
Royal Dutch Airlines "KLM"	Royal Dutch Airlines "KLM"	1
Economic issues (such as investments, production capacity, and world economy)	Dutch economy	1
Efficiency (including reorganization)	Efficiency (and inefficiency)	0
Employment	Employment	1
Return	Profitability	1
Mergers	Mergers and takeovers	0
Environment (including noise, new sustainable product)	Environment (pollution)	-1
Other: corporate issues, advertising, image, new product, service, delays, expansion Schiphol, strike, societal issues	Other: materials and lack of materials, fast connections, transport. tax-free shops, holiday trips, business trips	

Note. Strictly speaking, in the NET-method the airline "KLM" is an actor and not an issue.

Table F6. **Owned issues of the police**

Media issues	Associations of the respondents	Owned issue
Petty crime, capital crime, drug crimeCrime	Crime	0
Safety, traffic safety	Safety	1
Employment	Employment	1
Societal issues	Standards and values	0
Efficiency (including reorganization)	Efficiency (and inefficiency)	0
Other: corporate issues, strike, image, new product, police fraud	Other: "police on the streets," maintaining order, weak actions by police officers, tough actions by police officers	

Table F7. **Owned issues of the agricultural sector**

Media issues	Associations of the respondents	Owned issue
Economic issues (such as investments)	Dutch economy	1
Efficiency (including reorganization)	Efficiency (and inefficiency)	0
Employment (including salary and buy-out of farmers)	Employment	1
Strike	Protest actions	0
Environment	Environment (pollution of soil and water)Nature and landscape	-1 -1
Other: corporate issues, swine fever, BSE, cattle/harvest, societal issues, new product, image	Other: profitability, fresh products (milk, meat, and vegetables), EU and government subsidies	

APPENDIX G – THE SURVEY

The survey was conducted during the summers of 1998, 1999, and 2000. The questionnaires were returned during the periods July 24, 1998 to July 30, 1998, August 13, 1999 to August 16, 1999, and July 21, 2000 to July 25, 2000.

For each of the organizations and sectors, the respondents were asked to indicate what they think of first when they think about the organization in question. Subsequently, the respondents were asked to indicate whether they agree with the policy of the organization with regard to their first association, as mentioned in Chapter 3. Then the respondents were asked to state whether they think the policy of the organization is a success with regard to their first association. These questions were repeated with regard to the second association of the respondents. This appendix gives the answer categories offered to the respondents with regard to their first and second associations.

What do you think of first when you think of Shell?
/What do you think of second when you think of Shell?
1. Gas stations
2. The Dutch economy
3. Mergers and takeovers
4. Raw materials and scarcity of raw materials
5. Environmental pollution
6. Mini-supermarkets in gas stations
7. Standards and values
8. Advertising
9. Employment
10. Profitability
11. Efficiency
12. Inefficiency
13. Other
14. Do not know
15. Do not know Shell

What do you think of first when you think of BP?
/What do you think of second when you think of BP?
1. Gas stations
2. The Dutch economy
3. Mergers and takeovers
4. Raw materials and scarcity of raw materials
5. Environmental pollution
6. Mini-supermarkets in gas stations
7. Standards and values
8. Advertising

9. Employment
10. Profitability
11. Efficiency
12. Inefficiency
13. Other
14. Do not know
15. Do not know BP

What do you think of first when you think of ABN AMRO?
/What do you think of second when you think of ABN AMRO?
1. Bank accounts and bank cards
2. The Dutch economy
3. Financial services
4. Mergers and takeovers
5. Raw materials and scarcity of raw materials
6. Environmental pollution
7. Standards and values
8. Advertising
9. Employment
10. Profitability
11. Efficiency
12. Inefficiency
13. Other
14. Do not know
15. Do not know ABN AMRO

What do you think of first when you think of Rabobank?
/What do you think of second when you think of Rabobank?
1. Bank accounts and bank cards
2. The Dutch economy
3. Financial services
4. Mergers and takeovers
5. Raw materials and scarcity of raw materials
6. Environmental pollution
7. Standards and values
8. Advertising
9. Employment
10. Profitability
11. Efficiency
12. Inefficiency
13. Other
14. Do not know
15. Do not know Rabobank

What do you think of first when you think of Albert Heijn?
/What do you think of second when you think of Albert Heijn?
1. The Dutch economy
2. The opening hours
3. The service
4. Mergers and takeovers
5. Raw materials and scarcity of raw materials
6. The price
7. The wide selection of products
8. Environmental pollution
9. Standards and values
10. Advertising
11. Employment
12. Profitability
13. Efficiency
14. Inefficiency
15. Other
16. Do not know
17. Do not know Albert Heijn

What do you think of first when you think of Super de Boer?
/What do you think of second when you think of Super de Boer?
1. The Dutch economy
2. The opening hours
3. The service
4. Mergers and takeovers
5. Raw materials and scarcity of raw materials
6. The price
7. The wide selection of products
8. Environmental pollution
9. Standards and values
10. Advertising
11. Employment
12. Profitability
13. Efficiency
14. Inefficiency
15. Other
16. Do not know
17. Do not know Super de Boer

What do you think of first when you think of Schiphol?
/What do you think of second when you think of Schiphol?
1. KLM
2. The Dutch economy
3. Mergers and takeovers
4. Raw materials and scarcity of raw materials
5. Environmental pollution
6. Standards and values
7. Fast connections
8. Transport
9. Tax-free shops
10. Holiday trips
11. Profitability
12. Employment
13. Business trips
14. Efficiency
15. Inefficiency
16. Other
17. Do not know
18. Do not know Schiphol

What do you think of first when you think of the Dutch Railways (NS)?
/What do you think of second when you think of the Dutch Railways (NS)?
1. The Dutch economy
2. Environment (noise, air)
3. Advertising
4. Service in trains and stations
5. Strikes
6. Transport
7. Delays
8. Employment
9. Profitability
10. Efficiency
11. Inefficiency
12. Other
13. Do not know
14. Do not know the Dutch Railways (NS)

What do you think of first when you think of the police?
/What do you think of second when you think of the police?
1. More police on the streets (translation of the Dutch "Blauw op straat")
2. Crime
3. Maintaining authority (taking care of law and order)
4. Weak actions by police
5. Tough actions by police
6. Safety
7. Employment
8. Standards and values
9. Efficiency
10. Inefficiency
11. Other
12. Do not know
13. Do not know the police

What do you think of first when you think of the agricultural sector?
/What do you think of second when you think of the agricultural sector?
1. The Dutch economy
2. Environment (pollution of soil and water)
3. Nature and landscape
4. Protest actions
5. European Union and government subsidies
6. Fresh products (milk, meat, and vegetables)
7. Profitability
8. Employment
9. Efficiency
10. Inefficiency
11. Other
12. Do not know
13. Do not know the agricultural sector

APPENDIX H – CONSTRUCTION OF THE DATASETS

The dataset to test hypotheses 1 to 3 and 7
First of all, a selection was made of the news. As in a previous publication on the effects of media coverage on reputation (Oegema et al., 2000), not all the news was taken into account. In order to test hypotheses 2 and 3, the focus was on support & criticism news in which the focal organization received support or criticism, and success & failure news in which the focal organization was attributed success or failure. This means that news in which the focal actor gives support or criticism is not taken into account when hypothesis 2 is tested.

Subsequently, the media coverage was assigned to the respondents at the individual level. For example, if a respondent read NRC Handelsblad and watched the NOS news, only media coverage of these two media titles were assigned to the respondent. If a respondent did not use a certain medium title, a "o" was entered for the media variables of this medium. Respondents who did not use any of the focal media titles were excluded from the datasets (see endnote [28]).

One difficulty in the assignment of the media coverage to the respondents is how to weight the media content of the different media used by the respondent. There are several possibilities for doing this. One medium was chosen in the publication referred to in the previous subsection (Oegema et al., 2000). Priority was given to media coverage of newspapers and newspapers with a low circulation. This meant that if someone was watching television news and reading two newspapers, e.g., De Telegraaf and Trouw, only the news from the newspaper with the lowest circulation would be taken into account, in this case Trouw, while the news in De Telegraaf and the television news would be ignored. In this study, two separate variables were made, one variable for television and one variable for newspapers, so that the possible difference between the impact of television newspapers can be studied. Moreover, if respondents were reading two or more newspapers, the amount of media coverage of these newspapers was added up and the direction of the news weighted. For example, if NRC Handelsblad contained 10 assertions of support & criticism news about Shell with an average direction of −1, while de Volkskrant contained 20 assertions about Shell with an average direction of −0.5 of this type of news, the average direction of someone who was reading the NRC Handelsblad and the Volkskrant was calculated as follows: $(10 \times -1 + 20 \times -0.5) / (10 + 20) = -0.67$. The calculation of the average direction of news, that is, the direction weighted for the frequency of news per medium title, can be formulated as follows:

$$\text{Average direction of news: } \frac{f_1 q_1 + f_2 q_2 + f_3 q_3 + f_4 q_4 + f_5 q_5}{f_1 + f_2 + f_3 + f_4 + f_5}$$

Where f_1 is the frequency of news of medium A, q_1 is the direction of news in medium A, f_2 is the frequency of news in medium B, q_2 is the direction of news in medium B, f_3 is the frequency of news in medium C, q_3 is the direction of news in medium C and so on. Since five different newspapers are studied, f1 and q1 range up to 5. We considered making separate variables for each medium but that would have resulted in a lot of variables, which would not have stimulated the parsimony of a regression model.

It should be remarked that the surveys were returned on several days of the week. Most of the surveys were returned on a Friday, Saturday, or Sunday. For pragmatic reasons, it is assumed that all of the respondents have read the Saturday newspaper, which is not correct for the respondents who returned the survey on Friday. There was relatively little media coverage on the three Saturdays in the response period: there was no media coverage of Shell, BP, Rabobank, Albert Heijn, or Super de Boer. The police received the most media coverage on Saturday August 25, 1998 (14 assertions). After the datasets are merged, the structure of the final dataset (with fictitious data) with a limited selection of the variables looks like this:

FIGURE H.1. **The year organization respondent dataset to test hypotheses 1 to 3 and 7,** **N = 16,662**

	year	org	resp	medium	amtv	agree	mark	mark_lag	var	v
1	1998	Shell	1.00	NOS	300	.50	7			
2	1998	Shell	2.00	Trouw	0	1.00	6			
3	1998	BP	1.00	NOS	10	.00	7			
4	1998	BP	2.00	Trouw	0	-1.00	8			
5	1998	BP	3.00	Volksk	0	.75	9			
6	1998	ABN	1.00	NOS	157	.25	8			
7	1999	ABN	2.00	Trouw	0	.50	9			
8	1999	ABN	3.00	Volksk	0	.50	7			
9	1999	ABN	4.00	RTL	125	-.50	8			
10	2000	Rabo	5.00	NRC	0	.75	8	7		
11	2000	Rabo	6.00	Telegra	0	.00	6	6		
12	2000	Rabo	7.00	AD	0	.00	7	7		
13	2000	Super	7.00	AD	0	1.00	6	5		
14	2000	Super	8.00	Telegra	0	.75	7	8		
15	2000	AH	7.00	AD	0	.25	9	8		
16	2000	AH	8.00	Telegra	0	.75	8	8		
17	2000	Schiph	9.00	RTL	250	1.00	7	7		

Labels in figure: **Year × org × resp**, **Variables 1998, 1999 or 2000**, **Report mark previous year**

The ten organizations and the three years in which the reputation data were gathered are listed in a column in a *pooled dataset*. The number of cases in the pooled dataset may seem relatively high (N = 16,662). This is due to the fact that respondents can be included more than once in the dataset, namely if they answered the questions about several organizations and if they took part in more than one year. Starting with the first row, Figure H.1 shows that respondent 1 answered the questions about the reputation of Shell in 1998. The respondent watched the NOS news, which contained in total 300 assertions about Shell ("amtv") in the year before the public opinion polling took place (in this case, July 24, 1997 to July 25, 1998). The respondent slightly agreed (.5) with the policy of Shell related to his association with Shell. The second respondent reads *Trouw*, so the amount of *television* news is set to zero. The respondent fully agreed (1) with the policy of Shell and gave a report mark of 6.

Apart from the media variables of the period that preceded the year before the measurement of public opinion, the lagged report mark was listed in the columns. Since report mark is only measured twice (in 1999 and in 2000), the lagged report mark is only present for respondents who participated in 2000 and in 1999.

The advantage of the pooling of the years is that the analyses of 1998, 1999, and 2000 can be performed in one model. No separate analyses have to be conducted for each year.

The dataset to test hypotheses 4, 5, and 6
In order to test the hypotheses 4 to 6, a dataset is needed which lists the associations mentioned by the respondents and the issues that appeared in the media. The dataset is construed as follows. First of all, a variable that appears in the media coverage dataset and in the public opinion dataset – a "key variable" – is created. The key variable consists of the year, the focal organization, and the number of the association. Subsequently, in the media dataset the total amount of issue news and the total amount of owned issue news are calculated per medium. In the corporate reputation dataset, the three years, the ten organizations, and the different associations are arranged in a column. This results in a large number of cases ($N = 214,958$). The two files are merged by using the key variable constructed earlier. Subsequently, the amount of issue news is calculated by using the media-use variables of the respondents. The following shows the final dataset with fictitious data and some of the media variables:

FIGURE H.2. **The year organization association respondent dataset to test hypotheses 4 to 6,**
         $N = 214,958$

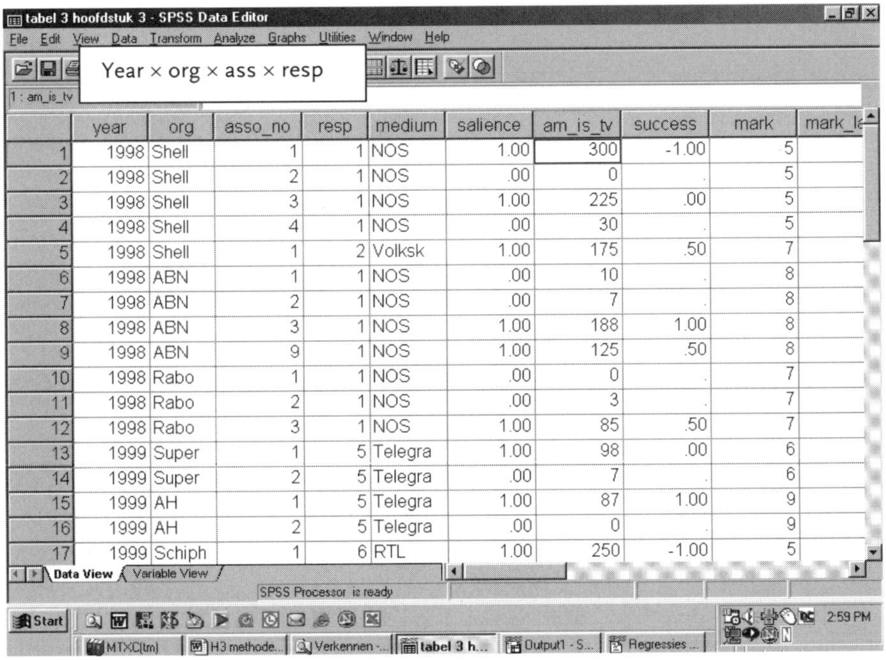

	year	org	asso_no	resp	medium	salience	am_is_tv	success	mark	mark_l
1	1998	Shell	1	1	NOS	1.00	300	-1.00	5	
2	1998	Shell	2	1	NOS	.00	0		5	
3	1998	Shell	3	1	NOS	1.00	225	.00	5	
4	1998	Shell	4	1	NOS	.00	30		5	
5	1998	Shell	1	2	Volksk	1.00	175	.50	7	
6	1998	ABN	1	1	NOS	.00	10		8	
7	1998	ABN	2	1	NOS	.00	7		8	
8	1998	ABN	3	1	NOS	1.00	188	1.00	8	
9	1998	ABN	9	1	NOS	1.00	125	.50	8	
10	1998	Rabo	1	1	NOS	.00	0		7	
11	1998	Rabo	2	1	NOS	.00	3		7	
12	1998	Rabo	3	1	NOS	1.00	85	.50	7	
13	1999	Super	1	5	Telegra	1.00	98	.00	6	
14	1999	Super	2	5	Telegra	.00	7		6	
15	1999	AH	1	5	Telegra	1.00	87	1.00	9	
16	1999	AH	2	5	Telegra	.00	0		9	
17	1999	Schiph	1	6	RTL	1.00	250	-1.00	5	

Starting with the first respondent, Figure H.2 shows that in 1998 respondent 1 associated Shell with the first association (e.g., the environment) of the answer categories. He considers the environmental policy of Shell to be unsuccessful (success = -1). The NOS news rendered 300 assertions about Shell and the environment in the period before the polling took place. It becomes clear from the second row of the data matrix that respondent 1 did not associate Shell with the second association (e.g., mergers); the value of salience is zero. The NOS did not broadcast news about Shell and mergers (am_is_tv = 0).

APPENDIX I – PEARSON'S CORRELATIONS

This appendix will present Pearson's correlations between the media variables on one hand and the reputation variables (report mark, agreement with organizational policy, and perceived success of organizational policy) on the other hand. Moreover, correlations between advertising intensity and reputation are shown. In addition, correlations between several media variables are presented, which give an indication of whether multicollinearity is present.

The appendix consists of three parts. In the first part, correlations are shown between the media variables, advertising, and report mark. These variables were also used in Chapter 5 to test hypotheses 1a, 1b, 2a, 3a, and 7. In the second part of the appendix, correlations are depicted between the media variables, advertising, and the perceived success of organizational policy and the agreement with organizational policy. These variables were used in Chapter 5 to test hypotheses 2b and 3b. In the third part of the appendix, correlations are depicted between issue news, the interaction between the frequency of issue news, and the salience of an issue and report mark. The salience of an issue is the dependent variable in hypothesis 4, and report mark is the dependent variable in hypotheses 5 and 6.

Part I – pooled data from 1999 and 2000 for the hypotheses 1a, 1b, 2a, 3a, and 7

Note. The numbers of cases in these correlation tables are based upon respondents who participated in 2000 (column C of Table 3.6), respondents who participated in 1999 (column D of Table 3.6), and respondents who participated in 1998-1999-2000 (column F of Table 3.6). The last group – the respondents who participated in three years – counts twice, since they gave a report mark in 1999 and in 2000. Note that the respondents did not give a report mark in 1998. Due to item non-response, the numbers of cases in the correlation tables cannot be calculated exactly on the basis of Table 3.6; this table provides only an upper limit for the actual numbers of cases in the correlation tables.

TABLE 11.   Shell

Variables	Mean	SD	1	2	3	4	5	6	7	8	9	10	11	12
1. Report mark	6.16	1.46												
2. Freq. TV news	185.76	101.01	.07*											
3. Freq. print news	32.14	47.92	.00	.17**										
4. director S&F TV	0.20	0.44	.02	-.47**	-.11**									
5. director S&F print	-0.03	0.52	.01	-.20**	-.45**	.15**								
6. director S&C TV	-0.29	0.15	-.05	.35**	.12**	-.84**	-.20**							
7. director S&C print	-0.06	0.13	-.05	-.14**	-.24**	.23**	-.41**	-.19**						
8. Fav. S&F TV	5.32	9.23	.03	-.42**	-.12**	.96**	.18**	-.88**	.22**					
9. Fav. S&F print	-1.06	2.36	.00	-.21**	-.91**	.16**	.69**	-.19**	.00	.18**				
10. Fav. S&C TV	-8.12	5.46	-.06*	.05	.07*	-.83**	-.14**	.85**	-.18**	-.92**	-.12**			
11. Fav. S&C print	-0.75	2.47	-.05	-.27**	-.61**	.32**	.16**	-.31**	.70**	.34**	.52**	-.27**		
12. Adv. print	152.79	134.94	.04	-.24**	.31**	.41**	.29**	-.43**	-.14**	.46**	-.06	-.43**	.07*	
13. Adv. TV	417.44	304.31	.06	-.09**	-.07*	.86**	.13**	-.86**	.19**	.94**	.12**	-1.00**	.27**	.42**

n report mark = 976; n media variables and advertising intensity = 988.

* Correlation is significant at the 0.05 level (2-tailed). **Correlation is significant at the 0.01 level (2-tailed).

TABLE 12.   BP

Variables	Mean	SD	1	2	3	4	5	6	7	8	9	10	11	12
1. Report mark	6.17	1.25												
2. Freq. TV news	7.21	9.71	.03											
3. Freq. print news	2.76	6.09	.01	.28**										
4. director S&F TV	-0.42	0.49	.00	.37**	.19**									
5. director S&F print	0.10	0.30	.01	.08*	.76**	.07*								
6. director S&C TV	0.66	0.47	.02	.50**	.07*	-.62**	.01							
7. director S&C print	0.05	0.27	.05	.37**	.48**	.19**	.26**	.13**						
8. Fav. S&F TV	-1.29	1.51	.00	.37**	.19**	1.00**	.07*	-.62**	.19**					
9. Fav. S&F print	0.10	0.33	.01	.22**	.76**	.13**	.89**	.08*	.33**	.13**				
10. Fav. S&C TV	2.27	2.66	.03	.98**	.27**	.24**	.09**	.62**	.35**	.24**	.23**			
11. Fav. S&C print	0.16	0.67	.03	.40**	.72**	.21**	.45**	.14**	.86**	.21**	.60**	.38**		
12. Adv. print	76.72	92.55	.09**	.71**	.34**	.36**	.09**	.25**	.36**	.36**	.22**	.67**	.44**	
13. Adv. TV	0.00	0.00	a	a	a	a	a	a	a	a	a	a	a	a

n report mark = 913; n media variables and advertising intensity in newspapers = 948; [a] No advertising expenditures on television.
* Correlation is significant at the 0.05 level (2-tailed). **Correlation is significant at the 0.01 level (2-tailed).

TABLE 13.   **Albert Heijn**

Variables	Mean	SD	1	2	3	4	5	6	7	8	9	10	11	12
1. Report mark	6.85	1.42												
2. Freq. TV news	34.96	26.35	.03											
3. Freq. print news	10.91	15.91	-.01	.18**										
4. director S&F TV	0.29	0.26	.03	.99**	.19**									
5. director S&F print	-0.05	0.20	-.01	-.08*	-.04	-.09**								
6. director S&C TV	-0.57	0.50	-.03	-.99**	-.19**	-1.00**	.09**							
7. director S&C print	-0.26	0.45	-.08*	.05	.18**	.06*	.31**	-.06*						
8. Fav. S&F TV	0.88	0.77	.03	.99**	.19**	1.00**	-.09**	-1.00**	.06*					
9. Fav. S&F print	-0.17	0.56	.02	-.14**	-.33**	-.15**	.86**	.15**	.00	-.15**				
10. Fav. S&C TV	-5.61	4.87	-.03	-.99**	-.19**	-1.00**	.09**	1.00**	-.06*	-1.00**	.15**			
11. Fav. S&C print	-0.14	0.50	-.07	.24**	.31**	.26**	.05	-.26**	.77**	.26**	-.30**	-.26**		
12. Adv. print	8728.31	5821.48	.09**	.19**	.16**	.18**	.03	-.18**	-.03	.18**	.01	-.18**	-.07*	
13. Adv. TV	1017.66	488.60	.01	.09**	-.07*	.00	.04	0.00	-.07*	.00	.08*	.00	-.17**	.14**

n report mark = 936; *n* media variables and advertising intensity = 957

* Correlation is significant at the 0.05 level (2-tailed). **Correlation is significant at the 0.01 level (2-tailed).

TABLE 14.   **Super de Boer**

Variables	Mean	SD	1	2	3	4	5	6	7	8	9	10	11	12
1. Report mark	6.33	1.35												
2. Freq. TV news	2.56	2.26	.02											
3. Freq. print news	0.36	0.85	.03	-.28**										
4. director S&F TV	0.56	0.50	.02	1.00**	-.28**									
5. director S&F print	-0.20	0.40	.07	.04	-.67**	.04								
6. director S&C TV	-0.56	0.50	-.02	-1.00**	.28**	-1.00**	-.04							
7. director S&C print	0.06	0.24	.00	.17	.00	.17**	-.8**	-.17**						
8. Fav. S&F TV	1.28	1.13	.02	1.00**	-.28**	1.00**	.04	-1.00**	.17**					
9. Fav. S&F print	-0.16	0.31	.08*	.00**	-.59**	.00	.99**	.00	-.20**	.00				
10. Fav. S&C TV	-1.28	1.13	-.02	-1.00**	.28**	-1.00**	-.04	1.00**	-.17**	-1.00**	.00			
11. Fav. S&C print	0.03	0.13	.00	.17**	.00	.17**	-.18**	-.17**	1.00**	.17**	-.20**	-.17**		
12. Adv. print	4745.98	3295.92	.04	.15**	.11**	.15**	-.05	-.15**	-.02	.15**	-.05	-.15**	-.02	
13. Adv. TV	546.91	374.20	.08*	.61**	-.23**	.61**	.11**	-.61**	.08*	.61**	.09**	-.61**	.08*	.16**

n report mark = 750; n media variables and advertising intensity = 973

* Correlation is significant at the 0.05 level (2-tailed). **Correlation is significant at the 0.01 level (2-tailed).

TABLE I5.   NS

Variables	Mean	SD	1	2	3	4	5	6	7	8	9	10	11	12
1. Report mark	5.63	1.42												
2. Freq. TV news	269.81	164.32	.07*											
3. Freq. print news	44.21	61.26	-.03	.22**										
4. director S&F TV	-0.03	0.19	.05	.69**	.17**									
5. director S&F print	-0.17	0.44	.02	-.32**	-.15**	-.15**								
6. director S&C TV	-0.13	0.23	-.02	-.18**	-.02	-.56**	-.02							
7. director S&C print	-0.10	0.35	.01	-.40**	-.42**	-.24**	.73**	.00						
8. Fav. S&F TV	1.11	4.92	.05	.74**	.20**	.91**	-.30**	-.49**	-.39**					
9. Fav. S&F print	-1.56	2.61	.03	-.37**	-.74**	-.23**	.69**	.02	.66**	-.35**				
10. Fav. S&C TV	-8.45	8.61	-.07*	-.91**	-.22**	-.89**	.30**	.41**	.39**	-.94**	.36**			
11. Fav. S&C print	-3.54	6.07	.03	-.33**	-.94**	-.25**	.28**	.04	.58**	-.33**	.84**	.34**		
12. Adv. print	1047.33	702.77	.01	.23**	.15**	.14**	-.02	.00	-.03	.13**	-.11**	-.20**	-.12**	
13. Adv. TV	2225.50	907.33	.06	.69**	.09**	.16**	-.08*	-.03	-.10***	.08*	-.11**	-.41**	-.10***	.21**

n report mark =968; *n* media variables and advertising intensity = 976 .
* Correlation is significant at the 0.05 level (2-tailed). **Correlation is significant at the 0.01 level (2-tailed).

TABLE 16.   Schiphol

Variables	Mean	SD	1	2	3	4	5	6	7	8	9	10	11	12
1. Report mark	6.70	1.47												
2. Freq. TV news	251.73	154.32	.04											
3. Freq. print news	52.18	72.57	-.04	.22**										
4. director S&F TV	0.13	0.16	.07*	-.62**	-.19**									
5. director S&F print	-0.12	0.18	-.07*	-.25**	-.36**	.31**								
6. director S&C TV	-0.12	0.11	-.06	.09**	.01	.14**	.11**							
7. director S&C print	0.03	0.12	-.08*	-.29**	-.05	.44**	.15**	.14**						
8. Fav. S&F TV	2.78	8.13	.08*	-.64**	-.20**	.97**	.29**	.17**	.42**					
9. Fav. S&F print	-1.60	3.27	-.03	-.39**	-.76**	.44**	.64**	.11**	.38**	.44**				
10. Fav. S&C TV	-3.83	4.10	-.03	-.90**	-.22**	.78**	.32**	.22**	.38**	.79**	.47**			
11. Fav. S&C print	0.36	3.69	-.06	-.31**	-.22**	.43**	.22**	.11**	.82**	.41**	.59**	.38**		
12. Adv. print	143.13	174.82	.02	.43**	.81**	-.44**	-.69**	-.09**	-.16**	-.44**	-.89**	-.49**	-.29**	
13. Adv. TV	0.00	0.00	a	a	a	a	a	a	a	a	a	a	a	a

n report mark = 966; n media variables and advertising intensity print news = 979; [a] The company did not have any advertising expenditures on television.

* Correlation is significant at the 0.05 level (2-tailed). **Correlation is significant at the 0.01 level (2-tailed).

TABLE 17. **ABN AMRO**

Variables	Mean	SD	1	2	3	4	5	6	7	8	9	10	11	12
1. Report mark	6.40	1.42												
2. Freq. TV news	146.23	60.25	.06											
3. Freq. print news	35.49	42.21	-.03	.06										
4. director S&F TV	-0.20	0.11	-.02	-.41**	-.03									
5. director S&F print	-0.02	0.24	.03	-.06	-.18**	.19**								
6. director S&C TV	-0.23	0.36	-.04	.17**	.01	-.40**	-.39**							
7. director S&C print	0.02	0.10	-.04	.05	-.01	-.23**	-.63**	.57**						
8. Fav. S&F TV	-3.30	1.78	-.04	-.86**	-.05	.71**	.23**	-.58**	-.29**					
9. Fav. S&F print	-0.02	1.90	.05	-.06	-.13**	.24**	.83**	-.52**	-.73**	.28**				
10. Fav. S&C TV	-2.78	4.92	-.06	-.02	-.01	-.58**	-.38**	.92**	.54**	-.46**	-.52**			
11. Fav. S&C print	0.24	1.19	-.03	.05	.19**	-.23**	-.55**	.51**	.84**	-.27**	-.77**	.49**		
12. Adv. print	1212.06	980.85	-.03	.11**	.62**	-.07*	.15**	.13**	.12**	-.14**	.06	.10**	.26**	
13. Adv. TV	3272.61	1581.05	.04	.94**	.06	-.61**	-.18**	.47**	.22**	-.98**	-.23**	.32**	.21**	.13**

n report mark = 893; *n* media variables and advertising intensity = 929
* Correlation is significant at the 0.05 level (2-tailed). **Correlation is significant at the 0.01 level (2-tailed).

TABLE 18.   Rabobank

Variables	Mean	SD	1	2	3	4	5	6	7	8	9	10	11	12
1. Report mark	6.44	1.48												
2. Freq. TV news	73.34	30.59	.05											
3. Freq. print news	18.56	20.11	.01	.07*										
4. director S&F TV	-0.46	0.47	.00	.36**	.09**									
5. director S&F print	-0.12	0.27	-.02	.14**	-.01	.27**								
6. director S&C TV	0.33	0.16	.01	.15**	-.06	-.50**	-.12**							
7. director S&C print	0.00	0.27	.02	-.10**	-.28**	-.31**	-.41**	.16**						
8. Fav. S&F TV	-1.93	4.93	.02	.22**	.08	.89**	.27**	-.63**	-.32**					
9. Fav. S&F print	-0.45	0.91	-.01	.18**	-.31**	.45**	.68**	-.23**	-.38**	.46**				
10. Fav. S&C TV	4.00	1.58	.04	.86**	.02	-.10**	-.01	.50**	.07*	-.30**	-.07*			
11. Fav. S&C print	-0.23	1.32	.02	-.20**	-.41**	-.45**	-.54**	.23**	.77**	-.45**	-.64**	.05		
12. Acv. print	1529.01	1102.30	.07*	.33**	.52**	.45**	.14**	-.21**	-.24**	.42**	.02	.10**	-.38**	
13. Acv. TV	1749.80	899.34	.05	.95**	.08*	.61**	.21**	.02	-.19**	.49**	.30**	.68**	-.32**	.41**

n report mark = 877; n media variables and advertising intensity = 914
* Correlation is significant at the 0.05 level (2-tailed). **Correlation is significant at the 0.01 level (2-tailed).

TABLE 19.   Police

Variables	Mean	SD	1	2	3	4	5	6	7	8	9	10	11	12
1. Report mark	6.17	1.39												
2. Freq. TV news	600.54	251.24	.18**											
3. Freq. print news	87.41	105.75	-.05	.10**										
4. director S&F TV	0.24	0.13	.14**	.22**	-.10**									
5. director S&F print	0.19	0.24	.00	.03	.55**	.26**								
6. director S&C TV	-0.06	0.12	.03	.39**	.06	.21**	.17**							
7. director S&C print	-0.25	0.27	.06*	.03	-.43**	-.24**	-.43**	-.08**						
8. Fav. S&F TV	36.18	23.10	.16**	.55**	-.05	.85**	.25**	.32**	-.22**					
9. Fav. S&F print	8.97	13.79	.01	.08*	.82**	.14**	.78**	.14**	-.30**	.15**				
10. Fav. S&C TV	-3.44	4.16	-.01	-.15**	-.01	.14**	.12**	.66**	-.04	.03	.09**			
11. Fav. S&C print	-7.52	8.56	.06	-.03	-.80**	-.14**	-.47**	-.08*	.79**	-.15**	-.65**	.00		
12. Adv. print	440.25	302.84	.08*	.19**	.18**	-.08*	-.03	.02	.09**	-.01	.09**	-.06*	-.06*	
13. Adv. TV	72.98	61.17	.11**	.33**	-.08*	.79**	.20**	.08**	-.25**	.89**	.10**	-.23**	-.17**	-.06

n report mark = 949; n media variables and advertising intensity = 951
* Correlation is significant at the 0.05 level (2-tailed). **Correlation is significant at the 0.01 level (2-tailed).

TABLE 110. Agriculture

Variables	Mean	SD	1	2	3	4	5	6	7	8	9	10	11	12
1. Report mark	6.47	1.26												
2. Freq. TV news	202.31	95.98	.06											
3. Freq. print news	31.39	40.55	-.04	.13**										
4. director S&F TV	-0.53	0.22	-.02	-.33**	-.05									
5. director S&F print	-0.34	0.38	-.01	.08*	-.26**	.12**								
6. director S&C TV	-0.14	0.20	-.01	.22**	.11**	.42**	.33**							
7. director S&C print	-0.16	0.18	.06	.02	-.27**	.04	.83**	.17**						
8. Fav. S&F TV	-20.97	9.54	-.07*	-.80**	-.08**	.71**	.11**	.24**	.06					
9. Fav. S&F print	-3.06	4.36	-.04	.00	-.80**	.15**	.46**	.16**	.23**	.13**				
10. Fav. S&C TV	-5.10	9.42	-.04	.15**	.10***	.44**	.35**	.88**	.17**	.40**	.20**			
11. Fav. S&C print	-1.79	2.40	.07*	-.13**	-.87***	.01	.23**	-.13**	.46**	.07*	.61**	-.12**		
12. Adv. print	111.25	98.67	.05	.27**	.39**	.01	.02	.21**	-.12**	-.13**	-.15**	.20**	-.43**	
13. Adv. TV	316.29	234.66	.07*	.30**	-.06	-.33**	-.29**	-.75**	-.16**	-.62**	-.15**	-.84**	.07*	-.06

n report mark =929; *n* media variables and advertising intensity = 957
* Correlation is significant at the 0.05 level (2-tailed). **Correlation is significant at the 0.01 level (2-tailed).

Part II – pooled data from 1998, 1999, and 2000 for the hypotheses 2b and 3b

Note. The numbers of cases in these correlation tables are based upon respondents who participated in 1998 (column A of Table 3.6), respondents who participated in 2000 (column C of Table 3.6), respondents who participated in 1998 and in 1999 (column D of Table 3.6), and respondents who participated in 1998-1999-2000 (column F of Table 3.6). The respondents who participated twice, count twice. The respondents who participated three times, count three times.

TABLE I11.   **Shell**

Variables	Mean	SD	1	2	3	4	5	6	7
1. Success policy	0.39	0.73							
2. Agreement policy	0.08	0.35	.64**						
3. Fav. S&F TV	2.22	8.82	.00	-.01					
4. Fav. S&F print	-1.49	2.79	-.02	-.02	.22**				
5. Fav. S&C TV	-2.28	9.75	.04	.04	-.82**	-.24**			
6. Fav. S&C print	-0.90	2.39	-.03	-.05*	.29**	.32**	-.18**		
7. Adv. print	118.23	123.14	.01	.02	55**	.00	-.52**	.03	
8. Adv. TV	327.65	283.74	.01	.00	.92**	.18**	-.76**	.24**	.52**

n success policy = 909; *n* agreement policy= 1,429; *n* media variables and advertising intensity= 1,470
* Correlation is significant at the 0.05 level (2-tailed). **Correlation is significant at the 0.01 level (2-tailed).

TABLE I12.   **BP**

Variables	Mean	SD	1	2	3	4	5	6	7
1. Success policy	0.50	0.64							
2. Agreement policy	0.10	0.25	.59**						
3. Fav. S&F TV	-0.88	1.38	.03	-.02					
4. Fav. S&F print	0.07	0.28	.01	-.01	.05				
5. Fav. S&C TV	1.25	2.69	-.02	-.04	-.07*	.27**			
6. Fav. S&C print	0.21	0.60	.02	.00	.22**	.52**	.22**		
7. Adv. print	188.59	235.37	.01	.05	.41**	-.04	-.21**	.33**	
8. Adv. TV	92.00	149.93	.01	.03	.39**	-.15**	-.53**	.11**	.66**

n success policy = 505; *n* agreement policy= 1,306; *n* media variables and advertising intensity = 1,393
* Correlation is significant at the 0.05 level (2-tailed). **Correlation is significant at the 0.01 level (2-tailed).

TABLE I13. **Albert Heijn**

Variables	Mean	SD	1	2	3	4	5	6	7
1. Success policy	0.62	0.60							
2. Agreement policy	0.28	0.41	.67**						
3. Fav. S&F TV	0.30	1.13	-.06*	-.07*					
4. Fav. S&F print	-0.03	0.62	.04	.05	-.30**				
5. Fav. S&C TV	-3.49	5.08	.04	.05	-.85**	.27**			
6. Fav. S&C print	-0.49	1.08	-.09**	-.10**	.37**	-.18**	-.38**		
7. Adv. print	8885.33	5989.60	.08**	.07**	.05	.07**	-.08**	-.11**	
8. Adv. TV	797.56	518.35	.03	.03	.41**	-.14**	-.37**	.24**	.07**

n success policy = 1,163; n agreement policy= 1,298; n media variables and advertising intensity = 1,419
*Correlation is significant at the 0.05 level (2-tailed). **Correlation is significant at the 0.01 level (2-tailed).

TABLE I14. **Super de Boer**

Variables	Mean	SD	1	2	3	4	5	6	7
1. Success policy	0.59	0.63							
2. Agreement policy	0.24	0.36	.67**						
3. Fav. S&F TV	0.29	1.80	-.01	.04					
4. Fav. S&F print	-0.11	0.27	.02	.03	-.22**				
5. Fav. S&C TV	2.46	6.06	-.01	-.07*	-.92**	.25**			
6. Fav. S&C print	0.22	0.52	.05	.04	-.43**	.11**	.51**		
7. Adv. Print	3782.97	3132.01	.03	.05	.42**	-.16**	-.40**	-.20**	
8. Adv. TV	522.40	340.19	-.03	.02	.36**	.05	.12**	.01	.18**

n success policy = 679; n agreement policy= 890; n media variables and advertising intensity = 1,434
*Correlation is significant at the 0.05 level (2-tailed). **Correlation is significant at the 0.01 level (2-tailed).

TABLE I15. **NS**

Variables	Mean	SD	1	2	3	4	5	6	7
1. Success policy	-0.29	0.77							
2. Agreement policy	-0.05	0.41	.68**						
3. Fav. S&F TV	-1.80	6.28	-.06*	-.01					
4. Fav. S&F print	-1.64	2.49	.06*	.05	-.15**				
5. Fav. S&C TV	-8.94	7.77	.01	.01	-.47**	.29**			
6. Fav. S&C print	-2.53	5.31	.05	.01	-.39**	.73**	.27**		
7. Adv. print	1089.07	718.08	-.02	-.02	-.01	-.19**	-.19**	-.07**	
8. Adv. TV	2258.11	924.56	.06*	.03	-.19**	-.10**	-.43**	-.06*	.19**

n success policy = 1,069; n agreement policy= 1,378; n media variables and advertising intensity = 1,439
*Correlation is significant at the 0.05 level (2-tailed). **Correlation is significant at the 0.01 level (2-tailed).

TABLE I16. **Schiphol**

Variables	Mean	SD	1	2	3	4	5	6	7
1. Success policy	0.45	0.72							
2. Agreement policy	0.18	0.37	.66**						
3. Fav. S&F TV	-3.10	12.21	.02	.00					
4. Fav. S&F print	-2.29	3.76	.01	-.02	.37**				
5. Fav. S&C TV	-8.67	8.88	-.01	-.02	.85**	.35**			
6. Fav. S&C print	-0.72	4.48	.03	-.02	.43**	.63**	.40**		
7. Adv. print	143.87	163.70	-.03	.03	-.23**	-.82**	-.19**	-.38**	
8. Adv. TV	0.00	0.00	a	a	a	a	a	a	a

n success policy = 961; *n* agreement policy= 1,404; *n* media variables and advertising intensity = 1,444
* Correlation is significant at the 0.05 level (2-tailed). **Correlation is significant at the 0.01 level (2-tailed).
[a] Cannot be computed because at least one of the variables is constant.

TABLE I17. **ABN AMRO**

Variables	Mean	SD	1	2	3	4	5	6	7
1. Success policy	0.59	0.64							
2. Agreement policy	0.18	0.31	.55**						
3. Fav. S&F TV	-2.98	1.82	.07*	.08**					
4. Fav. S&F print	-0.45	2.11	.03	-.01	.10**				
5. Fav. S&C TV	-4.75	5.36	-.04	-.08**	-.33**	-.11**			
6. Fav. S&C print	0.42	1.28	.00	.01	-.12**	-.81**	.16**		
7. Adv. print	1439.47	1189.40	-.02	.07**	-.03	-.29**	-.14**	.40**	
8. Adv. TV	2887.61	1511.25	-.07	-.07**	-.94**	-.05	.31**	.08**	.00

n success policy = 734; *n* agreement policy= 1,262; *n* media variables and advertising intensity = 1,362
* Correlation is significant at the 0.05 level (2-tailed). **Correlation is significant at the 0.01 level (2-tailed).

TABLE I18. **Rabobank**

Variables	Mean	SD	1	2	3	4	5	6	7
1. Success policy	0.61	0.63							
2. Agreement policy	0.19	0.35	.63**						
3. Fav. S&F TV	0.38	5.37	.06	.03					
4. Fav. S&F print	-0.35	0.82	.03	.02	.42**				
5. Fav. S&C TV	7.87	7.00	.05	.00	.57**	.11**			
6. Fav. S&C print	0.24	1.63	-.05	-.05	.04	-.51**	.37**		
7. Adv. print	2374.16	2206.49	.03	.06*	.50**	.11**	.52**	.17**	
8. Adv. TV	1639.43	820.26	-.02	.00	.30**	.22**	.17**	-.26**	.09**

n success policy = 757; *n* agreement policy= 1,242; *n* media variables and advertising intensity = 1,354
* Correlation is significant at the 0.05 level (2-tailed). **Correlation is significant at the 0.01 level (2-tailed).

TABLE l19. **Police**

Variables	Mean	SD	1	2	3	4	5	6	7
1. Success policy	-0.42	0.75							
2. Agreement policy	-0.03	0.47	.54**						
3. Fav. S&F TV	28.85	24.76	.14**	.09**					
4. Fav. S&F print	10.59	14.74	-.02	.00	.01				
5. Fav. S&C TV	-19.31	26.34	.05	.05	.39**	-.17**			
6. Fav. S&C print	-8.27	9.14	.07*	.08**	-.04	-.75**	.13**		
7. Adv. print	347.53	288.24	.01	.00	.19**	.02	.39**	-.03	
8. Adv. TV	49.36	60.80	.12**	.07*	.80**	-.03	.46**	-.04	.22**

n success policy = 1,234; n agreement policy= 1,386; n media variables and advertising intensity – 1,406
* Correlation is significant at the 0.05 level (2-tailed). **Correlation is significant at the 0.01 level (2-tailed).

TABLE l20. **Agriculture**

Variables	Mean	SD	1	2	3	4	5	6	7
1. Success policy	0.17	0.77							
2. Agreement policy	0.13	0.37	.61**						
3. Fav. S&F TV	-16.07	10.75	.00	.02					
4. Fav. S&F print	-3.13	4.14	.03	.04	.08**				
5. Fav. S&C TV	-5.35	7.90	-.04	-.03	.28**	.18**			
6. Fav. S&C print	-2.62	3.47	.08	.06*	-.18**	.66**	-.04		
7. Adv. print	88.85	90.43	-.02	.00	-.33**	-.20**	.19**	-.23**	
8. Adv. TV	242.32	222.27	.05	.03	-.72**	-.10**	-.71**	.20**	.13**

n success policy = 985; n agreement policy= 1,324; n media variables and advertising intensity = 1,412
* Correlation is significant at the 0.05 level (2-tailed). **Correlation is significant at the 0.01 level (2-tailed).

Part III – pooled data from 1998, 1999, and 2000 for hypothesis 4, and pooled data 1999 and 2000 for hypotheses 5 and 6

Note with regard to the numbers of cases for hypothesis 4: The numbers of cases in these correlation tables are based upon the number of respondents in Table 3.7. The respondents who participated twice, count twice. The respondents who participated three times, count three times. This number of respondents should be multiplied by the number of associations, since an year × organization × association × respondent dataset is needed. Consider, for example, the number of cases that indicate the salience of an association with Shell. This number should be: 100 (see column A of Table 3.7) + 377 (see column B) + 69 × 2 (see column D) + 84 × 2 (see column E) + 229 × 3 (see column F) × 11 (number of associations). This is 1,670, which is precisely the number of cases.

Note with regard to the numbers of cases for hypotheses 5 and 6: The numbers of cases in these correlation tables are based upon the respondents who participated in 2000 (column C of Table 3.7), the respondents who participated in 1999 (column D of Table 3.7), and the respondents who participated in 1998-1999-2000 (column F of Table 3.7). The last group, the respondents who participated in three years, count twice, since they gave a report mark in 1999 and in 2000. Again, this number of respondents should be multiplied by the number of associations. Consider, for example, the cases with regard to the report mark of Shell. The number of cases should be: 976 (see Table I1) × 11 (number of associations). This is 10,736, which is precisely the number of cases.

There are considerably fewer cases with regard to the frequency of issue TV news × success of policy, the variable used to test priming. This is due to the fact that the respondents were asked to indicate the success of the policy with regard to their association. This means that the total number of cases that were found in Part II should be multiplied by 2 (the first and the second associations) instead of by 11. In the case of Shell, the number of respondents can be at most: 909 (see Table I11) × 2 = 1,818. This is the upper limit, since some respondents only gave their opinion about the success of the organizational policy with regard to one association only.

TABLE I21.   **Shell**

Variables	Mean	SD	1	2	3	4	5	6	7
1. Report mark	6.16	1.46							
2. Salience issue	0.16	0.37	.02						
3. Freq. owned issue TV news	4.15	13.79	.01	-.04**					
4. Freq. owned issue print news	0.65	3.49	.01	-.04**	.51**				
5. Freq. issues print news	1.53	3.69	.01	-.05**	.33**	.69**			
6. Freq. Issue TV news	8.92	14.97	.02*	-.08**	.70**	.33**	.38**		
7. Freq. Issue TV news × success policy	2.61	14.03	.15**	a	.63**	.47**	.34**	.52**	
8. Freq. Issue print news × success policy	.39	3.77	.08*	a	.47**	.84**	.52**	.36**	.58**

n report mark = 10,736; *n* salience issue, freq owned issue tv/print news, freq issue tv/print news = 16,170; *n* freq issue tv news × success policy = 1,428; *n* freq issue print news × success policy = 1,428
* Correlation is significant at the 0.05 level (2-tailed). **Correlation is significant at the 0.01 level (2-tailed).
[a] Cannot be computed because the salience of an issue is constant.

TABLE I22.   **BP**

Variables	Mean	SD	1	2	3	4	5	6	7
1. Report mark	6.17	1.25							
2. Salience issue	0.14	0.35	.04**						
3. Freq. owned issue TV news	0.01	0.15	.00	-.02*					
4. Freq. owned issue print news	0.02	0.26	-.02*	-.03**	.00				
5. Freq. issues print news	0.09	0.49	.00	-.05**	-.01	.31**			
6. Freq. issues TV news	0.14	1.50	-.01	-.03**	.09**	-.01	.28**		
7. Freq. issues TV × success policy	.04	.92	.05	a	a	.00	.70**	.97**	
8. Freq. issues print news × success policy	.03	.48	.05	a	a	.09*	.97**	.72**	.71**

n report mark = 10,043; *n* salience issue, freq owned issue tv/print news, freq issue tv/print news = 15,323; *n* freq issue tv news × success policy = 738; *n* freq issue print news × success policy = 738
* Correlation is significant at the 0.05 level (2-tailed). **Correlation is significant at the 0.01 level (2-tailed).
[a] Cannot be computed because at least one of the variables is constant.

TABLE 123. **Albert Heijn**

Variables	Mean	SD	1	2	3	4	5	6	7
1. Report mark	6.85	1.42							
2. Salience issue	0.13	0.34	.04**						
3. Freq. owned issue TV news	0.38	1.35	.01	.07**					
4. Freq. owned issue print news	0.26	1.09	-.01	.13**	.31**				
5. Freq. issues print news	0.32	1.11	-.01	.11**	.29**	.97**			
6. Freq. issues TV news	0.68	2.01	.01	-.01	.63**	.17**	.21**		
7. Freq. issues TV × success policy	.25	1.80	.08**	a	.31**	.03	.03	.31**	
8. Freq. issues print news × success policy	.44	1.81	.03	a	-.00	.66**	.67**	-.00	.29**

n report mark = 12,168; n salience issue, freq owned issue tv/print news, freq issue tv/print news = 18,447;
n freq issue tv news × success policy = 2,079; n freq issue print news × success policy = 2,079
* Correlation is significant at the 0.05 level (2-tailed). **Correlation is significant at the 0.01 level (2-tailed).
a Cannot be computed because the salience of an issue is constant.

TABLE 124. **Super de Boer**

Variables	Mean	SD	1	2	3	4	5	6	7
1. Report mark	6.33	1.35							
2. Salience issue	0.08	0.27	.07**						
3. Freq. owned issue TV news	0.13	0.75	a	-.04**					
4. Freq. owned issue print news	0.02	0.15	.01	-.03**	.37**				
5. Freq. issues print news	0.04	0.25	.02	-.01	.21**	.59**			
6. Freq. issues TV news	0.40	1.91	a	.04**	.37**	.13**	.35**		
7. Freq. issues TV × success policy	.24	2.34	a	a	.02	.03	.40**	.52**	
8. Freq. issues print news × success policy	.02	.21	.05	a	.06*	.22**	.73**	.39**	.42**

n report mark = 9,750; n salience issue, freq owned issue tv/print news, freq issue tv/print news = 18,642;
n freq issue tv news × success policy = 1,129; n freq issue print news × success policy = 1,129
* Correlation is significant at the 0.05 level (2-tailed). **Correlation is significant at the 0.01 level (2-tailed).
a Cannot be computed because at least one of the variables is constant.

TABLE 125.   **NS**

Variables	Mean	SD	1	2	3	4	5	6	7
1. Report mark	5.63	1.42							
2. Salience issue	0.18	0.38	.00						
3. Freq. owned issue TV news	1.15	3.22	.01	.05**					
4. Freq. owned issue print news	0.28	1.33	-.01	.05**	.46**				
5. Freq. issues print news	1.28	2.53	-.03**	-.01	.10**	.44**			
6. Freq. issues TV news	5.35	6.48	.03**	.06**	.27**	.09**	.34**		
7. Freq. issues TV × success policy	-3.72	7.66	.34**	a	.08**	.07**	-.18**	-.51**	
8. Freq. issues print news × success policy	-.81	2.53	.24**	a	.05**	-.08**	-.50**	-.20**	.49**

n report mark = 9,680; *n* salience issue, freq owned issue tv/print news, freq issue tv/print news = 14,390; *n* freq issue tv news × success policy = 1,768; *n* freq issue print news × success policy = 1,768
*Correlation is significant at the 0.05 level (2-tailed). **Correlation is significant at the 0.01 level (2-tailed).
[a] Cannot be computed because the salience of an issue is constant.

TABLE 126.   **Schiphol**

Variables	Mean	SD	1	2	3	4	5	6	7
1. Report mark	6.70	1.47							
2. Salience issue	0.13	0.34	.01						
3. Freq. owned issue TV news	-3.80	22.82	.00	-.01					
4. Freq. owned issue print news	-0.21	4.06	.00	.00	.54**				
5. Freq. issues print news	0.97	3.95	-.02*	.10**	-.47**	-.71**			
6. Freq. issues TV news	5.65	22.44	.01	.08**	-.97**	-.49**	.51**		
7. Freq. issues TV × success policy	-0.49	30.88	.32**	a	.78**	.53**	-.46**	-.75**	
8. Freq. issues print news × success policy	-0.06	6.99	.28**	a	.52**	.87**	-.68**	-.48**	.60**

n report mark = 13,524; *n* salience issue, freq owned issue tv/print news, freq issue tv/print news = 20,216; *n* freq issue tv news × success policy = 1,568; *n* freq issue print news × success policy = 1,568
*Correlation is significant at the 0.05 level (2-tailed). **Correlation is significant at the 0.01 level (2-tailed).
[a] Cannot be computed because the salience of an issue is constant.

TABLE 127. **ABN AMRO**

Variables	Mean	SD	1	2	3	4	5	6	7
1. Report mark	6.40	1.42							
2. Salience issue	0.16	0.36	.03*						
3. Freq. owned issue TV news	3.15	8.27	.02	-.09**					
4. Freq. owned issue print news	0.84	3.42	-.01	-.06**	.37**				
5. Freq. issues print news	1.54	4.30	-.01	-.07**	.23**	.75**			
6. Freq. issues TV news	5.83	10.73	.02	-.12**	.68**	.22**	.41**		
7. Freq. issues TV × success policy	1.77	10.13	.12**	a	.52**	.27**	.00	.38**	
8. Freq. issues print news × success policy	.40	3.95	.10**	a	.23**	.50**	.05	.01	.60**

n report mark = 9,823; *n* salience issue, freq owned issue tv/print news, freq issue tv/print news = 14,982;
n freq issue tv news × success policy = 1,159; *n* freq issue print news × success policy = 1,159
* Correlation is significant at the 0.05 level (2-tailed). **Correlation is significant at the 0.01 level (2-tailed).
[a] Cannot be computed because the salience of an issue is constant.

TABLE 128. **Rabobank**

Variables	Mean	SD	1	2	3	4	5	6	7
1. Report mark	6.44	1.48							
2. Salience issue	0.15	0.36	.03**						
3. Freq. owned issue TV news	0.79	3.08	.01	-.07**					
4. Freq. owned issue print news	0.18	0.63	-.01	-.07**	.33**				
5. Freq. issues print news	1.04	3.41	.00	-.07**	.00	.11**			
6. Freq. issues TV news	1.65	3.66	.01	-.13**	.78**	.21**	.09**		
7. Freq. issues TV × success policy	.37	2.21	.15**	a	.69**	.28**	.15**	.75**	
8. Freq. issues print news × success policy	.39	2.66	.02	a	.04	.09**	.85**	.15**	.21**

n report mark = 9,647; *n* salience issue, freq owned issue tv/print news, freq issue tv/print news = 14,894;
n freq issue tv news × success policy = 1,212; *n* freq issue print news × success policy = 1,212
* Correlation is significant at the 0.05 level (2-tailed). **Correlation is significant at the 0.01 level (2-tailed).
[a] Cannot be computed because the salience of an issue is constant.

Table 129. **Police**

Variables	Mean	SD	1	2	3	4	5	6	7
1. Report mark	6.17	1.39							
2. Salience issue	0.21	0.41	.02						
3. Freq. owned issue TV news	18.63	43.21	.04**	.17**					
4. Freq. owned issue print news	3.02	11.49	-.01	.11**	.58**				
5. Freq. issues print news	3.26	11.47	-.01	.11**	.58**	1.00**			
6. Freq. issues TV news	20.83	42.95	.04**	.15**	.98**	.57**	.57**		
7. Freq. issues TV × success policy	-17.91	60.84	.18**	a	-.62**	-.36**	-.36**	-.62**	
8. Freq. issues print news × success policy	-3.12	15.44	.12**	a	-.37**	-.59**	-.59**	-.37**	.60**

n report mark = 8,541; n salience issue, freq owned issue tv/print news, freq issue tv/print news = 12,654; n freq issue tv news × success policy = 2,257; n freq issue print news × success policy = 2,257
* Correlation is significant at the 0.05 level (2-tailed). **Correlation is significant at the 0.01 level (2-tailed).
[a] Cannot be computed because the salience of an issue is constant.

Table 130. **Agriculture**

Variables	Mean	SD	1	2	3	4	5	6	7
1. Report mark	6.47	1.26							
2. Salience issue	0.20	0.40	.02*						
3. Freq. owned issue TV news	-2.60	7.66	-.01	-.12**					
4. Freq. owned issue print news	-0.29	2.13	.01	-.08**	.53**				
5. Freq. issues print news	1.02	2.27	-.03*	-.04**	-.25**	-.36**			
6. Freq. issues TV news	4.43	7.09	.01	.06**	-.85**	-.38**	.34**		
7. Freq. issues TV × success policy	-.30	9.30	.28**	a	.04	.12**	-.14**	.05	
8. Freq. issues print news × success policy	-.20	2.22	.21**	a	.09**	.22**	-.23**	-.11**	.54**

n report mark = 8,361; n salience issue, freq owned issue tv/print news, freq issue tv/print news = 12,708; n freq issue tv news × success policy = 1,719; n freq issue print news × success policy = 1,719
* Correlation is significant at the 0.05 level (2-tailed). **Correlation is significant at the 0.01 level (2-tailed).
[a] Cannot be computed because the salience of an issue is constant.

Appendix J – Success and agreement with organizational policy as predictors of reputation

In Chapter 3 it was argued that corporate reputation can be predicted from the agreement of respondents with organizational policy and the perceived success of organizational policy. This was tested by means of regression analyses. The regression model for the pooled data was presented in Chapter 5, while in this appendix the regression models per organization are presented.

TABLE J1. **Stepwise regression analyses of the success and agreement with organizational policy as predictors of reputation**

Models per organizationa	Predictors (news and advertising, previous year)	Reputation[a]
Shell	Agreement with the policy of an organization	.51***
	Success of the policy of an organization	.15***
	Adjusted R^2	.38
	df	2, 590
	F	178.76***
	n	593
BP	Agreement with the policy of an organization	.28***
	Success of the policy of an organization	.28***
	Adjusted R^2	.24
	df	2, 343
	F	56.05***
	n	346
Albert Heijn	Agreement with the policy of an organization	.38***
	Success of the policy of an organization	.32***
	Adjusted R^2	.41
	df	2, 780
	F	275.01***
	n	783
Super de Boer	Agreement with the policy of an organization	.32***
	Success of the policy of an organization	.30***
	Adjusted R^2	.31
	df	2, 461
	F	104.06***
	n	464
NS	Agreement with the policy of an organization	.41***
	Success of the policy of an organization	.20***
	Adjusted R^2	.32
	df	2, 707
	F	164.64***
	n	710

TABLE J1.   (Continued)

Models per organizationa	Predictors (news and advertising, previous year)	Reputation[a]
Schiphol	Agreement with the policy of an organization	.51***
	Success of the policy of an organization	.17***
	Adjusted R^2	.40
	df	2, 624
	F	205.40***
	n	627
ABN AMRO	Agreement with the policy of an organization	.47***
	Success of the policy of an organization	.21***
	Adjusted R^2	.37
	df	2, 489
	F	146.76***
	n	492
Rabobank	Agreement with the policy of an organization	.47***
	Success of the policy of an organization	.27***
	Adjusted R^2	.44
	df	2, 494
	F	196.96***
	n	497
Police	Agreement with the policy of an organization	.30***
	Success of the policy of an organization	.28***
	Adjusted R^2	.26
	df	2, 815
	F	146.53***
	n	818
Agricultural sector	Agreement with the policy of an organization	.36***
	Success of the policy of an organization	.20***
	Adjusted R^2	.26
	df	2, 652
	F	116.78***
	n	655

Note. Cell entries are Betas from Ordinary Least Squares regression.
* $p \leq .05$; ** $p \leq .01$; *** $p \leq .001$
[a] Cannot be computed because the salience of an issue is constant.

Table J1 shows that in all ten models, the reputation of the organization can be predicted from the agreement with the policy of the organization and the success of the policy of the organization.

Appendix K – The effects of the amount of news, the tenor of success & failure news, and advertising on reputation controlled for educational level

Table K1 presents per organization the results of the non-autoregressive models of the effects of the amount of news, the favorability and the direction of success & failure news, and the advertising expenditures on reputation, controlled for the educational level of the respondents.

Table K1. Stepwise regression analyses of the effects of the amount of news, tenor of success & failure news, and advertising on reputation, controlled for educational level

Models per organization[a]	Predictors (news and advertising, previous year)	
Super de Boer	Educational level	-.16***
	Adjusted R^2	.02
	df	1, 743
	F	18.47***
	n	745
NS	Amount of TV news	.07*
	Educational level	-.08*
	Adjusted R^2	.01
	df	2, 959
	F	5.07**
	n	962
Schiphol	Amount of TV news	.14***
	Favorability of success & failure TV news	.20***
	Direction of success & failure print news	-.09*
	Educational level	-.08*
	Adjusted R^2	.03
	df	4, 959
	F	8.21***
	n	964
Rabobank	Advertising intensity newspapers	.09*
	Educational level	-.09**
	Adjusted R^2	.01
	df	2, 868
	F	6.21**
	n	871
Police	Amount of TV news	.16***
	Direction of success & failure TV news	.11***
	Adjusted R^2	.04
	df	2, 942
	F	20.99***
	n	945

TABLE K1.   (Continued)

Models per organization[a]	Predictors (news and advertising, previous year)	
Agricultural sector	Favorability of success & failure TV news	-.08*
	Educational level	-.14***
	Adjusted R²	.02
	df	2, 921
	F	10.95***
	n	924

[a] Cell entries are Betas from Ordinary Least Squares regression. For ABN AMRO, no significant effects were found.

$* p \leq .05; ** p \leq .01; *** p \leq .001$

The cross-sectional models of BP and Albert Heijn did not change after controlling for the educational level of the respondents. These models are not presented in Table K1. Apart from some changes in the regression coefficients, the models of the NS, Rabobank, the police, and the agricultural sector did not change when the models were controlled for education. In the case of the NS, Rabobank, and the agricultural sector, the education variable was added to the model.

In the model of Schiphol, two media variables – the amount of television news and the favorability of success & failure television news – remained significant.[13]

The models of Shell and Super de Boer changed the most after controlling for education. In the model of Shell, the media variables and the advertising variables were no longer significant[14] after controlling for the educational level of respondents. When the model of Super de Boer was controlled for education, the amount of print news and the favorability of success & failure print news was no longer significant due to a small decrease in the number of cases.[15] This means that the non-autoregressive model of Super de Boer is relatively unstable.

Table K2 presents the results of the non-autoregressive pooled model of the effects of the amount of news, the favorability and the direction of success & failure news and the advertising expenditures on reputation, controlled for the educational level of the respondents.

TABLE K2   **Pooled confirmatory regression analysis of the impact of the amount of news, the tenor of success & failure news and advertising on corporate reputation, controlled for the educational level**

Predictors (news and advertising, previous year)	Reputation
Key variables	
Amount of TV news	.11***
Amount of print news	-.03*
Favorability of success & failure TV news	.07***
Favorability of success & failure print news	.01
Direction of success & failure TV news	-.004
Direction of success & failure print news	.01
Advertising intensity newspapers	.06***
Advertising intensity TV	.03
Control variables	
Educational level of respondents	-.05***
Shell	-.1***
BP	-.06***
Albert Heijn	.03
Super de Boer	-.05**
NS	-.23***
Schiphol	.01
ABN AMRO	-.05*
Rabobank	-.03
Police	-.20***
Adjusted R^2	.06
df	18, 9089
F	33.21***
N	9,108

Note. Cell entries are Betas from Ordinary Least Squares regression. The agricultural sector was the base category.
* $p \leq .05$; ** $p \leq .01$; *** $p \leq .001$

When the pooled model was controlled for the educational level of the respondents, the model did not change except for the education variable that was added to the model ($ß = -.05$). The negative direction of Beta indicates that the higher the educational level of the respondents, the more negative their attitude towards the organizations.[16]

The autoregressive pooled model was not controlled for education, because it did not contain any significant media or advertising variables.

APPENDIX L – THE EFFECTS OF THE AMOUNT OF NEWS, THE TENOR OF SUPPORT & CRITICISM NEWS, AND ADVERTISING ON REPUTATION CONTROLLED FOR EDUCATIONAL LEVEL

Table L1 displays the effects of the amount of news, the tenor of support & criticism news, and advertising on reputation controlled for the educational level of respondents.

In the case of three models (Albert Heijn, BP, Schiphol), there was only a minor change in the regression coefficients after controlling the non-autoregressive models for the educational level of the respondents. These models are not presented in Table L1. The model of Rabobank hardly changed either, except for the education variable that was entered in the model.

TABLE L1.   Stepwise regression analyses of the effects of the amount of news, the tenor of support & criticism news, and advertising on report mark, controlled for educational level

Models per organization[a]	Predictors (news and advertising, previous year)	Reputation
Shell	Favorability of support & criticism TV news	-.07*
	Adjusted R^2	.004
	df	(1, 969)
	F	4.56*
	n	971
Super de Boer	Educational level	-.16***
	Adjusted R^2	.02
	df	(1, 743)
	F	18.47***
	n	745
NS	Favorability of support & criticism TV news	-.07*
	Educational level	-.08*
	Adjusted R^2	.009
	df	(2, 959)
	F	5.17**
	n	962
Rabobank	Advertising intensity newspapers	.09*
	Educational level of the respondents	-.09**
	Adjusted R^2	.01
	df	(2, 868)
	F	6.21**
	n	871

TABLE L1.   (Continued)

Models per organization[a]	Predictors (news and advertising, previous year)	Reputation
Police	Amount of TV news	.18***
	Adjusted R^2	.03
	df	(1, 943)
	F	31.16***
	n	945
Agricultural sector	Amount of TV news	.07*
	Educational level	-.13***
	Adjusted R^2	.02
	df	(2, 921)
	F	10.14***
	n	924

[a] Cell entries are Betas from Ordinary Least Squares regression. For ABN AMRO, no significant effects were found.

* $p \leq .05$; ** $p \leq .01$; *** $p \leq .001$

A more detailed comparison between the cross-sectional models presented in Table 5.4 of Chapter 5 and the non-autoregressive models after controlling for education in Table L1 is provided below.

When the model of Shell was controlled for the educational level of the respondents, the amount of television news and the direction of support & criticism TV news were no longer significant, due to the minor change in the number of cases. Instead, the favorability of support & criticism TV news became significant ($ß = -.068$, $p < .05$). The direction of the regression coefficient was in agreement with the direction of the regression coefficient of support & criticism that was significant in Table 5.4. The more the NGOs criticize Shell, the better its reputation will be.[17]

In the model of Super de Boer, the advertising expenditures on TV variable, which was significant in the non-autoregressive model of Table 5.4, was not significant when the model was controlled for the educational level of the respondents. The educational level was the only significant variable[18] in Table L1.

When the model of the NS was controlled for the educational level of the respondents, this control variable became significant ($ß = -.079$, $p < .05$). The higher the education of the respondents, the more negative their attitude towards the NS. The amount of news was no longer significant, while the favorability of support & criticism TV news became significant ($ß = -.074$, $p < .05$). As in the model of Shell, the direction of the regression coefficient was negative. This indicates that the more the NS is criticized by the labor unions, the better its reputation will be.

In the model of the police, the amount of print news no longer has a significant influence on the reputation of the police when controlled for the educational level of the respondents. This was due to the small decrease in the number of cases. This was discovered by estimating the model of Table 5.4 with the same number of cases as in Table L1, while omitting the educational level of the respondents. In that model, the amount of print news was no longer significant either. Furthermore, it became clear that the educational level of the respondents did not change the model of the police, because it was not significant.

In the model of the agricultural sector, the advertising expenditures and the favorability of support & criticism print news were no longer significant after controlling the model for the educational level of the respondents. Instead, the amount of television news became significant ($ß = .066$, $p < .05$). The more television news about the agricultural sector, the better its reputation. Furthermore, the educational level of the respondents became significant when this variable was added to the model. The higher the educational level of the respondents, the more negative their attitude towards the agricultural sector ($ß = -.134$, $p < .001$).

In sum, in four models there was only a slight change in the regression coefficients of the key variables (Albert Heijn, BP, Schiphol, Rabobank). The models of Shell and the police changed solely due to the decrease in the number of respondents. The models of Super de Boer, the NS, and the agricultural sector changed due to the addition of the education variable. In the model of Super de Boer, multicollinearity was present.

Table L2 presents the results of the pooled model of the effects of the amount of news, the favorability and the direction of support & criticism news, and advertising on reputation, controlled for the educational level of the respondents.

When the pooled model was controlled for the educational level of the respondents, the amount of television news and the advertising expenditures in newspapers remained significant, as in the non-autoregressive pooled model without controlling for education (Table 5.5). The only difference was that when the model was controlled for education, the amount of print news was nearly significant ($ß = -.027$, $p = .086$), while it was significant in Table 5.5.

TABLE L2.   **Pooled confirmatory regression analysis of the impact of the amount of news, the tenor of support & criticism news, and advertising on corporate reputation, controlled for educational level**

Predictors (news and advertising, previous year)	Reputation
Key variables	
Amount of TV news	.11***
Amount of print news	-.03
Favorability of support & criticism TV news	-.01
Favorability of support & criticism print news	.00
Direction of support & criticism TV news	-.02
Direction of support & criticism print news	.00
Advertising intensity newspapers	.05***
Advertising intensity TV	.03
Control variables	
Educational level of respondents	-.05***
Shell	-.07***
BP	-.02
Albert Heijn	.05**
Super de Boer	-.03
NS	-.21***
Schiphol	.05**
ABN AMRO	-.03
Rabobank	.01
Police	-.12***
Adjusted R^2	.06
df	(18, 9089)
F	32.49***
N	9,108

Note. Cell entries are Betas from Ordinary Least Squares regression. The agricultural sector was the base category.
* $p \leq .05$; ** $p \leq .01$; *** $p \leq .001$

When the autoregressive pooled model was controlled for education, none of the media and advertising variables was significant, as in the autoregressive pooled model in which education was omitted. The educational level of respondents was not significant, either, when the autoregressive pooled model was controlled for education.

Appendix M – Testing causality

This appendix tested several pooled models to examine the causality. In this sub-section, it will be tested whether corporate reputation influences media coverage. In order to do so, several media variables were entered successively in the models as the dependent variables: the amount of television news, the amount of print news, the direction and the weight of success & failure television news, the direction and the weight of success & failure print news, the direction and the weight of support & criticism television news and the direction and the weight of support & criticism print news, the frequency of issue television news, and the frequency of issue print news.

Bearing in mind the scope of this appendix, it was decided to report only the full models that examined the effects of past reputation on the direction of the news, and not the models that examined the effects of past reputation on the weight of the news. Nevertheless, these analyses were conducted as well. It appeared that past reputation did not influence the weight of success & failure TV news, the weight of success & failure print news, the weight of support & criticism television news, or the weight of support & criticism print news. However, the past salience of an issue did influence the frequency of issue news.

TABLE M1.   **Pooled confirmatory regression analysis of the impact of past media coverage and past reputation on the amount of television news**

Predictors (news and advertising, previous year)	Amount of TV news
Key variables	
Amount of TV news, $_{t-1}$	.88***
Amount of print news, $_{t-1}$	-.00
Direction of success & failure TV news, $_{t-1}$	.03**
Direction of success & failure print news, $_{t-1}$	.00
Reputation, $_{1999}$	.01
Control variables	
Shell	-.08***
BP	-.01
Albert Heijn	.07***
Super de Boer	.01
NS	.44***
Schiphol	-.19***
ABN AMRO	.09***
Rabobank	-.01
Police	.16***
Adjusted R^2	.95
df	(14, 2023)
F	2540***
N	2,038

Note. Cell entries are Betas from Ordinary Least Squares regression. The agricultural sector was the base category.
$^\wedge$ p $\leq$ .05; $^{\wedge\wedge}$ p $\leq$ .01; *** p $\leq$ .001

Table M1 presents the results of the pooled model, which examines whether the prediction of the amount of television coverage based on past media influence can be improved by past reputation.

As shown in Table M1, past reputation does not influence the amount of television news. The amount of television news in the previous year is a strong predictor of the amount of television news in the current year. This is probably due to the fact that the media use of respondents was taken into account.

Table M2 presents the results of the pooled model, which examines whether the prediction of the amount of print news based on past media influence can be improved by past reputation.

TABLE M2. **Pooled confirmatory regression analysis of the impact of the past media coverage and past reputation on the amount of print news**

Predictors (news and advertising, previous year)	Amount of print news
Key variables	
Amount of TV news, $_{t-1}$	.01
Amount of print news, $_{t-1}$	.93***
Direction of success & failure TV news, $_{t-1}$	-.02
Direction of success & failure print news, $_{t-1}$	.02*
Reputation, $_{1999}$	-.01
Control variables	
Shell	-.05***
BP	-.02
Albert Heijn	.03*
Super de Boer	-.01
NS	.18***
Schiphol	-.08***
ABN AMRO	.04**
Rabobank	.01
Police	-.02
Adjusted R^2	.85
df	(14, 2023)
F	845,56***
N	2,038

Note. Cell entries are Betas from Ordinary Least Squares regression. The agricultural sector was the base category.
* $p \leq .05$; ** $p \leq .01$; *** $p \leq .001$

Past reputation does not influence the amount of print news. As in the previous subsection, the lagged dependent variable (in this case the amount of print news of the previous research period) strongly influences the dependent variable (the amount of print news). This is probably due to the fact that the media use of respondents was taken into account.

Table M3 presents the results of the pooled model, which examines whether the prediction of the direction of success & failure television news based on past media influence can be improved by past reputation. The variable "weight of success & failure news" was not included in the models in order to avoid multi-collinearity.

TABLE M3.   **Pooled confirmatory regression analysis of the impact of the past media coverage and past reputation on the direction of success & failure television news**

Predictors (news and advertising, previous year)	Direction of S&F TV news
Key variables	
Amount of TV news, $_{t-1}$	.09***
Amount of print news, $_{t-1}$	-.00
Direction of success & failure TV news, $_{t-1}$	-.02
Direction of success & failure print news, $_{t-1}$	.01
Reputation, $_{1999}$	-.01
Control variables	
Shell	.62***
BP	.08***
Albert Heijn	.62***
Super de Boer	.76***
NS	.39***
Schiphol	.46***
ABN AMRO	.22***
Rabobank	-.04
Police	.48***
Adjusted R^2	.78
df	(14, 2023)
F	524,93***
N	2,038

Note. Cell entries are Betas from Ordinary Least Squares regression. The agricultural sector was the base category.
* $p \leq .05$; ** $p \leq .01$; *** $p \leq .001$

Table M3 shows that past reputation does not influence the direction of success & failure television news. Contrary to the two models presented in the previous subsections, the lagged dependent variable (the direction of success & failure television news in the previous research period) does not have a significant impact on the dependent variable (the direction of success & failure television news).

Table M4 presents the results of the pooled model, which examines whether the prediction of the direction of success & failure print news based on past media influence can be improved by past reputation. As was the case in the previous subsection, the weight of success & failure news variable was not included in the models in order to avoid multicollinearity.

TABLE M4.   **Pooled confirmatory regression analysis of the impact of the past media coverage and past reputation on the direction of success & failure print news**

Predictors (news and advertising, previous year)	Direction of S&F print news
Key variables	
Amount of TV news, $_{t-1}$	.01
Amount of print news, $_{t-1}$	-.01
Direction of success & failure TV news, $_{t-1}$	.04
Direction of success & failure print news, $_{t-1}$	.02
Reputation, $_{1999}$	.01
Control variables	
Shell	.35***
BP	.32***
Albert Heijn	.23***
Super de Boer	.13***
NS	.03
Schiphol	.22***
ABN AMRO	.19***
Rabobank	.11**
Police	.42***
Adjusted R^2	.19
df	(14, 2023)
F	34,45***
N	2,038

Note. Cell entries are Betas from Ordinary Least Squares regression. The agricultural sector was the base category.
* $p \leq .05$; ** $p \leq .01$; *** $p \leq .001$

Past reputation does not influence the direction of success & failure print news. Again, the direction of success & failure (print) news, is not influenced by the past direction of success & failure (print) news.

Table M5 presents the results of the pooled model, which examines whether the prediction of the direction of support & criticism television news based on past media influence can be improved by past reputation. The weight of support & criticism news variable was not included in the models in order to avoid multi-collinearity.

TABLE M5   **Pooled confirmatory regression analysis of the impact of the past media coverage and past reputation on the direction of support & criticism TV news**

Predictors (news and advertising, previous year)	Direction of S&C TV news
Key variables	
Amount of TV news, $_{t-1}$	.02
Amount of print news, $_{t-1}$	.00
Direction of support & criticism TV news, $_{t-1}$	-.09***
Direction of support & criticism print news, $_{t-1}$	.01
Reputation, $_{1999}$	.00
Control variables	
Shell	-.11***
BP	.52***
Albert Heijn	-.37***
Super de Boer	-.34***
NS	.05**
Schiphol	.07***
ABN AMRO	.09***
Rabobank	.38***
Police	.11***
Adjusted R²	.73
df	(14, 2023)
F	391,45***
N	2,038

Note. Cell entries are Betas from Ordinary Least Squares regression. The agricultural sector was the base category.
* $p \leq .05$; ** $p \leq .01$; *** $p \leq .001$

As shown in Table M5, past reputation does not influence the direction of support & criticism TV news. This indicates that support & criticism influences corporate reputation and not vice versa. In addition, it appears that the lagged dependent variable (the direction of support & criticism TV news in the previous research period) influences the dependent variable.

Table M6 presents the results of the pooled model, which examines whether the prediction of the direction of support & criticism print news based on past media influence can be improved by past reputation. The variable "weight of support & criticism news" was not included in the models in order to avoid multicollinearity.

TABLE M6.   **Pooled confirmatory regression analysis of the impact of past media coverage and past reputation on the direction of support & criticism print news**

Predictors (news and advertising, previous year)	Direction of S&C print news
Key variables	
Amount of TV news, $_{t-1}$	.02
Amount of print news, $_{t-1}$	-.12
Direction of support & criticism TV news, $_{t-1}$	-.08
Direction of support & criticism print news, $_{t-1}$	.26***
Reputation, $_{1999}$	-.03
Control variables	
Shell	.12***
BP	.09
Albert Heijn	-.05
Super de Boer	.18***
NS	-.19***
Schiphol	.22***
ABN AMRO	.17***
Rabobank	.22***
Police	-.10*
Adjusted R^2	.27
df	(14, 2023)
F	55.55***
N	2,038

Note. Cell entries are Betas from Ordinary Least Squares regression. The agricultural sector was the base category.
* $p \leq .05$; ** $p \leq .01$; *** $p \leq .001$

As shown in Table M6, past reputation does not influence the direction of support & criticism print news. This indicates that support & criticism influences corporate reputation and not vice versa. In addition, it appears that the lagged dependent variable (the direction of support & criticism print news in the previous research period) influences the dependent variable. Also, there is a negative relationship between the amount of print news and the direction of support & criticism print news. This indicates that more print news in a given year results in more negative support & criticism print news in the next year. It is not clear why this effect occurs.

The last tables in this appendix, Table M7 and Table M8, present the results of
two pooled models that examine whether the prediction of the amount of issue
news based on past media influence can be improved by the salience of that issue
in the past. In other words, it is examined whether news about an organization
and an issue (such as environment) can be predicted by the number of people
who associated that company with that issue in the past. Table M7 presents a
model that estimates the impact of past media coverage and public opinion on
the frequency of issue TV news.

TABLE M7.   **Pooled confirmatory regression analysis of the impact of past media coverage
and past salience of an issue on the frequency of issue television news**

Predictors (news and advertising, previous year)	*Freq. of issue TV news*
Key variables	
Frequency of issue TV news, $_{t-1}$	.59***
Frequency of issue TV news, $_{t-2}$	.26***
Frequency of issue print news, $_{t-1}$	.02***
Frequency of issue print news, $_{t-2}$	.03***
Salience of an issue, $_{t-1}$	.02***
Salience of an issue, $_{t-2}$	.03***
Control variables	
Shell	-.13***
BP	-.05***
Albert Heijn	-.03***
Super de Boer	-.05***
NS	.05***
Schiphol	-.11***
ABN AMRO	-.03***
Rabobank	-.04***
Police	.13***
Adjusted R^2	.79
df	(15, 18119)
F	4664.92***
N	18,135

Note. Cell entries are Betas from Ordinary Least Squares regression. The agricultural sector was the base category.
* $p \leq .05$; ** $p \leq .01$; *** $p \leq .001$

Table M7 shows that if an issue was salient in the past there will be more television
news about that issue. This indicates there is a simultaneous relationship between
the frequency of issue TV news and the salience of an issue. The frequency of issue
TV news in the past is by far the most important predictor of issue TV news. Again,
this is probably due to the fact that the media use of respondents was taken into
account.

Table M8 presents a model that estimates the impact of past media coverage and public opinion on the frequency of issue print news.

TABLE M8.   **Pooled confirmatory regression analysis of the impact of the past media coverage and past salience of an issue on the frequency of issue print news**

Predictors (news and advertising, previous year)	Freq. of issue print news
Key variables	
Frequency of issue TV news, $_{t-1}$	-.06***
Frequency of issue TV news, t-2	.04***
Frequency of issue print news, t-1	.79***
Frequency of issue print news, t-2	.16***
Salience of an issue, t-1	.01***
Salience of an issue, t-2	.02***
Control variables	
Shell	-.08***
BP	-.02***
Albert Heijn	-.00
Super de Boer	-.01**
NS	.04***
Schiphol	-.01*
ABN AMRO	.02**
Rabobank	.01
Police	.02***
Adjusted R^2	.81
df	(15, 18119)
F	5009.80***
N	18,135

Note. Cell entries are Betas from Ordinary Least Squares regression. The agricultural sector was the base category.
* $p \le .05$; ** $p \le .01$; *** $p \le .001$

The results in the case of issue print news are in agreement with the results of issue television news. Table M8 shows that if an issue was salient in the past, there will be more print news about that issue. This indicates there is a simultaneous relationship between the frequency of issue print news and the salience of an issue. The frequency of issue print news in the past is by far the most important predictor of issue print news. Surprisingly, there is a negative effect of past frequency of issue TV news at one lag on the frequency of issue print news.

NOTES

1 – INTRODUCTION

1   Enron was once a major American energy company but went bankrupt in 2001 because of an accounting scandal. For publications about the Enron Corporation fiasco, see Hotaling and Lippitt (2003), Petrick and Scherer (2003), and Sims and Brinkmann (2003).

2   Lazarsfeld et al. (1944) did give attention to the partisanship of the newspapers, radio, and magazines. In addition, they also discussed the main theme of the campaign news, which was the campaign itself. The content of the media coverage was not linked to the media use of the respondents, however, so that attitude shifts resulting from exposure to the news could not be traced.

3   These mediating factors are: individual predispositions and selective perception processes, group norms, message dissemination via interpersonal channels, opinion leadership, and the free-enterprise nature of the mass media in some societies.

4   On the contrary, Klapper devoted a chapter to conversion of attitudes and described findings of research in which conversion did occur. For example, three quarters (71%) of a test audience for the British film "Naples is a Battlefield" changed their pre-exposure opinion that the U.S. should not send food to Italy (op.cit. Klapper 1960). He stated that conversions have been unintentionally hidden by many reports, which tell only of the "net effects." Klapper gave the example of a study of the effects of a campaign to improve attitudes towards the oil industry, which reported a net shift of 4% in the direction intended. In actuality, 22% of the respondents had changed their opinions, 13% in the direction intended and 9% in the opposite direction.

5   The *California Management Review*, the *Journal of Marketing* and the *Journal of Marketing Research* were examined electronically. The content of these journals from the period 1958-1970 was searched using the commands "content analysis," "media coverage," and "media content." This resulted in one article about content analysis (see Kranz, 1970). The *Harvard Business Review* from the period 1958-1970 was examined manually. No articles were found about media coverage, media content, or content analysis (Issues 4 and 5 of volume 46 were missing, however).

6   There are definitions without this common element of overall evaluation. For example, Bromley (2000, p. 241) defined corporate reputation as, "The way key external stakeholders' groups or other interested parties actually conceptualize that organization."

7   In this regard, Van Riel (1997) used the term "corporate image" for the image at the level of the holding (such as Ahold), the term "company image" for the image of a subsidiary (such as Albert Heijn) and the term "store image" for the image at the level of the store (such as the image of a particular neighborhood Albert Heijn).

8   Verčič (2000) criticized corporate reputation researchers for not testing these types of models longitudinally.

9 *Fortune* magazine's annual survey of corporate reputation is regularly used in corporate reputation research (Black et al., 2000; Fombrun & Shanley, 1990; Jones et al., 2000). In *Fortune*'s survey, each company is rated relative to its leading competitors on eight characteristics using an 11-point scale (0 = poor, 10 = excellent). The eight characteristics are: innovation, financial soundness, employee talent, use of corporate assets, long-term investment value, social responsibility, quality of management, and quality of products and services. To arrive at each company's final score – which determines its ranking in its industry group – the scores that survey respondents gave were averaged on these eight criteria. The respondents consist of 10,000 executives, directors, and securities analysts. They did not rate all the firms, but only those in their own industry or economic sector.

10 For a more elaborated discussion on bilateral causality, see for example Gujarati (1995).

2 – EFFECTS OF NEWS AND ADVERTISING ON REPUTATION

1 Part of this subsection has been translated from an article by Meijer and Kleinnijenhuis (2001).

2 Stone et al. (1966) did not provide references when they stated that "traditional research into product image has concentrated mainly on the image of the product as conveyed through mass media." Furthermore, they used the term "product image" and not "corporate image." These concepts do have a lot in common, however.

3 The "neutral category" was not used, since raters tended to either over- or underuse that category (Fombrun & Shanley, 1990).

4 Fombrun and Shanley (1990) reported that a varimax factor analysis of the eight attributes extracted a single factor. On basis of their empirical findings, Fryxell and Wang (1994) concluded that this single factor appears to be predominantly financial. According to them, the usefulness of the Fortune reputation data is limited to measuring the extent to which a firm is perceived as being financially successful.

5 Wartick (1992) studied 29 companies, which he divided into three subgroups: organisations with a good, an average, and a bad reputation. He also distinguished three types of changes in corporate reputation: 1) the magnitude of change in corporate reputation, 2) the direction of change in corporate reputation, and 3) the total movement of change in corporate reputation, which is continuous from largest positive change to largest negative change.

6 Verčič (2000) used the term *trust* people place in companies, rather than the concept of *corporate reputation*.

7 The concept was operationalized in two variables: one expressing the relative and one the absolute number of negative stories about a focal company. The changes in the percentages of negative stories are symmetrical to the changes of positive stories (e.g., a 5% increase in negative articles is a 5% decrease in positive articles, and vice versa).

8 According to Green's (1992) rule of thumb, the number of cases in regression analysis should be more or equal to "50 cases + 8 × number of independent variables." Green (1992) also proposed a slightly more complex rule of thumb that estimates minimum sample size on the basis of effect size as well as on the number of predictors.

9 Zajonc was primarily concerned with the effects of repeated exposure to the attitude objects themselves (such as the Chinese-like characters) and not with the effects of repeated exposure to *information* about the attitude object (Cacioppo & Petty, 1979).

10 This is called *judgement extrapolation* (see Van Cuilenburg, Kleinnijenhuis & De Ridder 1988).

11 In American publications the term "horse race coverage" is used. The following publications describe horse race coverage in American politics: Brady and Johnston (1987), Broh (1977), and Patterson (1980).

12 Van Praag and Van der Eijk (1998) found higher percentages of horse race news in campaign news during the Dutch elections of 1994 than Kleinnijenhuis et al. (2003). Van Praag and Van der Eijk analyzed the news of only two television channels (NOS and RTL). According to their results, 30% of the news on both television channels consisted of horse race news in 1994. Kleinnijenhuis et al. (2003), however, did not report the percentage of horse race news in the 1994 elections per medium.

13 In my opinion, it is important to note that it doesn't matter whether others are actually asking for more or less of the product, but whether people *think* that others are asking for more or less of the product.

14 The experiments 1 to 3 were originally designed to test the agenda-setting hypothesis. For the purpose of testing the priming hypothesis, they had some shortcomings, such as the inclusion of too few participants in each condition. For this reason, another experiment was designed by Iyengar & Kinder to overcome these limitations.

15 In categorical priming a person's judgment about a person or product is influenced by the constructs that are activated in an earlier task (Herr, 1989).

16 Although Budge and Farlie (1983) mentioned that parties are widely perceived as "owning" certain issue types, they usually use the concept of *Saliency Theory* rather than *issue ownership*. The term saliency theory stresses the attempts of the political parties to make their favorable issues salient.

17 Though *wearout* generally refers to the loss of effectiveness of an ad, several authors (Craig et al., 1976; Calder & Sternthal, 1980; Nordhelm, 2002) use the term to refer to the process by which commercials are initially effective, while subsequent exposures causes effectiveness to level off and ultimately decline.

18 Bornstein (1989) mentioned that the common procedure in mere exposure studies is to present stimuli a maximum of 10-50 times, with the mean ceiling in the number of stimulus presentations being 20.95 ($SD = 32.28$).

3 – METHOD

1 Respondents who did not use any of the focal media are included in these numbers.

2 The NIPO Telepanel was replaced by CASI (computer-assisted self-interviewing), a database consisting of approximately 40,000 households.

3 Van Cuilenburg et al. (1988) used the term *meaning object* instead of *attitude object*, since the term meaning object is broader and can also be used in non-evaluative texts.

4 For a description of the types of predicates, see also Kleinnijenhuis et al. (1997).

5 As mentioned by Van Cuilenburg et al. (1992), the concepts "attitude-object," "common meaning term," "verbal connector," and "assertion" of Osgood et al. (1954) are important building blocks of evaluative text analysis.

6 Therefore, in her study of the NET-method from a linguistic perspective, Vis (2002) recommended using the term "actor" instead of subject and the term "undergoer" instead of the term object. This recommendation was not applied, however, and the current use of the term actor is used for organizations or persons that can also fill the role of object.

7 The terminology as used by De Ridder (1994a) differs slightly from Van Cuilenburg et al.'s (1988) publication. Van Cuilenburg et al. used the concepts "arrows," "links," "chains," and "bundles," while De Ridder used the concepts "arrows," "paths," and "bundle." The concept "chains" is replaced by de Ridder with the concept "paths," while the concept "bundle" is used in the same respect. Van Cuilenburg et al. explicitly distinguished between "arrows" and "links," a distinction that was not made by De Ridder.

8 The NET-method departs from the representational model of communication (Van Cuilenburg et al., 1988), which postulates that a text represents what the author means.

9   The transitivity axiom is discussed by Abelson (1968). He showed that the transitivity ax-
    iom (or "evaluative induction hypothesis" as he called it) can lead to peculiar conclusions.
    Van Cuilenburg et al. (1988) admitted that the transitivity axiom can lead to peculiar con-
    clusions if two actors are after the same *scarce* object. If two men are in love with the same
    woman, it does not necessarily mean the two men also love each other, as would follow
    from the transitivity axiom. The solution is to create two separate meaning objects: "love of
    the woman for man A" and "love of the woman for man B." The relationship between
    these two meaning objects is negative. For a more detailed explanation, see Chapter 6 of
    Van Cuilenburg et al. (1988) and Abelson (1968).

10 FactLANE has been owned by LexisNexis since January 2003, and has been included in the
    LexisNexis database since 2004.

11 LexisNexis did not indicate the page numbers of the articles from *De Telegraaf*. Therefore,
    the average page number of *De Telegraaf* articles selected manually is taken to be the page
    number of *De Telegraaf* articles retrieved from the LexisNexis database.

12 With regard to *De Telegraaf* articles selected manually, the coders wrote down the size of
    the column instead of the number of words. In a phone call (on May 21, 2002), the edito-
    rial staff of *De Telegraaf* stated that 100 words are 80 millimeters long, and a column is 550
    millimeters. This means that a column without a picture contains approximately 687
    words. Therefore, an average of 600 words per column was used. Those smaller than a
    quarter of a column were counted as 100 words.

13 With regard to news about the agricultural sector in *De Telegraaf*, articles about BSE and
    swine fever were not selected during the period from July 24, 1997 to January 1, 1999.
    Nevertheless, it is not expected this will influence the testing of the hypotheses that are
    related to issues (hypotheses 4, 5, and 6). Hypothesis 4 tests the effects of issue news on
    salience. Animal diseases belong to the category "other associations" (see Appendix F),
    since this association was not mentioned by the respondents. The same applies to hypoth-
    esis 6. Animal diseases belong to the category "other associations" because this associa-
    tion was not mentioned by the respondents. Hypothesis 6 tests the effects of news about
    owned issues on reputation. Swine fever and BSE were not perceived to be owned issues
    for the agricultural sector, however.

14 Common meaning terms can also belong to "the list of meaning objects" used by the in-
    vestigator to code the text.

15 A total of 21 headlines (20 headlines on Shell and 1 on BP) were coded by the researcher by
    mistake. This is 2% of the total number of headlines or news items about Shell and BP.

16 More precisely, for each issue an estimate was made of the magnitude and direction of
    change it would produce in support for the political party in question (Budge & Farlie,
    1983).

17 Andreassen and Lanseng (1997) used the term "corporate image" instead of the term "cor-
    porate reputation."

18 Although this is a different example, the structure of the example is based on the one given
    by Ajzen and Fishbein (1980) in describing a woman's beliefs about using birth control
    pills. However, in that example a unipolar scale was used and the woman emitted the be-
    liefs herself.

19 It might be questioned whether the belief strength component can take a value of -1. By
    using the term "likely," Fishbein and Ajzen (1980) suggested that they, like Huygens
    (1998), used a probability measure for belief strength. This would indicate that in the case
    that something is extremely unlikely, the value of belief strength should be 0 instead of -1,
    since a chance cannot be negative.

20 Although in marketing research a weight or importance measure is used, this is not the
    case here. Sheth and Talarzyk (1972) found that this component somewhat lowered the
    predictive power of the model. That is why Fishbein argued that salient beliefs should not

be weighted for their importance in determining attitude (Ajzen & Fishbein, 1980; Cohen et al., 1972; Fishbein & Ajzen, 1975). He stated that all salient beliefs are important, and any differences in their degree of importance are included by the measures of belief strength and outcome evaluation. To put it in another way, importance will "tend to be related to *polarity* of evaluation." For a more complete discussion about the use of the importance measure, see Fishbein and Ajzen (1975).

21 Respondents who did not use any of the focal media were excluded from this number.

22 On average over the three years, this was 4% of the respondents per sample per year.

23 A low level of education means the respondents attended one of the following Dutch secondary school types: LO or LBO. A moderate level of education means the respondents attended one of the following Dutch secondary school types: MAVO or MBO. A high level of education means the respondents attended one of the following Dutch secondary school types: HAVO, HBO, or WO.

24 Since more moderately educated respondents participated in this sample than those with higher levels of education, there are more newspaper readers with moderate than higher educational levels. On the average of the three years, nearly three quarters (74%) of the more highly educated respondents subscribed to a newspaper; for respondents with moderate and lower levels of education, this percentage was 54% and 37% respectively.

25 Strictly speaking, regression analysis is a statistical tool that requires both the dependent variable and the independent variables to be metric (Hair et al., 1998); this means that the variables should be measured at an interval or a ratio scale. In this study, it can be argued that the frequently used reputation measure "report mark" is an ordinal or ratio scale. The variables "success of the policy of the organization" and "agreement with the policy of the organization" are both ordinal. As mentioned by Tacq (1997), the number of techniques in which the ordinal character of the variables is respected is very small. He stated that contributions to develop an ordinal equivalent for multiple regression analysis are very complicated in arithmetic, and their ins and outs have not yet been elaborated from an inferential point of view (i.e., for generalization from the sample to the population). In line with the dominant social science research tradition, regression analysis was employed despite the ordinal nature of some of the dependent variables. See also the comparable study of Verčič (2000) in which time series regression analysis was used as well, even though the dependent variable was ordinal. The variables such as "success of the policy of the organization" and "agreement with the policy of the organization" were measured with two associations, with 3 categories per association. This led to 6 different scores.

26 Heteroscedasticity is one of the most common assumption violations (Hair et al., 1998). As mentioned by Gujarati (1995), heteroscedasticity is generally expected if data with heterogeneous units (small, medium-sized, and large firms) are sampled together. In this study, data of the same size (the respondents) were the unit of analysis, and because of this, heteroscedasticity is not expected to be a problem here.

27 As mentioned by Gujarati (1995), strictly speaking, multicollinearity refers to the presence of more than one exact linear relationship, and collinearity refers to the existence of a single linear relationship. This distinction is rarely maintained in practice, and multicollinearity refers to both cases.

28 As is mentioned by Gujarati (1995), some authors distinguish between the terms autocorrelation and serial correlation. Whereas autocorrelation is the lag correlation of a given series with itself, lagged by a number of time units, serial correlation is the lag correlation between two different time series. Nevertheless, Gujarati (1995) stated that, as he does himself, it is common practice to treat the terms autocorrelation and serial correlation synonymously. This was also done in this study.

29 The number of cases in these analyses was based upon the respondents who participated in 2000 (column C of Table 3.7), the respondents who participated in 1999 (column D of

Table 3.7), and the respondents who participated in 1998-1999-2000 (column F of Table 3.7). The latter group – the respondents who participated in three years – count twice, because they gave report marks in 1999 and in 2000.

30   There is no unique critical value that will lead to the rejection of the acceptance of the null hypothesis that there is no first-order autocorrelation in the disturbances u_i (Gujarati, 1995).

5 – RESULTS

1   This variable became significant after the removal of the variable advertising on television news. The variable advertising expenditures on television news was removed from the model of Shell, since it was highly correlated with the variable favorability of success & failure TV news ($r = .94$, $p < .01$, see Table F1), and suppressed its effects.

2   This variable became significant after the removal of the variable advertising in television news. If this variable was included in the confirmatory specification model, the variable "Favorability of success & failure TV news" had a VIF value of 18.18, which was far above the threshold of 10. This indicates a high degree of multicollinearity among the independent variables. After removal of the control variable, the VIF values of the media variables all reached an acceptable level (beneath the threshold of 10).

3   After the inclusion of the lagged dependent variable, all VIF values were beneath the threshold of 10.0.

4   Like the model of Shell, the model of Super de Boer contained high VIF values (the highest VIF value is 733) if the model was controlled for the educational level of respondents. The educational level is unknown for five of the respondents.

5   Strictly speaking, it was not necessary to include the dummy variables of the organizations in the model, since the lagged dependent variable was included in the model. However, the dummy variables were included in the model for reasons of comparison.

6   The variable "advertising expenditures television news" had a VIF value of 10.4 and thus slightly exceeded the threshold of 10.0. This did not suppress the effects of the media variables, however.

7   Strictly speaking, it was not necessary to include the dummy variables of the organizations in the model, because the lagged dependent variable was included. However, the dummy variables were included in the model for reasons of comparison.

8   In the case of Schiphol, the *favorability* of success & failure print news was significant in the model that consisted only of the amount of news and success & failure news, whereas in the model that consisted of the amount of news, success & failure news, and support & criticism news, the *direction* of success & failure print news was significant.

9   For the agricultural sector, support & criticism news is in the model with the two types of news no longer significant.

10   Furthermore it should be remarked that if the non-autoregressive model of ABN AMRO is estimated with the same number of cases as in the autoregressive model of ABN AMRO, the direction of success & failure print news becomes significant ($ß = .17$, $p < .05$). This influence disappears due to the inclusion of the lagged dependent variable into the autoregressive model.

11   Because there were two points in time and hence the assumption of sphericity will not be violated, Maucly's test of sphericity will not be reported.

12   In the model of Super de Boer, the educational level of five respondents was unknown; in the model of the police the educational level of four respondents was unknown.

13   In the case of the model of Shell (Table 5.2), the significant "television advertising expenditures" variable was excluded from the model because of its high correlation with the favorability of success & failure TV news (see endnote 1 of this chapter).

14   In the model of the agricultural sector (Table 5.2), the variable "television advertising expenditures" was excluded from the model due to the high VIF value of the variable "Favorability of success & failure TV news" (see endnote 2 of this chapter).

6 – Conclusions and discussion

1   The pseudo R^2 is comparable with the R^2 measure in multiple regression (Hair et al., 1998).
2   Part of this subsection was translated from Meijer and Kleinnijenhuis (2001).

Appendices

1   In a bridging strategy, the organization tries to create relationships with the external stakeholders, while keeping the stakeholders at a distance in the more defensive buffering strategy (Van den Bosch & Van Riel, 1998). To put it briefly, Van den Bosch and Van Riel concluded that Shell adopted a buffering strategy first, ignoring the perspective of external stakeholders such as Greenpeace, who argued that sinking the Brent Spar would establish a precedent. Later on the corporation seemed to move to a bridging strategy by admitting it should have a greater external focus and that it has to communicate more, both internally and externally.
2   In *Fortune* magazine's ranking of the largest corporations in 2001, BP was ranked fourth and Royal Dutch/Shell group was ranked eighth.
3   Due to the merger between Exxon and Mobile in 1999, the alliance between BP and Mobile had to be dismantled (ExxonMobil website, accessed 19 November 2003).
4   For individuals and small and medium-sized enterprises.
5   For individuals and institutional investors.
6   For major international corporations.
7   Because the Coöperatieve Centrale Raiffeisen-Bank was founded on June 12, 1898 (Rabo-bank website, accessed 20 November 2003), Rabobank celebrated its 100-year anniversary during the research period, on June 12, 1998.
8   Rabobank Nederland supports the independent local Rabobanks. The local Rabobanks are members of Rabobank Nederland (Rabobank website, accessed 20 November 2003).
9   Aldi ranks second, and 55% of the Dutch population bought something at Aldi at least once in 2002. C1000 ranks third, and 53% of the Dutch population bought something at C1000 at least once in 2002 (GFK, 2002).
10   Lovers Rail operated the Amsterdam-Haarlem and Amsterdam-Keukenhof railway lines.
11   The coders received the coding instructions in Dutch. For the benefit of this dissertation the coding instructions have been translated into English. As a result of this translation, the abbreviation of meaning objects might seem illogical.
12   Krippendorff (2004) warned against placing agreement coefficients on a conservative-liberal continuum. He argued that statistical coefficients are mathematical functions that must be understood *before* selecting one coefficient over another.
13   The amount of print news was no longer significant when the model was controlled for the educational level of the respondents. In addition, the variable "favorability of success & failure print news" was replaced by the variable "direction of success & failure print news" ($ß = -.09, p < .05$).
14   As mentioned earlier, the intercorrelation between the variable advertising expenditures on television news and variable "favorability of success & failure TV news" was very high ($r = .94, p < .01$, see Table F1). In this model with high levels of multicollinearity, the "favorability of success & failure TV news" remained significant after controlling for the educational level of the respondents, because the sign of the regression coefficient changed from positive to negative ($ß = -.22, p < .05$). The "amount of television news" was

no longer significant and the variable "advertising intensity TV" was added to the model. In order to avoid the multicollinearity problem, another stepwise model was estimated in which the variable "advertising intensity TV" was excluded beforehand. In that model, no significant media effects were found. In other words, introducing the educational level of respondents led to extraordinarily high VIF values in the model of Shell. As mentioned by Hair et al. (1998), high degrees of multicollinearity can lead to results in regression coefficients being incorrectly estimated and even exhibiting the wrong signs.

15   Like the model of Shell, the model of Super de Boer contained high VIF values (the highest VIF value was 733), when the model was controlled for the educational level of respondents. The educational level was unknown for five of the respondents.

16   The VIF values stayed beneath the threshold of 10.

17   The levels of multicollinearity were extremely high. When the variable "advertising expenditures TV" was omitted, multicollinearity reached an acceptable level, without changing the level of significance of the favorability of support & criticism TV news and without changing the value of the regression coefficient.

18   Due to their low tolerance values, SPSS excluded three variables (amount of TV news, direction support & criticism print news, favorability support & criticism print news) from the model, if the degree of multicollinearity was examined in a model that was estimated by using confirmatory specification.

References

Abelson, R. P. (1968). Psychological implication. In R. P. Abelson (Ed.), *Theories of cognitive consistency* (pp. 112-139). Chicago: Rand McNally.

ABN AMRO (2004). *About ABN Amro*. Retrieved October 29, 2004 from http://www.abnamro.com.

Abratt, R. (1989). A new approach to the corporate image management process. *Journal of Marketing Management, 5*(1), 45-57.

Achen, C. H. (2000, July). *Why lagged dependent variables can surpress the explanatory power of other independent variables*. Paper presented at the Annual Meeting of the Political Methodology Section of the American Political Science Association, University of California, Los Angeles. Retrieved December 19, 2003, from http://polmeth.wustl.edu/working00.brief.html

Ahold (2004). *Historie* [History]. Retrieved October 29, 2004 from http://www.ahold.nl.

Aiken, L. S., & West, S. G. (1991). *Multiple regression: Testing and interpreting interactions* (2nd ed.). Newbury Park, CA: Sage Publications.

Aitkin, M., Anderson, D., & Hinde, J. (1981). Statistical modeling of data on teaching styles. *Journal of the Royal Statistical Society, 144*(4), 419-461.

Ajzen, I. (1988). *Attitudes, personality and behavior*. Milton Keynes, UK: Open University Press.

Ajzen, I., & Fishbein, M. (1980). *Understanding attitudes and predicting social behavior*. Englewood Cliffs, NJ: Prentice-Hall.

Albert, S., & Whetten, D. A. (1985). Organizational identity. *Research in Organizational Behavior, 7*, 263-295.

Allen, C. T., & Madden, T. J. (1985). A closer look at classical conditioning. *Journal of Consumer Research, 12*(2), 301-315.

Anand, P., & Sternthal, B. (1990). Ease of message processing as a moderator of repetition effects in advertising. *Journal of Marketing Research, 27*(3), 345-353.

Andreassen, W., & Lanseng, E. (1997). The principal's and agents' contribution to customer loyalty within an integrated service distribution channel. *European Journal of Marketing, 31*(7/8), 487-503.

Atteveldt, W. van (in press). *iNet: Input and visualization for relational content analysis* (Tech. Rep.). Amsterdam: Vrije Universiteit Amsterdam, Department of Communication Science.

Atwood, L. E., Sohn, A. B., & Sohn, H. (1978). Daily newspaper contributions to community discussion. *Journalism Quarterly, 55*, 570-576.

Bakker, P., & Scholten, O. (2003). *Communicatiekaart in Nederland: Overzicht van media en communicatie* [Communication map in the Netherlands: Overview of media and communication]. Alphen aan den Rijn, the Netherlands: Kluwer.

Balmer, J. M. T. (1995). Corporate branding and connoisseurship. *Journal of General Management, 21*(1), 24-46.

Balmer, J. M. T. (2001). Corporate identity, corporate branding and corporate marketing – Seeing through the fog. *European Journal of Marketing, 35*(3/4), 248-291.

Bandura, A. (1994). Social cognitive theory of mass communication. In J. Bryant & D. Zillmann (Eds.), *Media effects: Advances in theory and research* (pp. 61-90). Hillsdale, NJ: Lawrence Erlbaum Associates.

BBC. (2004). Reclametoppers [Top commercials]. *Adformatie, 32*(5), 12.

Bennett, N. (1976). *Teaching styles and pupil progress.* London: Open books Ltd.

Berelson, B., Lazarsfeld, P. F., & McPhee, W. (1954). *Voting.* Chicago: University of Chicago Press.

Berendse, A. C. (2003). *LNV in-druk: Onderzoek naar het Ministerie van Landbouw, Natuurbeheer en Visserij in de landelijke dagbladen en de invloed van persberichten van het LNV op de media* [LNV under pressure: Study on the Ministry of Agriculture and Fisheries in the daily newspapers, and the influence on the media of Ministry press releases]. Unpublished master's thesis, Vrije Universiteit Amsterdam, the Netherlands.

Berens, G., & Riel, C. B. M. van (2004). Corporate associations in the academic literature: Three main streams of thought in the reputation measurement literature. *Corporate Reputation Review, 7*(2), 161-178.

Berg, H. van den, & Veer, K. van der. (2000). Computerondersteunende tekstanalyse. Het geval CETA [Computer-assisted text analysis. The CETA case]. *Tijdschrift voor communicatiewetenschap, 28*(1), 26-40.

Berger, I. E., & Mitchell, A. A. (1989). The effect of advertising on attitude accessibility, attitude confidence and the attitude-behavior relationship. *Journal of Consumer Research, 16,* 269-279.

Berlyne, D. E. (1970). Novelty, complexity, and hedonic value. *Perception and Psychophysics, 8*(November), 279-286.

Biddle, J. (1991). A bandwagon efffect in personalized license plates? *Economic Inquiry, 29,* 375-388.

Bierley, C., McSweeney, F. K., & Vannieuwkerk, R. (1985). Classical conditioning of preferences for stimuli. *Journal of Consumer Research, 12*(3), 316-323.

Birkigt, K., & Stadler, M. M. (1995). *Corporate identity, grundlagen, funktionen, fallspielen* [Corporate identity, basics, functions, roles]. Landsberg am Lech, Germany: Verlag Moderne Industrie.

Black, E. L., Carnes, T., & Richardson, V. J. (2000). The market valuation of corporate reputation. *Corporate Reputation Review, 3*(1), 31-42.

Boer, C. de, & Brennecke, S. I. (1999). *Media en publiek: Theorieën over media-impact* [Media and audience: Theories about media impact]. Amsterdam: Boom.

Bornstein, R. F. (1989). Exposure and affect: Overview and meta-analysis of research, 1968-1987. *Psychological Bulletin, 106*(2), 265-289.

Bosch, F. A. J. van den, & Riel, C. B. M. van (1998). Buffering and bridging as environmental strategies for firms. *Business Strategy and Environment, 7,* 24-31.

Boulding, K. E. (1956). *The image.* Ann Arbor: University of Michigan Press.

Bouwman, M. Y. (1998). *Op zoek naar het merkimago: Mogelijkheden en beperkingen van verschillende onderzoekstechnieken* [In search of the brand image: Possibilities and limitations of different research techniques]. Amsterdam: Stichting Wetenschappelijk Onderzoek Commerciële Communicatie.

BP (2004). *Over BP Groep* [About BP Group]. Retrieved October 29, 2004 from http://www.bpnederland.nl.

Brady, H. E., & Johnston, R. (1987). What's the primary message: Horse race or issue journalism? In G. R. Orren & N. W. Polsby (Eds.), *Media and Momentum*. Chatham, NJ:
Chatham House.

Broh, C. A. (1977). Horse-Race Journalism: Reporting the polls in the 1976 presidential
election. *Public Opinion Quarterly, 41*, 514-529.

Bromley, D. B. (1993). *Reputation, image and impression management.* Chichester, UK: Wiley.

Bromley, D. B. (2000). Psychological aspects of corporate identity, image and reputation.
Corporate Reputation Review, 3(3), 240-252.

Brown, T. J. (1998). Corporate associations in marketing: Antecedents and consequences.
Corporate Reputation Review, 1(3), 215-233.

Bucholz, R. A., Evans, W., D., & Wagley, R. A. (1985). *Management response to public issues.*
Upper Saddle River, NJ: Prentice Hall.

Budge, I., & Farlie, D. J. (1983). *Explaining and predicting elections: Issue effects and party
strategies in twenty-three democracies.* London: George Allen & Unwin.

Cacioppo, J. T., & Petty, R. E. (1979). Effects of message repetition and position on cognitive
response, recall, and persuasion. *Journal of Personality and Social Psychology, 37*(January), 97-109.

Cacioppo, J. T., & Petty, R. E. (1985). Central and peripheral routes to persuasion: The role
of message repetition. In L. F. Alwitt & A. A. Mitchell (Eds.), *Psychological processes and
advertising effects: Theory, research and applications.* Hillsdale, NJ: Lawrence Erlbaum
associates.

Calder, B. J., & Sternthal, B. (1980). Television commercial wearout: An information processing view. *Journal of Marketing Research, 17*(May), 173-186.

Carroll, C. E. (2004). *How the mass media influence perceptions of corporate reputation:
Exploring agenda-setting effects within business news coverage.* Unpublished doctoral dissertation, The University of Texas, Austin.

Carroll, C. E., & McCombs, M. (2003). Agenda-setting effects of business news on the public's images and opinions about major corporations. *Corporate Reputation Review, 16*(1),
36-46.

Centraal Bureau voor de Statistiek (2002). *CBS-landbouwtelling veehouderij, akkerbouw,
tuinbouw en arbeid* [Statistics Netherlands CBS, agricultural census of livestock farming,
arable farming, market gardening, and labor]. Doetinchem, the Netherlands: Reed
business information.

Centraal Bureau voor de Statistiek (2003). *Landbouwtelling met indelingen naar bedrijfstypen
per regio* [Statistics Netherlands CBS, 'Agricultural census classified according to farm
types per region']. Retrieved November 22, 2003 from http://statline.cbs.nl.

Centre for Multilevel Modelling, Institute of Education, University of London (2004).
Introduction to Multilevel Modelling. Retrieved January 12, 2004, from
http://multilevel.ioe.ac.uk/intro/index.html

Christensen, L. T., & Askegaard, S. (2001). Corporate identity and corporate image revisited – A semiotic perspective. *European Journal of Marketing, 35*(3/4), 292-315.

Cohen, B. (1963). *The press and foreign policy.* Princeton: Princeton University Press.

Cohen, J. (1978). Partialed products are interactions; Partialed owers are curve components. *Psychological Bulletin, 85*, 858-866.

Cohen, R., Fishbein, M., & Ahtola, O. T. (1972). The nature and uses of expectancy-value
models in consumer attitude research. *Journal of Marketing Research, 9*, 456-460.

Collins, A. M., & Loftus, E. F. (1975). A spreading activation theory of semantic processing.
Psychological Review, 82(6), 407-428.

Craig, C. S., Sternthal, B., & Leavitt, C. (1976). Advertising wear-out: An experimental analysis. *Journal of Marketing Research, 13*(November), 365-372.

Cramwinckel, M. (2000). De mediamonitor van Oegema c.s.: Toch meer aandacht voor de ontvanger gewenst? [The media monitor of Oegema & co.: Should more attention be paid to the receiver?]. In D. Oegema, M.M. Meijer, & J. Kleinnijenhuis (Eds.), *De mediamonitor: Zijn effecten van persaandacht meetbaar?* Alphen aan den Rijn, the Netherlands: Samsom.

Cuilenburg, J. J. van, Kleinnijenhuis, J., & Ridder, J. A. de (1988). *Tekst en betoog: Naar een computergestuurde inhoudsanalyse van betogende teksten* [Text and argument: Towards a computer-assisted content analysis of argumentative texts]. Muiderberg, the Netherlands: Dick Coutinho.

Cuilenburg, J. J. van, Scholten O., & Noomen G.W. (1992). *Communicatiewetenschap* (3rd ed.). [Communication Science]. Muiderberg, the Netherlands: Dick Coutinho.

Dalton, R. J., Beck, P. A., & Huckfeldt, R. (1998). Partisan cues and the media: Information flows in the 1992 presidential election. *The American Political Science Review, 92*(1), 111-126.

Davies, G., Chun, R., Da Silva, R. V., & Roper, S. (2001). The personification metaphor as a measurement approach for corporate reputation. *Corporate Reputation Review, 4*(2), 113-127.

Dearing, J. W., & Rogers, E. (1996). *Agenda-setting.* Thousand Oaks, CA: Oaks.

Deephouse, D. L. (2000). Media reputation as a strategic resource: An integration of mass communication and resource-based theories. *Journal of Management, 26*(6), 1091-1112.

Deephouse, D. L., Carroll, C., Meijer, M. M., Kleinnijenhuis, J., Verčič, D., & McCombs, M. (2001, May). *Convergencies and controversies on the role of the media in corporate reputation research.* Paper presented at the Conference on Corporate Reputation, Identity, and Competitiveness, Paris.

DeFleur, M. L., Davenport, L., Cronin, M., & DeFleur, M. (1992). Audience recall of news stories presented by newspaper, computer, television and radio. *Journalism Quarterly, 69*(4), 1010-1022.

Domke, D. (2001). Racial cues and political ideology: An examation of associative priming. *Communication Research, 28*(6), 772-881.

Domke, D., Shah, D. V., & Wackman, D. (1998). Media priming effects: Accessibility, association, and activation. *International Journal of Public Opinion Research, 1,* 51-74.

Dowling, G. R. (1994). *Corporate reputations: strategies for developing the corporate brand.* London: Kogan Page.

Dunbar, R. L. M., & Schwalbach, J. (2000). Corporate reputation and performance in Germany. *Corporate Reputation Review, 3*(2), 115-123.

Dutton, J. E., & Dukerich, J. M. (1991). Keeping an eye on the mirror: Image and identity in organizational adaption. *Academy of Management Journal, 34*(3), 517-554.

Eagly, A. H., & Chaiken, S. (1976). Why would anyone say that? Causal attribution of statements about the Watergate scandal. *Sociometry, 39,* 236-243.

Eagly, A. H., & Chaiken, S. (1993). *The psychology of attitudes.* Orlando, FL: Harcourt.

Eells, R. (1959). The corporate image in public relations. *California management review, 1*(4).

Entman, R. M. (1991). Framing U.S. coverage of international news: Contrasts in narratives of the KAL and Iran air incidents. *Journal of Communication, 41,* 6-25.

ExxonMobile (2004). *Ruim een eeuw Mobil* [Mobile: Over a century]. Retrieved October 29, 2004 from http://www.essobenelux.com

Faber, H. (1996, December 14). De straat op, weg van de zappende kijker [Onto the street, away from the zapping viewer]. *de Volkskrant,* p. 43.

Fan, D., & Tims, A. R. (1990). The impact of the news media on public opinion: American presidential election 1987-1988. *International Journal of Public Opinion Research, 1*, 151-163.

Fazio, R. H. (1989). On the power and functionality of attitudes: The role of attitude and accessiblity. In A.R. Pratkanis, S.J. Breckler, & A.G. Greenwald (Eds.), *Attitude structure and function* (pp. 153-179). Hillsdale, NJ: Erlbaum.

Festinger, L. (1957). *A theory of cognitive dissonance.* New York: Row Peterson.

Finkel, S. E. (1995). *Causal analysis with panel data.* London: Sage.

Fishbein, M. (1963). An investigation of the relationships between beliefs about an object and the attitude toward that object. *Human Relations, 16*, 233-240.

Fishbein, M. (1967). A behavior theory approach to the relations between beliefs about an object and the attitude toward that object. In M. Fishbein (Ed.), *Readings in attitude theory and measurement* (pp. 389-400). New York: Wiley.

Fishbein, M., & Ajzen, I. (1975). *Belief, attitude, intention and behavior: An introduction to theory and research.* Reading, MA: Addison-Wesley.

Fishbein, M., & Ajzen, I. (1980). Predicting and understanding consumer behavior: Attitude-behavior correspondence. In I. Ajzen & M. Fishbein (Eds.), *Understanding attitudes and predicting social behavior* (pp. 148-172). Upper Saddle River, NJ: Prentice-Hall.

Fleitas, D. W. (1971). Bandwagon and underdog effects in minimal information elections. *American Political Science Review, 65*, 434-438.

Fombrun, C. J. (1996). *Reputation: Realizing value from the corporate image.* Boston, MA: Harvard Business School Press.

Fombrun, C. J., Gardberg, N. A., & Sever, J. M. (2000). The Reputation QuotientSM: A multi-stakeholder measure of corporate reputation. *The Journal of Brand Management, 7*(4), 241-255.

Fombrun, C. J., & Ginsberg, A. (1990). Shifting gears: Enabling and disabling forces on change in corporate posture. *Strategic Management Journal, 11*(4), 297-308.

Fombrun, C. J., & Riel, C. B. M. van (1997). The reputational landscape. *Corporate Reputation Review, 1*(1/2), 5-13.

Fombrun, C. J., & Riel, C. B. M. van (2004). *Fame & fortune: How successful companies build winning reputations.* Upper Saddle River, NJ: FT Prentice Hall.

Fombrun, C. J., & Rindova, V. P. (2000). The road to transparency: Reputation management at Royal Dutch/Shell. In M. Schultz, M.J. Hatch, & M.H. Larsen (Eds.), *The expressive organization: linking identity, reputation and the corporate brand* (pp. 77-96). New York: Oxford University Press.

Fombrun, C. J., & Shanley, M. (1990). What's in a name? Reputation building and corporate strategy. *Academy of Management Journal, 33*(2), 233-258.

Fryxell, G. E., & Wang, J. (1994). The *Fortune* corporate reputation index: Reputation for what? *Journal of Management, 20*(1), 1-14.

Fuente Sabate, J. M. de la, & Quevedo Puento, E. de (2003). Empirical analysis of the relationship between corporate reputation and financial performance: A survey of the literature. *Corporate Reputation Review, 6*(2), 161-177.

Galtung, J., & Ruge, M. H. (1965). The structure of foreign news. *Journal of Peace Research, 2*, 64-91.

GFK PanelServices Benelux BV (2002). *Ranking meeste klanten* [Ranking of most customers]. Dongen, the Netherlands: Author.

Gioia, D. A., Schultz, M., & Corley, K. G. (2000). Organizational identity, image and adaptive instability. *Acadamy of Management Review, 25*(1), 63-81.

Glass, G. C., & Hopkins, K. (1996). *Statistical methods, in education and psychology*. Englewood Cliffs, NJ: Prentice Hall.

Global 500: The world's largest corporations (Europe edition, 2003, July 21). *Fortune*, p. F-1.

Goidel, R. K., Shields, T. G., & Peffley, M. (1997). Priming theory and RAS models: Toward an intergrated perspective of media influence. *American Politics Quarterly, 25*, 287-318.

Granger, C. W. J. (1980). Testing for causality: A personal viewpoint. *Journal of Economics Dynamics and Control, 2*, 329-352.

Green, S. B. (1992). How many subjects does it take to do a regression analysis? *Multivariate Behavioral Research, 26*(3), 499-510.

Gruning, J. E., & Hunt, T. (1984). *Managing public relations*. New York: Holt, Rinehart & Winston.

Gujarati, D. N. (1995). *Basic econometrics* (3rd ed.). Singapore: McGraw-Hill international editions.

Hair, J. F., Anderson, R. E., Tatham, R. L., & Black, W. C. (1998). *Multivariate data analysis* (5th ed.). Upper Saddle River, NJ: Prentice Hall.

Hampton, J. (2003). *Repeated measures factors/designs*. Retrieved November 19, 2003, from City University London, Department of Psychology website: http://www.staff.city.ac.uk/hampton/stats/lecture07/

Harcup, T., & O'Neill, D. (2001). What Is News? Galtung and Ruge revisited. *Journalism Studies, 2*(2), 261-280.

Hatch, M. J., & Schultz, M. (2000). Scaling the tower of Babel: relational differences between identity, image and culture in organizations. In M. Schultz & M. J. Hatch & M. H. Larsen (Eds.), *The expressive organization: linking identity, reputation and the corporate brand* (pp. 11-35). Oxford: Oxford University Press.

Heath, R. L., & Nelson, R. A. (1986). *Issues management: corporate public policymaking in an information society*. California: Sage publications.

Heeler, R. M., Kearney, M. J., & Mehaffey, B. J. (1973). Modeling supermarket product selection. *Journal of Marketing Research, 10*, 37-44.

Heider, F. (1946). Attitudes and cognitive organization. *The Journal of Psychology, 21*, 107-112.

Heider, F. (1958). *The psychology of interpersonal relations*. New York, N.Y.: Wiley.

Heijting, K., & Scheeringa, S. (2000). Het imago van de mediamonitor [The image of the media monitor]. In D. Oegema, M.M. Meijer, & J. Kleinnijenhuis (Eds.), *De mediamonitor: Zijn effecten van persaandacht meetbaar?* Alphen aan den Rijn, the Netherlands: Samsom.

Herr, P. M. (1989). Priming price: prior knowledge and context effects. *Journal of Consumer Research, 16*(June), 323-340.

Hertog, J. K., & Fan, D. P. (1995). The impact of press coverage on social beliefs: The case of HIV transmission. *Communication Research, 22*(5), 545-574.

Himmelfarb, S. (1993). The measurement of attitudes. In A.H. Eagly & S. Chaiken (Eds.), *The psychology of attitudes* (pp. 23-88). Orlando: Harcourt, Brace, Jovanovich.

Hotaling, A., & Lippitt, J. (2003). Reconstructing financial reporting following the accounting fraud disclosure at Enron energy company. *International Journal of Management, 20*(4), 464-469.

Huberts, L. W. J. C. (1998). *Blinde vlekken in de politiepraktijk en de politiewetenschap: Over politie, wetenschap, macht, beleid, integriteit en communicatie* [Blind spots in police practice and police science: About the police, science, power, policy, integrity, and communication]. Arnhem, the Netherlands: Gouda Quint.

Huberts, L. W. J. C. (2001). Persoonlijke integriteit als kompas voor de politie [Personal integrity as guiding principle for the police]. In L. G. Moor (Ed.), *Rechtvaardigheid en barmhartigheid.* Dordrecht, the Netherlands: SMVP.

Huygens, C. (1998). *Van rekeningh in spelen van geluck De ratiociniis in ludo aleae* (On reasoning or computing in games of chance, W. Kleijne, Trans.). Utrecht, the Netherlands: Epsilon uitgaven. (Original work published in 1660).

Irwin, G. A., & Holsteyn, J. van (1988). Opiniepeilingen en stemgedrag [Opinion polls and voting behavior]. In R. B. Andeweg (Ed.), *Tussen steekproef en stembus* (pp. 43-63). Leiden, the Netherlands: DSWO Press.

Iyengar, S., & Kinder, D. R. (1987). *News that matters.* Chicago: University of Chicago Press.

Iyengar, S., & Simon, A. (1997). News coverage of the gulf crisis and public opinion. In S. Iyengar & R. Reeves (Eds.), *Do the media govern?: Politicians, voters, and reports in America.* Thousand Oaks, CA: Sage.

Jo, E., & Berkowitz, L. (1994). A priming effect analysis of media influences: An update. In J. Bryant & D. Zillmann (Eds.), *Media effects: Advances in theory and research.* Hillsdale, NJ: Lawrence Erlbaum Associates.

Jones, G. H., Jones, B. H., & Little, P. (2000). Reputation as reservoir: Buffering against loss in times of economic crises. *Corporate Reputation Review, 3*(1), 21-29.

Katrichis, J. (1992). The conceptual implications of data centering in interactive regression models. *Journal of Market Research Society, 35,* 183-192.

Kerrick, J. (1958). The effect of relevant and non-relevant sources on attitude change. *Journal of Social Psychology, 47,* 15-20.

Kim, J., Allen, C. T., & Kardes, F. R. (1996). An investigation of the mediational mechanisms underlying attitudinal conditioning. *Journal of Marketing Research, 33*(3), 318-328.

Kiousis, S., Bantimaroudis, P., & Hyun, B. (1999). Candidate image attributes: Experiments on the substantive dimension of second level agenda setting. *Communication Research, 26*(4), 414-428.

Kiousis, S., & McCombs, M. (2004). Agenda-setting effects and attitude strength: Political figures during the 1996 presidential election. *Communication Research, 31*(1), 36-57.

Klapper, J. T. (1960). *The effects of mass communication.* New York: Free Press.

Kleinnijenhuis, J. (1990). *Op zoek naar nieuws: Onderzoek naar journalistieke informatieverwerking en politiek* [Looking for news: A study into journalistic information-processing and politics]. Amsterdam: Vrije Universiteit.

Kleinnijenhuis, J. (1998). Effecten van vrije publiciteit op het imago [Effects of free publicity on image]. In V. M. G. Damoiseaux & A. A. van Ruler (Eds.), *Effectiviteit in Communicatie-management: Zoektocht naar criteria voor professioneel succes* (pp. 55-76). Deventer, the Netherlands: Samsom.

Kleinnijenhuis, J. (2001). Organisatiereputatie en publiciteit [Organizational reputation and publicity]. In C. B. M. van Riel (Ed.), *Corporate communication: Het managen van reputatie* (3rd ed., pp. 131-154). Alphen aan den Rijn, the Netherlands: Kluwer.

Kleinnijenhuis, J. (2004). *Media, mores en macht* (Talmalezing) [Media, mores, and power]. Amsterdam: Vrije Universiteit, Faculty of Social Sciences.

Kleinnijenhuis, J., Oegema, D., Ridder, J. de, Hoof, A. van, & Vliegenthart, R. (2003). *De puinhopen in het nieuws: De rol van de media bij de Tweede Kamerverkiezingen van 2002* [The chaos in the news: The role of the media in the 2002 Dutch parliamentary elections]. Alphen aan den Rijn, the Netherlands: Kluwer.

Kleinnijenhuis, J., Oegema, D., Ridder, J.A. de, Ruigrok, P. C. (1998). *Paarse polarisatie. De slag om de kiezer in de media* [Polarization in the Dutch coalition government: The battle for the voter in the media]. Alphen aan den Rijn, the Netherlands: Samson.

Kleinnijenhuis, J., & Ridder, J. de (1997, March). *Effects of strategic news framing on party preferences*. Paper presented at the Annual Meeting of the American Political Science Association, Washington D.C.

Kleinnijenhuis, J., Ridder, J. de, Oegema, D., & Bos, H. (1995). *De democatie op drift: Een evaluatie van de verkiezingscampagne van 1994* [Democracy adrift: An evaluation of the 1994 Dutch general election campaign]. Amsterdam: VU Uitgeverij.

Kleinnijenhuis, J., Ridder, J. de, & Rietberg, E. M. (1997). Reasoning in economic discourse: An application of the network approach to the Dutch press. In C. W. Roberts (Ed.), *Text analysis for the Social Sciences: Methods for drawing statistical inferences from texts and transcripts* (pp. 191-207). New York: Erlbaum.

Knight, M. G. (1999). Getting past the impasse: Framing as a tool for public relations. *Public Relations Review, 25*(3), 381-398.

Kranz, P. (1970). Content analysis by word group. *Journal of Marketing Research, 7*(3), 377-380.

Krippendorff, K. (1986). *Content Analysis: An introduction to its methodology* (6th ed.). London: Sage Publications.

Krippendorff, K. (2004). Reliability in content analysis: Some common misconceptions and recommendations. *Human Communication Research, 30*(3), 411-433.

Kromrey, J. D., & Foster-Johnson, L. (1998). Mean centering in moderated multiple regression: Much ado about nothing. *Educational and Psychological Measurement, 58*, 42-68.

Krosnick, J. A., & Kinder, D. R. (1990). Altering the foundations of support for the president through priming. *American Political Science Review, 84*, 497-512.

Lang, G., & Lang, K. (1981). Mass communication and public opinion: Strategies for research. In M. Rosenberg & R. H. Turner (Eds.), *Social psychology: Sociological perspectives* (pp. 653-682). New York: Basic Books.

Laponce, J. A. (1966). An experimental method to measure the tendency to equibalance in a political system. *American Political Science Review, 60*, 982-993.

Lasthuizen, K., Huberts, L., & Kaptein, M. (2002). Integrity problems in the police organization: Police officers' perceptions reviewed. In M. Pagon (Ed.), *Policing in Central and Eastern Europe: Deviance, violence, and victimization* (pp. 25-37). Leicester, UK, and Ljubljana, Slovenia: Scarman Centre University of Leicester & Ljubljana, College of Police and Security Studies.

Lazarsfeld, P. F., Berelson, B., & Gaudet, H. (1944). *The people's choice: How the voter makes up his mind in a presidential campaign* (3rd ed.). New York: Duell, Sloan and Pearce.

Leibenstein, H. (1950). Bandwagon, snob, and veblen effects in the theory of consumers' demand. *The Quarterly Journal of Economics, 64*, 183-207.

Livesey, S. M. (2001). Eco-identity as discursive struggle: Royal Dutch/Shell, Brent Spar, and Nigeria. *The Journal of Business Communication, 38*, 58-91.

Lombard, M., Snyder-Duch, J., & Campanella Bracken, C. (2002). Content analysis in mass communication: Assessment and reporting intercoder reliability. *Human Communication Research, 28*(4), 587-604.

Lombard, M., Snyder-Duch, J., & Campanella Bracken, C. (2004). A call for standardization in content analysis reliability. *Human Communication Research, 30*(3), 434-437.

LTO (2003). *De sector in kengetallen* [Basic information on the sector]. Retrieved, November 22, 2003 from http://www.lto.nl

Maathuis, O. J. M. (1999). *Corporate Branding: The value of the corporate brand to customers and managers*. Unpublished doctoral dissertation, Erasmus University Rotterdam, the Netherlands.

MacKenzie, S. B. (1986). The role of attention in mediating the effect of advertising on attribute importance. *Journal of Consumer Research, 13*, 174-195.

Mandel, N., & Johnson, E. J. (2002). When web pages influence choice: Effects of visual primes on experts and novices. *Journal of Consumer Research, 29*, 235-245.

Markham, V. W. R. (1972). *Planning the corporate reputation*. London: Allen and Unwin.

Marsh, C. (1984). Do polls affect what people think? In C. F. Turner & E. Martin (Eds.), *Surveying subjective phenomena* (Vol. 2, pp. 565-591). New York: Sage.

Marsh, C., & O'Brien, J. (1989). Opinion bandwagons in attitudes towards the Common Market. *Journal of Marketing Research Society, 31*(3), 295-305.

Martineau, P. (1958). Sharper focus for the corporate image. *Harvard Business Review, 36*(6), 49-58.

McCombs, M. (1992). Explorers and surveyors: Expanding strategies for agenda setting research. *Journalism Quarterly, 69*(4), 813-824.

McCombs, M. (1994). News influence on our pictures of the world. In J. Bryant & D. Zillmann (Eds.), *Media effects: Advances in theory and research* (pp. 1-16). Hillsdale, NJ: Lawrence Erlbaum Associates.

McCombs, M., Llamas, J. P., Lopez-Escobar, E., & Rey, F. (1997). Candidate images in Spanish elections: Second-level agenda-setting effects. *Journalism & Mass Communication Quarterly, 74*(4), 703-717.

McCombs, M., Lopez-Escobar, E., & Llamas, J. P. (2000). Setting the agenda of attributes in the 1996 Spanish general election. *Journal of Communication, 50*(2), 77-92.

McCombs, M., Shaw, D. L., & Weaver, D. (1997). *Communication and democracy: Exploring the intellectual frontiers in agenda-setting theory*. Mahwah, NJ: Erlbaum.

McCombs, M. E., & Shaw, D. L. (1972). The agenda-setting function of mass media. *Public Opinion Quarterly, 36*, 176-187.

McGuire, W. J. (1985). Attitudes and attitude change. In G. Lindzey & E. Aronson (Eds.), *Handbook of social psychology* (3rd ed., Vol. 2, pp. 233-346). New York: Random House.

McKeone, D. (1995). *Measuring your media profile*. Hampshire, England: Gower.

McKoon, G., & Ratcliff, R. (1995). Conceptual combinations and relational contexts in free association and in priming in lexical decision and naming. *Psychonomic Bulletin and Review, 2*, 527-533.

McNamara, T. P. (1992). Priming and constraints it places on theories of memory and retrieval. *Psychological Review, 99*, 650-662.

McQuail, D. (1994). *Mass communication theory: An introduction*. London: Sage.

McQuail, D. (2000). *McQuail's mass communication theory* (4th ed.). London: Sage.

Meijer, M. M., & Kleinnijenhuis, J. (2001). Mediamonitor als kompas? Systematische nieuwsanalyse ten behoeve van organisaties en sectoren [Media monitor as precept? Systematic news analysis for the benefit of organizations and sectors]. *Tijdschrift voor communicatiewetenschap, 29*(4), 243-263.

Meijer, M. M., Oegema, D., & Kleinnijenhuis, J. (2000, June). *Exploring media effects on corporate image: Towards the development of a media monitor to forecast corporate reputation*. Poster presented at the 50th Annual conference of the International Communication Association, Acapulco, Mexico.

Miller, R. L. (1976). Mere exposure, psychological reactance and attitude change. *The Public Opinion Quarterly, 40*(2), 229-233.

Miller, A. H., Goldenberg, E. N., & Erbring, L. (1979). Type-set politics: Impact of newspapers on public confidence. *The American Political Science Review, 73*(1), 67-84.

Miller, J. M., & Krosnick, J. A. (2000). News media impact on the ingredients of presidential evaluations: Politically knowledgeable citizens are guided by a trusted source. *American Journal of Political Science, 44*(2), 301-315.

Moingeon, B., & Ramanantsoa, B. (1997). Understanding corporate identity: The French school of thought. *European Journal of Marketing, 31*(5/6), 383-395.

Mutz, D. C. (1995a). Effects of horse-race coverage on campaign coffers: Strategic contributing in presidential primaries. *The Journal of Politics, 57,* 1015-1042.

Mutz, D. C. (1995b). Media, momentum and money: Horse race spin in the 1988 republican primaries. In P. L. Lavrakas & M. W. Traugott & P. V. Miller (Eds.), *Presidential polls and the news media* (pp. 229-245). Boulder, CO: Westview Press.

Neuendorf, K. A. (2002). *The content analysis guidebook.* Thousand Oaks, CA: Sage Publications.

Neuman, W. R., Just, M. R., & Crigler, A. N. (1992). *Common knowledge, news and the construction of political meaning.* Chicago: The University of Chicago Press.

Noelle-Neumann, E. (1973). Return to the concept of the powerfull mass media. *Studies of Broadcasting, 9,* 67-112.

Nord, W. R., & Peter, J. P. (1980). A behavior modification perspective on marketing. *Journal of Marketing, 44* (Spring), 36-47.

Nordhelm, C. L. (2002). The influence of level of processing on advertising repetition effects. *Journal of Consumer Research, 29,* 371-382.

NS (2003). *Jaarverslag 2002* [Annual report 2002]. Retrieved, November 21, 2003 from http://www.ns.nl.

NS (2004). *Archief NS* [NS archive]. Retrieved October 29, 2004 from http://www.ns.nl.

Oegema, D., Haan, M. de, & Leur, B. van (1998). Shell en de publiciteit over de Brent Spar [Shell and the publicity on the Brent Spar]. In V. M. G. Damoiseaux & A. A. van Ruler (Eds.), *Effectiviteit in communicatiemanagement* (pp. 77-93). Deventer/Diegem, the Netherlands: Samsom.

Oegema, D., Meijer, M. M., & Kleinnijenhuis, J. (2000). *De mediamonitor: Zijn effecten van persaandacht meetbaar?* [The media monitor: Can the effects of press attention be measured?] Alphen aan den Rijn, the Netherlands: Samsom.

Osgood, C. E., Saporta, S., & Nunnally, J. C. (1954). *Evaluative assertion analysis: A method for studying the semantic content of messages.* University of Illinois, Institute of Communications Research.

Osgood, C. E., & Tannenbaum, P. H. (1955). The principle of congruity in the prediction of attitude change. *Psychological Review, 62*(1), 42-55.

Page, B. I., Shapiro, R. Y., & Dempsey, G. R. (1987). What moves public opinion? *American Political Science Review, 81,* 23-43.

Pan, Z., & Kosicki, G. M. (1997). Priming and media impact on the evaluation of the President's performance. *Communication Research, 24,* 3-30.

Patterson, T. (1993). *Out of order.* New York: Knopf.

Patterson, T. E. (1980). *The Mass Media Election.* New York: Praeger.

Pennings, P., Keman, H., & Kleinnijenhuis, J. (1999). *Doing research in political science: An introduction to comparative methods and statistics.* London: Sage Publications.

Peter, J. (2003). *Why European TV news matters: A cross-nationally comparative analysis of TV news about the European Union and its effects.* Unpublished doctoral dissertation. University of Amsterdam.

Petrick, J. A., & Scherer, R. F. (2003). The Enron scandal and the neglect of management integrity capacity. *Mid - American Journal of Business, 18*(1), 37-49.

Petrocik, J. R. (1996). Issue ownership in presidential elections, with a 1980 case study. *American Journal of Political Science, 40*(3), 825-850.

Petty, R. E., & Cacioppo, J. T. (1986). *Communication and persuasion: Central and peripheral routes to attitude change.* New York: Springer Verlag.

Pligt, J. van der, & Eiser, J. R. (1984). Dimensional salience, judgment and attitudes. In J. R. Eiser (Ed.), *Attitudinal judgment* (pp. 162-177). New York: Springer-Verlag.

Pligt, J. van der, & Vries, N. K. de (1995). *Opinies en attitudes: Meting, modellen en theorie* [Opinions and attitudes: Measurement, models, and theory]. Amsterdam: Boom.

Police (2003). *Organisatie [Organization].* Retrieved November 22, 2003 from http://www.politie.nl

Police (2004). *Kerngegevens Nederlandse politie 2002 [Basic information on the Dutch police, 2002].* Retrieved October 29, 2004 from http://www.politie.nl

Popping, R. (2000). *Computer-assisted text analysis.* London: Sage.

Praag, P. J. van, & Eijk, C. van der (1998). News content and effects in a historic campaign. *Politcal Communication, 15*(2), 165-183.

Pras, B., & Summers, J. (1975). A comparison of linear and nonlinear evaluation process models. *Journal of Marketing Research, 12,* 276-281.

Pratkanis, A. R. (1989). The cognitive representation of attitudes. In A.R. Pratkanis, S.J. Breckler, & A.G. Greenwald (Eds.), *Attitude structure and function* (pp. 71-98). Hillsdale, NJ: Lawrence Erlbaum.

Price, V. E. (1992). *Communication concepts 4: Public opinion.* Newbury Park, CA: Sage.

Price, V. E., Tewksbury, D., & Powers, E. (1997). Switching trains of thought: The impact of news frames on readers' cognitive responses. *Communication Research, 24*(5), 481-506.

Pruyn, A. T. (2002). *All eyes are on Kwatta: On the dangers of a myopic view on marketing communication.* Unpublished inaugural lecture, University of Twente, the Netherlands.

Publilius Syrus (1969). *Die Sprüche des Publilius Syrus* [The maxims of Publilius Syrus]. H. Beckby (Ed. & Trans.). München: Ernst Heimeran Verlag. (Original work dates from approximately 100 B.C.)

Rabobank (2003). *Korte geschiedenis van de Rabobank* [Short history of the Rabobank]. Retrieved November 20, 2003 from http://www.rabobankgroep.nl

Rabobank (2004). *Profiel van de Rabobank Groep* [Profile of the Rabobank Group]. Retrieved October 29, 2004 from http://www.rabobankgroep.nl

Rekom, J. van (1998). *Corporate identity : development of the concept and a measurement method.* Unublished doctoral dissertation, Erasmus University Rotterdam, the Netherlands.

Reynolds, T. J., & Gutman, J. (1984). Advertising is image management. *Journal of Advertising Research, 24,* 27-37.

Ridder, J. de (1994a). *Van tekst naar informatie* [From text to information]. Unpublished doctoral dissertation, University of Amsterdam.

Ridder, J. de (1994b). Computer-aided Evaluative Text Analysis *(CETA2.1): Software + reference manual.* Groningen, the Netherlands: ProGamma.

Riel, C. B. M. van (1997). *Identiteit en imago: Grondslagen van corporate communication* [Identity and image: Foundations of corporate communication, 2nd ed.]. Schoonhoven, the Netherlands: Academic Service.

Riel, C. B. M. van (Ed.). (2001). *Corporate communication: Het managen van reputatie* [Corporate communication: The management of reputation]. Alphen aan den Rijn, the Netherlands: Kluwer.

Riel, C. B. M. van, & Balmer, J. M. T. (1997). Corporate identity: The concept, its measurement and management. *European Journal of Marketing, 31*(5/6), 342-357.

Riel, C. B. M. van, Stroeker, N. E., & Maathuis, O. J. M. (1998). Measuring corporate images. *Corporate Reputation Review, 1*(4), 313-326.

Riffe, D., & Freitag, A. A. (1997). A content analysis of content analyses: Twenty-five years of Journalism Quarterly. *Journalism & Mass Communication Quarterly, 74*, 873-882.

Riffe, D., Lacy, S., & Fico, F. (1998). *Analyzing media messages: Using quantitative content analysis in research.* Mahwah, NJ: Lawrence Erlbaum Associates.

Riley, J. W., Jr, & Levy, M. F. (1963). The image in perspective. In J. W. Riley Jr. (Ed.), *The corporation and its publics* (pp. 176-189). New York: John Wiley & Sons.

Robson, M. J., & Dunk, M. A. J. (1999). Case study developing a pan-European co-marketing alliance: The case of BP-Mobil. *International Marketing Review, 16*(3), 216-230.

Rohlfs, J. H. (2001). *Bandwagon effects in high-technology industries.* Cambridge, MA: MIT Press.

Rokeach, M., & Rothman, G. (1965). The principle of belief congruence and the congruity principle as models of cognitive interaction. *Psychological Review, 72*(2), 128-142.

Ruler, A. A. van (2000). Communication management in the Netherlands. *Public Relations Review, 26*(4), 403-423.

Ruth, J. A., & York, A. (2004). Framing information to enhance corporate reputation: The impact of message source, information type, and reference point. *Journal of Business Research, 57*(1), 14-20.

Salama, A., Holland, W., & Vinten, G. (2003). Challenges and opportunities in mergers and acquisitions: Three international case studies – Deutsche Bank-Bankers Trust; British Petroleum-Amoco; Ford-Volvo. *Journal of European Industrial Training, 27*(6/7), 313-321.

Sawyer, A. (1981). Repetition, cognitive responses, and persuasion. In R. E. Petty & T. M. Ostrom & T. C. Brock (Eds.), *Cognitive responses in persuasion* (pp. 237-261). Hillsdale, NJ: Lawrence Erlbaum Associates.

Scheufele, D. A. (1999). Framing as a theory of media effects. *Journal of Communication, 49*, 103-122.

Scheufele, D. A. (2000). Agenda-setting, priming, and framing revisited: Another look at cognitive effects of political communication. *Mass communication and society, 3*(2-3), 297-316.

Schiphol Group (2003). *Profiel Schiphol Group* [Profile Schiphol Group]. Retrieved November 21, 2003 from http://www.schiphol.nl

Shell (2004). *About the Royal Dutch/Shell Group of Companies.* Retrieved October 29, 2004 from http://www.shell.com.

Shoemaker, P. J., & Reese, S. D. (1990). Exposure to what? Integrating media content and effects studies. *Journalism Quarterly, 67*, 649-652.

Schoenbach, K., & Lauf, E. (2002). The "trap" effect of television and its competitors. *Communication Research, 29*(5), 564-583.

Schumann, D. W., Petty, R. E., & Scott Clemons, D. (1990). Predicting the effectiveness of different strategies of advertising variation: A test of the repetition-variation hypotheses. *Journal of Consumer Research, 17*(2), 192-202.

Semetko, H. A., & Valkenburg, P. M. (2000). Framing European politics: A content analysis of press and television news. *Journal of Communication, 50*(2), 93-109.

Shah, D. V., Watts, M. D., Domke, D., & Fan, D. P. (2002). News framing and cueing of issue regimes. *Public Opinion Quarterly, 66*, 339-370.

Sheth, J. N., & Talarzyk, W. W. (1972). Perceived instrumentality and value importance as determinants of attitudes. *Journal of Marketing Research, 9*, 6-9.

Simon, H. A. (1979). *Models of thought* (Vol. 1). New Haven: Yale University Press.

Sims, R. R., & Brinkmann, J. (2003). Enron ethics (or: Culture matters more than codes). *Journal of Business Ethics, 9*(4), 243-256.

Smits, J. (2000). Turven is belangrijk, maar het toeval ook [Keeping score is important, but so is coincidence]. In D. Oegema, M. M. Meijer, & J. Kleinnijenhuis (Eds.), *De media-monitor: Zijn effecten van persaandacht meetbaar* (pp. 67-69). Alphen aan den Rijn, the Netherlands: Samsom.

Snijders, T., & Bosker, R. (1999). *Multilevel analysis: An introduction to basic and advanced multilevel modeling.* London: Sage Publications.

Spector, A. J. (1961). Basic dimensions of the corporate image. *Journal of Marketing, 25*(6), 47-51.

Stevens, J. P. (2002). *Applied multivariate statistics for the social sciences* (4th ed.). Mahway, NJ: Lawrence Erlbaum Associates.

Stigler, G. J. (1962). Information in the labor market. *The Journal of Political Economy, 70*(5), 94-105.

Stone, G. C., & McCombs, M. E. (1981). Tracing the time lag in agenda setting. *Journalism Quarterly, 58*, 51-55.

Stone, P. J., Dunphy, D. C., & Bernstein, A. (1966). The analysis of product image. In P. J. Stone, D. C. Dunphy, M. S. Smith, & D. M. Ogilvie (Eds.), *The General Inquirer: A computer approach to content analysis* (pp. 619-627). Cambridge, MA: MIT. Press.

Tabachnick, B. G., & Fidell, L. S. (2001). *Using multivariate statistics* (4th ed.). Toronto, Canada: Allyn and Bacon.

Tacq, J. (1997). *Multivariate analysis techniques in social science research: From problem to analysis.* London: Sage.

Tannenbaum, P. H. (1968). The congruity principle: Retrospective reflections and recent research. In R. P. Abelson, E. Aronson, W. J. McGuire, T. M. Newcomb, M. J. Rosenberg, & P. H. Tannenbaum (Eds.), *Theories of cognitive consistency: a sourcebook* (pp. 52-72). Chicago: Rand McNally.

Tiemeijer, W. (2000). Zééééér interessant...maar heb ik er ook wat aan? De bruikbaarheid van de mediamonitor voor een politieke organisatie [Veeeery interesting ... but what can I do with it? The usefullness of the media monitor for a political organization]. In D. Oegema, M. M. Meijer, & J. Kleinnijenhuis (Eds.), *De mediamonitor: Zijn effecten van persaandacht meetbaar?* (pp. 86-100) Alphen aan den Rijn, the Netherlands: Samsom.

Timmerman, T. (2001). *Researching brand images: The nature and activation of brand repre-sentations in memory.* Unpublished doctoral dissertation, University of Amsterdam.

Trumbo, C. (1995). *Longitudinal modeling of public issues: An Application of the agenda-setting process to the issue of global warming. Journalism & Mass Communication Monographs, No. 152.*

Vakratsas, D., & Ambler, T. (1999). How advertising works: What do we really know? *Journal of Marketing, 63*(January), 26-43.

Valkenburg, P. M., Semetko, H. A., & Vreese, C.H. de (1999). The effects of news frames on readers' thoughts and recall. *Communication Research, 26*(5), 550-569.

Verčič, D. (2000). *Trust in organisations: A study of the relations between media coverage, public perceptions and profitability.* Unpublished doctoral dissertation, The London School of Economics and Political Science, London.

Vis, L. (2002). *The conflict in Northern Ireland: A methodological study of the influence of ideology on the representation of news events.* Unpublished master's thesis, Vrije Universiteit, Amsterdam.

Vreese, C. H. de (2003). *Framing Europe: Television news and European integration.* Amsterdam: Aksant.

Wanta, W., & Foote, J. (1994). The president-news media relationship: A time series analysis of agenda-setting. *Journal of Broadcasting and Electronic Media, 38,* 437-448.

Wartick, S. L. (1992). The relationship between intense media exposure and change in corporate reputation. *Business & Society, 31,* 33-49.

Wartick, S. L., & Heugens, P. P. M. A. R. (2003). Future directions for issues management. *Corporate Reputation Review, 6*(1), 7-18.

Waterman, R. W., Jenkins-Smith, H. C., & Silva, C. L. (1999). The expectation gap hypothesis: Public attitudes toward an incumbent president. *Journal of Politics, 61*(4), 944-966.

Wilkie, W. L., & Pessemier, E. A. (1973). Issues in marketing's use of multi-attribute attitude models. *Journal of Marketing Research, 10,* 428-441.

Winters, L. C. (1986). The effect of brand advertising on company image: Implications for corporate advertising. *Journal of Advertising Research, 26*(April/May), 54-59.

Winters, L. C. (1988). Does it pay to advertise to hostile audiences with corporate advertising? *Journal of Advertising Research, 28*(June/July), 11-18.

Zajonc, R. (1960). The concepts of balance, congruity, and dissonance. *Public Opinion Quarterly, 24,* 280-296.

Zajonc, R. (1968). Attitudinal effects of mere exposure. *Journal of Personality and Social Psychology, 9*(Monograph Suppl. 2), 1-32.

Zaller, J. R. (1992). *The nature and origins of mass opinion.* Cambridge, UK: Cambridge University Press.

Zaller, J. R. (1996). The myth of massive media impact revived: New support for a discredited idea. In D. C. Mutz, P. M. Sniderman, & R. A. Brody (Eds.), *Political persuasion and attitude change* (pp. 17-78). Ann Arbor: University of Michigan Press.

Zoch, L. M., & VanSlyke Turk, J. (1998). Women making news: Gender as a variable in source selection and use. *Journalism & Mass Communication Quarterly, 75*(4), 762-775.

Index

SUBJECTS